Sewage Treatment Plants: Economic Evaluation of Innovative Technologies for Energy Efficiency

Integrated Environmental Technology Series

Sewage Treatment Plants: Economic Evaluation of Innovative Technologies for Energy Efficiency

Edited by

Katerina Stamatelatou and
Konstantinos P. Tsagarakis

IWA PUBLISHING

Published by **IWA Publishing**
Republic - Export Building
1 Clove Crescent
London E14 2BA, UK
Telephone: +44 (0)20 7654 5500
Fax: +44 (0)20 7654 5555 Email:
publications@iwap.co.uk
Web: www.iwaponline.com

First published 2015

British Library Cataloguing in Publication Data
A CIP catalogue record for this book is available from the British Library

ISBN 9781780405018 (Hardback)
ISBN 9781789065473 (Paperback)
ISBN 9781780405025 (eBook)

Contents

About the Editors

Katerina Stamatelatou is an assistant professor in the Democritus University of Thrace in Greece (Department of Environmental Engineering) and member of the part time academic staff of Hellenic Open University, included in the post graduate course "Environmental Design of Infrastructure Works MSc". She is a chemical engineer and obtained her diploma and Ph.D. in the Department of Chemical Engineering of University of Patras. Her field of research focuses on the study and development of anaerobic digestion processes, design and operation of bioreactors and bioprocess modeling, targeting in waste and biomass valorisation. She has been chair and member in Committees of International Conferences. Her published work enumerates more than 40 papers in peered review international journals and chapters in collective volumes. She is currently an associate editor in "Water Science and Technology" and reviewer in international scientific journals.

Konstantinos P. Tsagarakis is Associate Professor in the Department of Environmental Engineering at the Democritus University of Thrace. He holds a degree from the department of Civil Engineering of the Democritus University of Thrace (Greece), a BA degree from the Department of Economics of the University of Crete (Greece) and a Ph.D. Degree from the School of Civil Engineering from the University of Leeds (UK). His research interests include: wastewater management, environmental economics, energy economics, environmental policy, public awareness and behavior, quantitative methods, and techno-economic project analysis. His research work has been published in more than 50 papers in prestigious refereed journals. He is Associate Editor in Water Policy journal. He is the Director of Postgraduate Master Course "The Technologies of Environmental Legislation".

About the Editors

Preface

Sewage treatment plants (STPs) have been evolved over time to adapt to the growth of cities, the environmental changes (including climate change), the economic conditions and, finally, the requirements of society under the influence of both environment and economy. Initially, the goal of STPs was to simply release the water of the drains from the pollutants before discharging it back to the environment. As a result, the STPs were designed on the principle of the activated sludge process, which is energy consuming and does not take into account the potential of energy and nutrient recovery. The technological achievements in the fields of monitoring and control, the design of stable and efficient processes (both physicochemical and biological), the development of suitable benchmarking and economic tools have begun to change the philosophy of STP from treatment to valorisation facilities. This means that, sewage treatment should be incorporated into a more holistic management scheme, which aims at reducing the pollutants as well as enhancing nutrient, water and energy recycling in order to maintain the environment's integrity in an economic feasible but also efficient way.

In this respect, "Sewage Treatment Plants: Economic Evaluation of Innovative Technologies for Energy Efficiency" focuses on the novel, energy and/or economic efficient technologies or modification of the conventional, energy demanding treatment facilities towards the concept of energy streamlining and their economic impact. The book brings together knowledge from Engineering, Economics, Utility Management and Practice and helps to provide a better understanding of the real economic value with methodologies and practices about innovative energy technologies and policies in STP. It consists of two parts; the first part is dedicated to critical discussion of technologies aiming at enhancing the energy efficiency of STP including economic aspects as well, while the second part includes case studies demonstrating the economic impact of applying the energy efficient technologies at full scale.

The first two chapters are introductory. The first one briefly overviews novel, but well established technologies in a STP as well. The second one explains how the cost benefit analysis methodology can be used to assess the economic feasibility of a technology or change in the operation of a STP. Chapter three focuses on how strategic management, when regarding a STP as a whole, may lead to a better performance at a lower cost (from a total asset life cycle point of view). Chapter four presents the save in energy in the case of aerobic bioprocesses alternatives to the conventional activated sludge process and when advanced technologies of oxygen transfer are applied. More particularly, the nutrient removal technologies in energy efficient integrated systems are discussed in chapter five, while the promising aerobic granulation is the subject of chapter six. The application of anaerobic digestion and recent developments in the field of both sewage and sewage sludge treatment is presented in chapter seven. Focusing on the sewage sludge not only for energy but also for nutrient recovery is the subject of chapter eight. Besides liquid and solid effluents, STP produces gases that affect the atmospheric environment. In chapter 9, an energetic and economic efficiency analysis of common odour abatement technologies in STPs is performed. The advances in monitoring and control boosted the performance and improved the economics of the STP. This is examined in chapter 10, which also addresses the plant wide control. Although the Microbial Fuel Cell technology is still technically far from its full scale application, it deserves attention due to the rapid evolvements in this field (chapter 11).

Chapter 12 is the first case study presented in the second part of the book, based on the experience of two companies managing integrated water service in northeastern Italy and focusing on the energy savings in municipal STPs. Next, the concept of the energy factory for STP is introduced and case studies of implementing this approach in the Netherlands are presented (chapter 13). In the case studies of Austrian STPs, the energy consumption and costs are related to nitrogen removal efficiency and plant size (chapter 14). A methodology for evaluating sludge dewatering devices is presented in chapter 15 and a case study example of the implementation of this methodology is given. Chapter 16 proposes an enhanced nutrient removal process, which is necessary if anaerobic digestion becomes the core technology in STP, so that the nutrient rich anaerobic effluents are adequately treated. The subject of chapter 17 is the Sequencing Batch Reactor (SBR) technology and the potential for energy savings though aeration schemes, as has been demonstrated in pilot scale studies. Next, the cost impact of changing the end use of biogas and transform a STP in Norway to an energy supplier of the public transportation sector is presented. Chapter 19 finalizes the second and last part of the book with a study that shows how the alternative energy sources can be integrated into STP to contribute into cost reduction of the plant.

On the completion of this collected volume, we would like to thank the contributing authors for sharing their experience and perspective of future STPs.

Katerina Stamatelatou
Konstantinos P. Tsagarakis

Part I

Innovative technologies and economics in sewage treatment plants – an overview

Chapter 1

Reducing the energy demands of wastewater treatment through energy recovery

Efthalia Chatzisymeon

Institute for Infrastructure and Environment, School of Engineering, The University of Edinburgh, Edinburgh EH9 3JL, United Kingdom

1.1 INTRODUCTION

Wastewater treatment is a significant aspect of water industry that safeguards public health, natural environment and allow for a high quality of life and economic development. The rapid population growth in highly urbanized and industrialized societies has resulted to the production of large volumes of wastewater, which require energy and cost-intensive treatment to be sanitized and safely discharge into receiving water bodies. In order to meet discharge limits, existing wastewater treatment facilities utilize energy-intensive treatment techniques, although current scientific knowledge can provide the know-how to achieve energy saving and recovery in treatment plants. This chapter gives a brief overview of well-established as well as novel technologies that have the potential to reduce energy demands of existing, typical wastewater treatment facilities, either by energy recovery or saving during treatment, in order to reduce the environmental footprint and attain energy efficient treatment facilities.

1.1.1 Wastewater management

Water and wastewater management are highly important and interdependent tasks that can strongly affect human well-being and quality of life. If left untreated, wastewater can pollute surface and ground water reservoirs, thus posing serious threats onto public health and the environment. Hence, the role of water and wastewater industry is to provide reliable protection and safely discharge wastewater into the aquatic environment. However, rapid and localized population growth has led to large volumes of clean water being consumed daily and respectively large volumes of wastewater being produced, which stresses even more the existing wastewater facilities. On top of this a rapid deterioration of the quality of water reservoirs, mainly due to the increased urbanization, industrialization and farming activities, is observed. This is evident by the excess of organic pollutants and nutrients (N and P) loads in aquatic bodies. All the above indicate that more intensive water and wastewater treatment technologies, which are associated with high energy demands and costs, need to be adopted to safeguard public health and the natural environment.

Although estimations vary, on average the daily municipal water use per capita reaches 400 L in USA (USGS, 2014), while the mean municipal water consumption in Europe is half, about 200 L (EC, 2012), with substantial differentiations among EU countries. In developing countries the municipal water use per capita is substantially lower, reaching an order of magnitude less than the developed ones (UNDP, 2006). Used water is collected in sewage systems and then is led to treatment plants, as to be sanitized and safely discharged to environment and/or recycled for agriculture and other uses. In the UK about 625×10^3 km of sewers are used daily to collect over 11×10^6 m^3 of municipal and industrial wastewater (DEFRA, 2012). These vast quantities should be treated before ending up to receiving water bodies, but that is not always the case. For example, in Abbey Mills Pumping Stations, London around 16×10^6 t of raw wastewater is annually discharged to the River Lee, ending up to river Thames. In USA, in 2008 60.41×10^9 m^3 of municipal wastewater were produced, of which 47.2×10^9 m^3 were collected and finally only 40.89×10^9 m^3 were treated (FAO, 2014).

Wastewater treatment comprises various physical, chemical and biological processes, as well as their combination, in order to produce an effluent that can be safely disposed to environment without causing any short or long term adverse effects to humans or other living beings. Nonetheless, in order to meet wastewater discharge permits, high energy demands are required, leading to high operational costs and making wastewater management unsustainable. Therefore, more efficient and energy friendly treatment systems, that require lower to zero external amounts of energy to operate and hence lower operational costs, should be introduced in large scale.

1.1.2 Energy demands for wastewater treatment

Wastewater treatment has improved significantly over the past 20 years, with approximately 75% of UK surface waters now being in good biological and chemical quality (POST, 2007). However, the energy required to treat wastewater to this standard is high; with energy being used to collect, treat and discharge wastewater and manage sewage sludge. Insufficient data were available to assess accurately the actual energy intensity of each step of the water treatment.

However, there is no doubt that as our demand for clean water increases, so does the total amount of energy needed to safely discharge wastewater into the environment.

For example, over 10^{10} L of sewage are produced every day in England and Wales and it takes approximately 6.34 GWh of energy to treat this volume of sewage, which is almost 1% of the average daily electricity consumption of England and Wales (POST, 2007). Moreover, Shoener *et al.* (2014) reported that current energy-intensive approaches to wastewater treatment, which consume roughly 0.3–0.6 kWh m^{-3} (i.e., 3% of U.S. electricity demand), further contribute to climate change through greenhouse gas emissions from electricity production (Shoener *et al.*, 2014).

Nevertheless, to accurately estimate the actual energy demands of a wastewater treatment facility, treatment stages and utilized technologies should be taken into account. In addition, energy demands are strongly related to the physicochemical characteristics of sewage (i.e., organic load, total solids, etc.) and the desirable use of the final effluent (i.e., aquifer recharge, agriculture use, etc.), since these affect the degree of treatment intensity. Typically, a sewage treatment plant consists of five main stages, as described below (POST, 2007):

- *Pre-treatment*: includes bar screens to remove large objects, a flow equalization tank and a grit removal channel.
- *Primary treatment*: consists of a primary sedimentation tank where solids are physically settled out by gravity.
- *Secondary treatment*: typically is based on an activated sludge system, where bacteria are used to convert organic pollutants to carbon rich sludge.
- *Tertiary treatment*: might include UV irradiation, activated carbon filters or other advanced techniques to further remove non-biodegradable organic matter and/or disinfect the water.
- *Sludge treatment*: usually incineration, or sludge thickening and disposal is applied.

Table 1.1 presents a typical energy demands' breakdown for a common wastewater treatment facility. It is evident that the highest amount of energy, that is, 55.6%, is consumed in the activated sludge aeration process. The primary clarifier and sludge pumps is the second largest energy demanding stage, it consumes 10.3%, followed by heating for digesters (7.1%) and solids dewatering (7%). All the above stand for about 80% of the total energy demands of a common treatment facility. Since, as described above, conventional wastewater treatment processes are energy-intensive and hence not environmentally friendly, future strategies should focus on reducing energy demands and enabling zero to negative energy treatment requirements, as to create economic incentives and enable access to sustainable sanitation in both developed and developing communities (Shoener *et al.*, 2014).

Table 1.1 Typical energy demands for a wastewater treatment facility (Tchobanoglous *et al.*, 2003).

Stage	Energy demand (%)
Inlet pumping and headworks	4.9
Primary clarifier and sludge pumps	10.3
Activated sludge aeration	55.6
Secondary clarifier and RAS	3.7
Thickener and sludge pump	1.6
Effluent filters and process water	4.5
Solids dewatering	7.0
Tertiary treatment	3.1
Heating	7.1
Lighting	2.2
TOTAL	100

1.2 ENERGY RECOVERY

To take a step towards wastewater treatment facilities that have zero to negative net energy demands (i.e., energy produced during treatment is greater than the energy required for their operation), all potential energy saving and energy production steps in a typical treatment facility should be identified. Figure 1.1 illustrates how and where within the train system of wastewater treatment, the greatest potential for energy saving and recovery can be achieved.

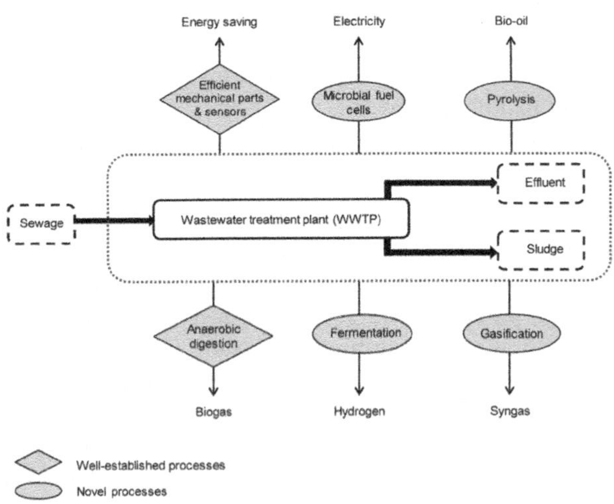

Figure 1.1 Processes that have the potential to save and/or recover energy within a wastewater treatment facility.

As shown in Figure 1.1, well-established processes exist, such as anaerobic digestion, that are already applied at industrial-scale treatment plants, as well as novel technologies, such as microbial fuel cells (MFCs) and gasification, that need further investigation and optimization for their application at larger scale. In specific, energy demands can be reduced either by decreasing energy consumption or by achieving energy recovery, by the means of the technologies described in the following paragraphs.

1.2.1 Use of efficient mechanical parts and sensors

Significant savings can be achieved in the most energy-intensive stages of a typical treatment facility, with the use of efficient mechanical parts. For example in aeration, the most energy demanding stage, it is possible to reduce energy consumption by about 30% (Caldwell, 2009). Therefore, replacement of aged machine parts, such as pumps, motors, and so on, with more efficient ones should be carefully considered, since they can significantly decrease energy consumption. A typical energy saving of 10–20% can be achieved through efficient blowers in the aeration process. Modern blowers are usually based on high speed, oil free turbo systems, which can be further improved if aeration supply and control is well designed. Efficient motors can be used to replace existing ones, hence achieving energy savings of about 5–15%. Furthermore, real-time monitoring with automatic instrumentation (i.e., sensors technology) of the treatment facility can significantly contribute to energy and cost savings, through the precise and rapid estimation of important operating parameters. Up to date, technological advances allow the real-time monitoring of several parameters, such as conductivity, pH, turbidity, dissolved oxygen, and other. This enables the optimization of the process by properly adjusting addition of chemicals, flowrates, retention time and other significant operating conditions, thus avoiding the excess of chemicals use and air sparging.

A successful case study of energy reduction through efficient mechanical parts and sensors technology was presented by an Anglian Water treatment facility. This was achieved by replacing the aeration system, which was approaching its treatment limit due to the increased load of wastewaters entering the plant with a higher efficiency one and by using a plate aerator that gave higher area coverage of the aeration zone and created smaller air bubbles (Caldwell, 2009). It is well-known that small bubbles rise more slowly and offer a larger surface area, which increases oxygen transfer into the wastewater from a typical 5.5% to more than 7.0% per meter of water depth. This reduces air requirements to treat the same load of wastewater by about 27% and allows more oxygen transfer in its volume. Moreover, the old blowers of the sewage plant were replaced with efficient oil free turbo blowers. These use 10–20% less energy to eject the same volume of air in the system. The combination of the use of efficient aerators and blowers with a new real-time monitoring system can significantly improve total energy and cost savings. A real-time control system, comprised of air flowmeters and pressure sensors, allowed air flowrates optimization, thus avoiding excess of air sparging consequently reducing its energy demands. The aeration capital cost for this solution was similar to conventional disc aerators, while energy and air requirements were reduced by 20% and 33%, respectively.

1.2.2 Anaerobic digestion

Anaerobic digestion (AD) is a well-known natural process where biodegradable materials are broken down by the action of microorganisms, in the absence of oxygen, which result to decreased organic loads and simultaneously to the production of biogas. Biogas is a mixture of gases that mainly consist of methane (typically 60–65%) and carbon dioxide. The process takes place in sealed anaerobic digesters under appropriate temperatures of about 30 to 38°C (mesophillic digestion) or about 49–57°C (thermophillic digestion), with the first being a more stable process (Reith *et al.*, 2003). AD can be divided into three main steps, (a) hydrolysis, where microorganisms split the organic matter to simpler forms in the presence of water, (b) volatile acid fermentation, which include acidogenesis – and acetogenesis, with end products being acetic acid, carbon dioxide, and hydrogen, and (c) methane formation, where products from the previous step are converted to methane and carbon dioxide. AD can take place either in a single stage, digestion is performed in a single tank at constant temperature, or in multiple stages, different tanks or different temperatures or both, are used. The latter finds favourable use in wastewater management since it allows AD facilities to optimize both organic removal and biogas production.

Wastewaters, as well as the sludge that is generated in the aeration stage, are rich in organic matter and therefore can be used to produce energy (biogas) and simultaneously reduce their organic load through AD. Depending on its quality and quantity, biogas can be used for heating purposes, electricity production or can be fed into a combined heat and power (CHP) system to provide heating to AD and power the high energy intensive processes within the treatment plant, such as the aeration blowers (Cao & Pawlowski, 2012). In general, anaerobic digesters are able to create enough biogas to maintain their own heating temperature and provide heat and/or electricity to other stages of the plant and to the building facilities on site (Caldwell, 2009).

AD is a well-established solution for energy recovery and organic load reduction, presenting both environmental and economic benefits, while its application is steadily increasing in wastewater treatment facilities throughout the world. For example, the volume of biogas captured and utilized in two Norwegian wastewater treatment facilities rose from 8.1×10^6 m^3 in 2000 to 14.6×10^6 m^3 in 2007 (Venkatesh & Elmi, 2013). Furthermore, in 2005–2006 the UK water industry generated 493 GWh from AD, while currently, with 110 AD facilities installed, it annually generates approximately 800 GWh through AD of sewage sludge treatment (Mills *et al.*, 2014; POST, 2007). To add, most of the regional sludge centres already produce enough energy to be able to provide for all of their internal processes and also are able to export electricity to the grid. Interestingly, the most recent UK strategy plan, after estimating the AD potential, sets out a goal that heat and electricity production could reach 3–5 TWh by 2020 (DEFRA, 2013).

Although AD is a widely applied technology, with significant contribution to treatment sustainability and wastewater and sewage sludge valorisation, there is also an increasing interest to enhance biogas quantity and quality (e.g., high ratio of methane to carbon dioxide) and further optimize the process. Hence, research efforts investigate possible alternatives, such as optimization of process conditions (e.g., sludge retention time and sludge loading rate), application of multi-stage process (e.g., temperature-phased and microorganism community-phased), and sludge pre-treatment to increase biodegradability (Cao & Pawlowski, 2012).

1.2.3 Fermentation

During fermentation specific microorganisms, in the absence of oxygen, follow a certain metabolic pathway and convert monomers (sugars) to acids. Anaerobic wastewater digestion to generate methane as a final product is a well-established technique. Nonetheless, if the growth of methanogenic bacteria is inhibited, thus preventing methane formation, and only hydrogen producing microorganisms are left to flourish, then acetogenesis will be the last step of AD, thus generating hydrogen (H_2), acetic acid and CO_2 (Reith *et al.*, 2003).

H_2 is a high energy density (122 KJ/g) fuel that produces zero CO_2 emissions when burned. Nonetheless the most common H_2 generation processes are steam reforming of natural gas and water electrolysis, which are extremely energy and cost-intensive (Su *et al.*, 2010; Argun & Kargi, 2011). Therefore, increasing research interest has been directed towards more sustainable and energy-efficient techniques for its production. Among them, anaerobic wastewater fermentation has proven to be a promising process that operates under mild conditions and requires low energy demands, since it achieves both waste reduction and clean energy production, namely H_2 (Chen *et al.*, 2008).

H_2 production through wastewater fermentation can be achieved (a) under the presence of light (photo-fermentation), where light provides metabolic energy, (b) under the absence of light (dark-fermentations), where organic compounds provide metabolic energy, or (c) by a combination of both techniques (combined-fermentation) (Su *et al.*, 2010; Argun & Kargi, 2011). The latter has been reported to provide higher H_2 yields and can also achieve higher reduction of the effluents' organic load (Chen *et al.*, 2008).

In photo-fermentation anaerobic photosynthetic bacteria, such as *Rhodobacter* and *Rhodopseudomonas*, catalyze organic acids, such as acetic and butyric acids and more simple ones, as glucose, fructose and sucrose, while in dark fermentation anaerobic bacteria, such as *Clostridium* and *Enterobacter*, can catalyze glucose, sucrose, starch and cellulosic materials to produce H_2 (Su *et al.*, 2010). Restriction factors of applying the process at large scale include low hydrogen yields (i.e., typically less than 15% of the maximum theoretical potential), high cost and the need for carbohydrate-rich wastewaters, thus this technology has yet to be effectively introduced at industrial scale.

1.2.4 Microbial fuel cells

Microbial fuel cells (MFCs) are an emerging sustainable technology that can achieve both the removal of organic pollutants and electricity generation (Ahn *et al.*, 2014). When utilizing MFCs for wastewater treatment, microorganisms use organic matter to produce electricity as well as water, CO_2 and other inorganic residue as by-products (Barua & Deka, 2010; Du *et al.*, 2007). MFCs are bioreactors that operate under anaerobic conditions and consist of two electrodes, an anode and a cathode separated by a positively charged ion membrane. On the anode organics (i.e., wastewater) are oxidized by microorganisms, thus generating CO_2, electrons and protons. Electrons are transferred to the cathode compartment through an external electric circuit, generating electricity, while protons are transferred to the cathode compartment through the membrane. Water is also produced by the combination of electrons and protons with oxygen, on the cathode (Oh & Logan, 2005; Rabaey & Verstraete, 2005).

The main benefits of MFCs are (a) their low-cost, since they use inexpensive catalysts; namely microorganisms present in wastewaters, (b) their high energy efficiency, theoretically energy can be recovered by far beyond 50%, and (c) their ability to operate under mild reaction conditions (Barua & Deka, 2010; Scott & Murano, 2007). Moreover, when MFCs are used significant lower amounts of solid needs to be disposed of, since they can achieve solids removal in the range of 50–90% and therefore potentially can reduce the energy required for the aeration treatment of wastewater by up to 50%.

Two different types of MFCs exist, the ones that require a mediator and the mediator less, with the latter showing significant potential for wastewater treatment applications (Oh & Logan, 2005). Also, various designs exist, with the single-chamber, air-cathode MFCs being promising for practical applications (Ahn *et al.*, 2014). For wastewater applications close electrode spacing is favorable, whilst very close electrode spacing can be achieved by placing a separator between the electrodes, as to avoid short-circuiting. The separator configuration can produce a 16% higher maximum power density but the separator-less closely spacing configuration requires significant less time for wastewater treatment and hence is better in terms of treatment efficiency (Ahn *et al.*, 2014).

1.2.5 Energy recovery from sewage sludge

Sewage sludge constitutes one of the most significant challenges in wastewater management, since large volumes are produced that apart from the high organic content, may contain hazardous substances, such as heavy metals and persistent micro-pollutants. Sewage sludge contains from 0.25–12% solids by weight, depending on the wastewater treatment technique that was adopted (Tchobanoglous *et al.*, 2003). Therefore, sludge management tradition handling routes, such as agricultural use, can be unsafe, while sludge incineration is associated with high energy demands and costs and landfill disposal faces various legislation restrictions, for example, Directive 2000/76/EEC and 2003/33/EEC (Manara & Zabaniotou, 2012).

Alternative management processes include the thermochemical treatment of sludge in the absence of oxygen or in oxygen-starved environments, as to prevent combustion. Under carefully controlled conditions and extreme temperatures (350–1000°C), sludge may undergo chemical reactions to produce fuels that can be used for heat and/or energy production and simultaneously achieve organic load removal. Processes include gasification, which produces syngas, and pyrolysis, which produces bio-oil. These are potential alternatives to sludge incineration, but similarly operational costs are still high, especially when using high temperatures. Also, special consideration should be given to the monitoring of operating conditions to avoid any formation of harmful by-products, such as hydrogen cyanide (Samolada & Zabaniotou, 2014).

1.2.5.1 Pyrolysis

During pyrolysis sewage sludge is thermally decomposed in an oxygen-free environment to gases (biogas), liquids (bio-oil) and solids (biochar). The major product obtained from this process is the bio-oil, which can be used as a fuel, the same stands for biogas, as well as a source of valuable chemical products. Biochar, a carbon-rich solid, can be used in various applications ranging from agriculture to adsorbent material for contaminants in soils, depending on its quality (Agrafioti *et al.*, 2013).

Pyrolysis, a rather endothermic process (100 kJ kg^{-1}), operates at temperatures ranging from 350°C to 1000°C. Pyrolysis by-product formation is affected by the process operating conditions, such as temperature and pressure as well as the initial sludge characteristics. Therefore, when bio-oil is the target, fast pyrolysis is employed, during which high heating rates, moderate temperatures (500°C) and short gas residence times (<2 s) are applied, whilst when biochar is the desired product, slow pyrolysis, characterized by mild temperatures (350–600°C) and heating rates, is applied (Leszczynski, 2006).

Pyrolysis is a 'greener' technology when compared to incineration, since the lower operating temperatures applied and the absence of oxygen result to toxic-free by-products. In contrast, oxygen utilization and high temperatures applied during incineration process can result to the formation of toxic substances, such as furans and dioxins. Although research has been focused on pyrolysis of sewage sludge for bio-oil production, thus recovering energy within the wastewater treatment facility, large scale applications of the technology is limited. This is due to the need for relatively complex and expensive equipment and the need for using drying feedstock (Samolada & Zabaniotou, 2014).

1.2.5.2 Gasification

The gasification process uses heat, pressure and steam to convert carbonaceous materials, in the presence of oxygen and/or steam, into a synthesis gas called syngas, which is a mixture of CO, H_2 as well as N_2 and traces of CO_2, CH_4 and other hydrocarbons and slag. Gasification mainly transforms organic materials to combustible gas or syngas, using between 20% and 40% of the oxygen required for total combustion, whereas pyrolysis is a thermochemical reaction carried out at elevated temperatures (500–1000°C) and theoretically in an oxygen-free environment.

Gasification has the advantage of reducing the volume of sewage sludge and toxic organic compounds; while simultaneously it generates syngas that can be used for heat (e.g., syngas from sewage sludge has a heating capacity of about 4 MJ m^{-3}) or electricity (i.e., in fuel cells) production (Dogru *et al.*, 2002; Judex *et al.*, 2012). In addition, problems commonly faced in incineration process, like the need for supplementary fuel and emissions of toxic by-products, such as SOx and NOx, heavy metals and fly ash, can be avoided by the gasification process.

Limitations of the technology include feedstock characteristics, such as moisture (>90% dry solids) content, and the complexity of the reactors design, such as design of the feeding system, mixing and separation of the feedstock. Also, the generated syngas must be cleaned and purified before its further use and the high cost of the initial set-up still prevents the wide application of this technology at large-scale.

1.3 CONCLUDING REMARKS

Reducing energy demands and increasing energy recovery in wastewater treatment facilities can be a feasible venture by means of current technological advances. If existing treatment facilities are upgraded as to achieve lower energy demands and simultaneously take advantage of energy harvesting techniques from wastewater and sewage sludge, then positive net energy facilities could exist, that will further benefit local or national communities by providing the excess heat and energy.

In this chapter the energy demands of a typical sewage treatment plant as well as options to reduce them were demonstrated. Furthermore, techniques that can achieve substantial energy recovery, within the various treatment stages, were presented. It is clear that scientific knowledge and the know-how to create energy, and thus save cost exist and can lead to the establishment of highly sustainable sewage treatment plants. These technologies are described and discussed in detail in this book, while emphasis is given to economic aspects of wastewater treatment facilities. Finally, successful case studies of energy recovery during wastewater treatment are demonstrated in Part II of this book.

1.4 REFERENCES

Agrafioti E., Bouras G., Kalderis D. and Diamadopoulos E. (2013). Biochar production by sewage sludge pyrolysis. *Journal of Analytical and Applied Pyrolysis*, **101**, 72–78.

Ahn Y., Hatzell M. C., Zhang F. and Logan B. E. (2014). Different electrode configurations to optimize performance of multi-electrode microbial fuel cells for generating power or treating domestic wastewater. *Journal of Power Sources*, **249**, 440–445.

Argun H. and Kargi F. (2011). Bio-hydrogen production by different operational modes of dark and photo-fermentation: an overview. *International Journal of Hydrogen Energy*, **36**, 7443–7459.

Barua P. K. and Deka D. (2010). Electricity generation from biowaste based microbial fuel cells. *International Journal of Energy, Information and Communications*, **1**(1), 77–92.

Caldwell P. (2009). Energy efficient sewage treatment can energy positive sewage treatment works become the standard design? In: *Proceedings of the 3rd European Water and Wastewater Management Conference*, 22–23 September.

Cao Y. and Pawłowski A. (2012). Sewage sludge-to-energy approaches based on anaerobic digestion and pyrolysis: brief overview and energy efficiency assessment. *Renewable and Sustainable Energy Reviews*, **16**, 1657–1665.

Chen C.-Y., Yang M.-H., Yeh K.-L., Liu C.-H. and Chang J.-S. (2008). Biohydrogen production using sequential two-stage dark and photo fermentation processes. *International Journal of Hydrogen Energy*, **33**, 4755–4762.

DEFRA (2012). Waste water treatment in the United Kingdom – 2012. Department for Environment, Food and Rural Affairs, Nobel House 17, Smith Square London SW1P 3JR.

DEFRA (2013). Anaerobic Digestion Strategy and Action Plan. Department for Environment, Food and Rural Affairs, Nobel House 17, Smith Square London SW1P 3JR.

Dogru M., Midilli A. and Howarth C. R. (2002). Gasification of sewage sludge using a throated downdraft gasifier and uncertainty analysis. *Fuel Processing Technology*, **75**, 55–82.

Du Z., Li H. and Gu T. (2007). A state of the art review on microbial fuel cells: a promising technology for wastewater treatment and bioenergy. *Biotechnology Advances*, **25**, 464–482.

EC (2012). Would you drink your wastewater? European Commission, Directorate-General for the Environment, Luxembourg: Publications Office of the European Union, ISBN: 978-92-79-22529-1.

FAO (2014). AQUASTAT database, food and agriculture organization of the United Nations (FAO). Website accessed 12 September 2014.

Judex J. W., Gaiffi M. and Burgbacher H. C. (2012). Gasification of dried sewage sludge: status of the demonstration and the pilot plant. *Waste Management*, **32**, 719–723.

Leszczynski S. (2006). Pyrolysis of sewage sludge and municipal organic waste. *Acta Metallurgica Slovaca*, **12**, 257–261.

Manara P. and Zabaniotou A. (2012). Towards sewage sludge based biofuels via thermochemical conversion – a review. *Renewable and Sustainable Energy Reviews*, **16**, 2566– 2582.

Mills N., Pearce P., Farrow J., Thorpe R. B. and Kirkby N. F. (2014). Environmental & economic life cycle assessment of current & future sewage sludge to energy technologies. *Waste Management*, **34**, 185–195.

Oh S. E. and Logan B. E. (2005). Hydrogen and electricity production from a food processing wastewater using fermentation and microbial fuel cell technologies. *Water Research*, **39**, 4673–4682.

POST (2007). Energy and Sewage, Postnote number 282. Parliamentary Office of Science and Technology, 7 Millbank, London, SW1P 3JA, UK.

Rabaey K. and Verstraete W. (2005). Microbial fuel cells: novel biotechnology for energy generation. *Trends Biotechnology*, **23**(6), 291–298.

Reith J. H., Wijffels R. H. and Barten H. (2003). Bio-methane & Bio-hydrogen: status and perspectives of biological methane and hydrogen production. Dutch Biological Hydrogen Foundation, The Netherlands.

Samolada M. C. and Zabaniotou A. A. (2014). Comparative assessment of municipal sewage sludge incineration, gasification and pyrolysis for a sustainable sludge-to-energy management in Greece. *Waste Management*, **34**, 411–420.

Scott K. and Murano C. (2007). Microbial fuel cells utilising carbohydrates. *J. Chem. Technol. Biotechnol.*, **82**, 92–100.

Shoener B. D., Bradley I. M., Cusick R. D. and Guest J. S. (2014). Energy positive domestic wastewater treatment: the roles of anaerobic and phototrophic technologies. *Environmental Science: Processes & Impacts*, **16**, 1204–1222.

Su H., Cheng J., Zhou J., Song W. and Cen K. (2010). Hydrogen production from water hyacinth through dark- and photo-fermentation. *International Journal of Hydrogen Energy*, **35**, 8929–8937.

Tchobanoglous G., Burton F. L., Stensel H. D. and Metcalf & Eddy (2003). Wastewater Engineering Treatment and Re-use, 4th edn, McGraw Hill, New York.

UNDP (2006). Human development report 2006. United Nations Development Programme, 1 UN Plaza, New York, 10017, USA, ISBN: 0-230-50058-7.

USGS (2014). Water Questions & Answers. U.S. Geological Survey, available online at: http://water.usgs.gov/edu/qa-home-percapita.html (accessed 12 September 2014).

Venkatesh G. and Elmi R. A. (2013). Economic-environmental analysis of handling biogas from sewage sludge digesters in WWTPs (wastewater treatment plants) for energy recovery: case study of bekkelaget WWTP in Oslo (Norway). *Energy*, **58**, 220–235.

Chapter 2

The principles of economic evaluation and cost-benefit analysis implemented in sewage treatment plants

María Molinos-Senante[1], Nick Hanley[2], Francesc Hernández-Sancho[3] and Ramón Sala-Garrido[1]

[1]Department of Mathematics for Economics, University of Valencia, Avda. Tarongers S/N, 46022 Valencia, Spain (maria.molinos@uv.es; ramon.sala@uv.es)

[2]Department of Geography and Sustainable Development, University of St Andrews, Scotland (ndh3@st-andrews.ac.uk)

[3]Department of Applied Economics, University of Valencia, Avda. Tarongers S/N, 46022 Valencia, Spain (francesc.hernandez@uv.es)

2.1 INTRODUCTION

Traditional wastewater treatment technologies, most of them based on activated sludge, have been widely implemented in the last decades over the world (Gavasci *et al.* 2010). However, growing public concern over environmental protection and increasing energy costs have led to the development of innovative technologies for energy saving. Improving energy efficiency is a challenge that should be taken into account in the construction of new wastewater treatment plants (WWTPs), in the renovation of the plants and in the operation of all facilities.

The development and implementation of innovative technologies for energy efficiency involves costs and benefits that should be assessed. Economic feasibility studies are an essential tool in the decision making process for the implementation of new technologies alternatives in the field of wastewater treatment (Molinos-Senante *et al.* 2012).

One of the most popular tools to assess the economic feasibility of any project is cost-benefit analysis (CBA) since it ensures the economic rationality of investments testing whether the benefits of action outweigh the costs. The approach followed in the performance of CBA in the evaluation of projects has been modified taken into account the objectives of the development policies. There are three stages (Molinos-Senante *et al.* 2010):

(1) *Traditional approach*: it is a financial analysis based on the comparison of incomes and costs generated during the life of the project, that is, what are known as internal or private impacts. It follows a clear economic approach aimed to increase the level of welfare in monetary terms, typically defined as profits.
(2) *Socio-economic approach*: this arises when the concept of social equity is incorporated. The aim is to achieve equitable income distribution, or at least to include some kind of income-related weights into the calculation of benefits and costs to different groups.
(3) *CBA involving environmental externalities valuation*: It results from the incorporation of environmental criteria in the decision-making process. This type of CBA originated in the 1980s and become more widespread in the 1990s (Pearce & Nash, 1981; Sudgen & Williams, 1988; Hanley & Spash, 1993; among others).

Wastewater treatment in general and innovative wastewater treatment technologies for energy efficiency in particular have important associated environmental benefits which are defined in economic terms as positive externalities. Hence, the assessment of the economic feasibility of wastewater treatment processes must be carried out through CBA instead of financial analysis. Otherwise, the environmental benefits of cutting pollution and to reduce energy consumption and consequently greenhouse gas (GHG) emissions would be underestimated since they are not accounted by the market (unless governments or the market offers payments for reductions in carbon emissions, for example through carbon trading).

Other reasons for selecting CBA as the preferred method to assess the economic feasibility are that: (i) it allows planners and decision-makers to take a long-term view of the project lifetime; (ii) it provides a project ranking, which, for all practical purposes, proves to be quite scientific and satisfactory (Molinos-Senante *et al.* 2013a) and; (iii) it clearly sets the impacts of a project in terms of who is affected, by how much, and when (Hanley & Barbier, 2009).

2.2 COST BENEFIT ANALYSIS METHODOLOGY

2.2.1 Cost benefit analysis basis

The objective of a CBA is to compare the economic feasibility (net social benefit) of several scenarios, including the 'do nothing' scenario, that is, maintain current conditions. CBA proposes various decision rules in the decision making process: (i) an intervention is only feasible if benefits are greater than costs; (ii) if alternative options are available, the best option is the one with the highest net present value; and (iii) time can be incorporated in the assessment through the use of discount rates.

Since CBA starts from the premise that a project should only be commissioned if all benefits exceed the aggregate costs, the benefits of each proposal are compared with their costs by using a common analytical methodology (Eq. (2.1)).

$$NP = B_1 + B_E \tag{2.1}$$

where *NP* is the net profit (total income – total costs); B_1 is the total internal benefit (internal income – internal costs); and B_E is total the external benefit (positive externalities – negative externalities). A project is economically feasible if, and only if, $NP > 0$. If the result of the calculation is $NP < 0$, then the project is not economically feasible. The best option offers the highest net profit (Benedetti *et al.* 2006; Chen & Wang, 2009). Moreover, total income can be divided by total costs to get a ratio which can be used to rank policies/project that are competing for scarce funds, with the option of having the highest benefit to cost ratio being the most preferred (De Anguita *et al.* 2011).

The implementation of an innovative wastewater treatment technology is a project whose life period is more than one year, and as a result, the internal and external benefits must be adjusted for the time they will occur. For this reason the *NP* must be expressed in present value terms. By means of a properly chosen discount rate the investor becomes indifferent regarding cash amounts receiving at different points of time. The net present value is calculated as shown in Eq. (2.2):

$$NPV = \sum_{t=1}^{T} \frac{NP_t}{(1+r)^t} \tag{2.2}$$

NPV is the net present value, NP_t is the net profit at time *t*; *r* is the discount rate and *T* is the project lifespan.

NPV results will determine the project's feasibility. As well as NP, a positive NPV means that the investment will be profitable and the project can be accepted. If NPV is negative, the investment is not economically feasible. Therefore, the decision rule is to select the option that will induce NPV optimisation. This NPV rule can be linked to overall social welfare by the Kaldor-Hicks principle, namely that a positive NPV implies that the gainers could compensate the losers and still be better off (Hanley & Barbier, 2009).

It should be highlighted that the selection of the lifespan of the technologies is always a controversial choice since it is well known that it depends on many factors including the maintenance and management of the facilities.

Regarding the discount rate, higher discount rates favours solutions that are weighted toward future spending, that is, those with relatively high operating costs and lower investment cost (Woods *et al.* 2013). There is much debate over which discount rate governments should use in public sector policy and project appraisal.

The opportunity cost of investments should be reflected in the discount rate because, when a particular project is invested in, it is assumed that this capital may not be invested elsewhere, in other words, there is an opportunity cost. In water reuse-projects, the opportunity cost generally refers to land on which the WWTP is placed (Seguí, 2007). For projects that last for a long time, it is recommended to use lower values of discount rate than in projects with a shorter lifespan (Termes-Rifé *et al.* 2013). When dealing with a long-term project, a declining value for the discount rate at different time periods may be used. When a decision is about environmental aspects, a very low value should be used, even near to zero, since environmental damages may have an impact that is likely to last for many years in the future. Pearce *et al.* (2003) stressed the need for modification of the traditional assumption in discounting rates using for example a declining discount rate which replaces the exponential discount factor with a hyperbolic function. The UK government uses a declining discount rate formula for long-term impacts. In particular, Her Majesty's Tresure (2003) recommends a discount rate of 3.5% for 1 to 30 years, a 3% rate for 31 to 75 years, a 2.5% rate for 76 to 125 years, a 2% rate for 125 to 200 years, 1.5% for 201 to 300 years, and 1% for longer periods. Similarly, in 2004, France replaced its constant discount rate of 8% with a 4% discount rate that decreases to 2% for longer maturities (MacLeod & Filion, 2012). Another approach suggested by Almansa and Martínez-Paz (2011) is the use of a dual-rate discount rate which involves the use of different discount factors for tangible and intangible goods.

A weakness of the CBA approach is that the final decision depends on the alternatives proposed. Thus, other alternatives not evaluated in the CBA may be considerably better.

Any investment project that is analysed through the CBA tool should follow a series of steps (see Figure 2.1).

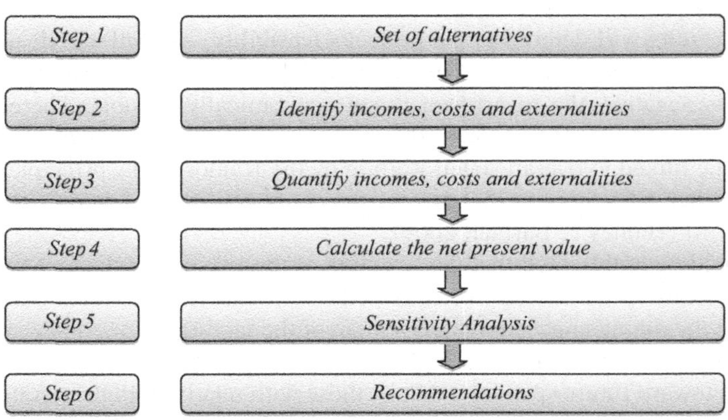

Figure 2.1 Steps of a cost benefit analysis.

(1) *Specify the set of alternatives to the project.* The CBA compares the NPV of investing resources in different projects or alternatives. In the majority of the situations, the alternative to be compared with the one proposed is the *status quo* situation, that is, the situation in which the project is not carried out.
(2) *Identify the incomes, costs and positive and negative externalities.* Once the alternatives to the project have been specified, the next step is to establish the internal and negative impacts of each one. Wastewater treatment creates a number of externalities, including negatives such as GHG emissions and biological and chemical risks if the treated water is reused and positive externalities as health benefits, education services, and especially environmental benefits.
(3) *Quantify the incomes, costs and positive and negative externalities.* Almost certainly, this is the most complicated phase of a CBA. On the one hand, internal impacts are those that have a price determined by the market and therefore, can be quantified directly. On the other hand, the quantification of the externalities is much more complex since they have not a price determined by the market. However, it does not mean that they do not have value since they contribute to improve people welfare. To quantify externalities, specific economic valuation methods are needed. Hence, it is possible to standardize all the units involved in the CBA.
(4) *Calculate the net present value.* As it is shown in Eq. (2.1), the NPV is defined as the addition of internal benefits and external benefits. Hence all the parameters involved to calculate the NPV must be expressed in the same units (monetary units).
(5) *Carry out a sensitivity analysis.* So far none of the steps described to apply a CBA has been taken into account the existence of uncertainty. Accounting for uncertainty is important in the development of any CBA since uncertainty could influence the ranking of selecting projects (Flores-Alsina *et al.* 2012). Xu and Tung (2008) reviewed a large number of methodologies applied to deal with uncertainty in the water and wastewater management including Monte Carlo simulations (Prat *et al.* 2012), fuzzy logic models (Kafetzis *et al.* 2010), Bayesian network models (Barton *et al.* 2005), statistical tolerances (Bonilla *et al.* 2004), among others.
(6) *Make a recommendation based on the NPV and sensitivity analysis.* As it has been pointed out, the alternative that generates the highest NPV will be chosen, assuming that some other alternatives have a positive NPV. Moreover, the sensitivity analysis could show that the project with the highest NPV is not the best option when uncertainty is considered.

2.2.2 Internal benefit

Internal benefit is the difference between internal costs and internal incomes (that is, private benefits minus private costs). The internal impacts are those directly linked with wastewater treatment or with the implementation of a technology to save energy in a WWTP. This can be calculated directly since both costs and incomes have market value.

2.2.2.1 Internal cost

In a wastewater treatment project, internal costs are composed by investment costs (IC) and operation and maintenance costs (OMC) of the facility. If we focus on cost assessment, both IC and OMC should be adjusted for the time they will occur. The cost estimation on an annual basis, that is, the total annualised equivalent cost (TAEC) can be calculated (Eq. (2.3)):

$$TAEC = \frac{r(1+r)^T}{(1+r)^T - 1} I + OMC \tag{2.3}$$

where *TAEC* is the total annualised equivalent cost in €/year; *IC* are the investments costs in €; *OMC* are the operation and maintenance costs in €/year; *r* is the discount rate; and *T* is the useful life-span of the project.

In the planning of a new investment, cost functions are a useful tool to quantify IC and OMC as they show the relationship between the dependent variable (cost) and independent variables (a set of representative variables of the process). Costs functions are also useful for comparing different treatment technologies from an economic point of view (Hernández-Sancho *et al.* 2011). Therefore, cost functions are widely used to predict IC and OMC of wastewater treatment projects (Panagiotakopoulos, 2004; Tsagarakis *et al.* 2003; Nogueira *et al.* 2007).

In the framework of 'water and wastewater economics', there are three main methodologies to develop costs functions (Molinos-Senante *et al.* 2013b):

(1) The facility is viewed as a system consisting of components or subsystems, each of which is simulated in detail (Panagiotakopoulos, 2004). Following an engineering approach, the design parameters are allowed to assume values within a wide but realistic range, thus simulating many alternative facility forms, each with its own estimation.

(2) In the so-called 'factor method', major cost drivers related to specific major cost parameters are known and they are directly estimated (Le Bozec, 2004). Though the use of conversion coefficients for the cost drivers, estimates from one region or country can be transferred to another.

(3) Statistical and mathematical methods are often used when cost figures (actual or estimates) are available. These figures might relate to set-up cost and/or operating cost to the main variables of the facilities.

Previous studies (Sipala *et al.* 2005; Gonzalez-Serrano *et al.* 2006) illustrated that the statistical method is the most common approach for developing cost functions. Steps from the collection of the raw data to the generation of the costs functions are shown in Figure 2.2.

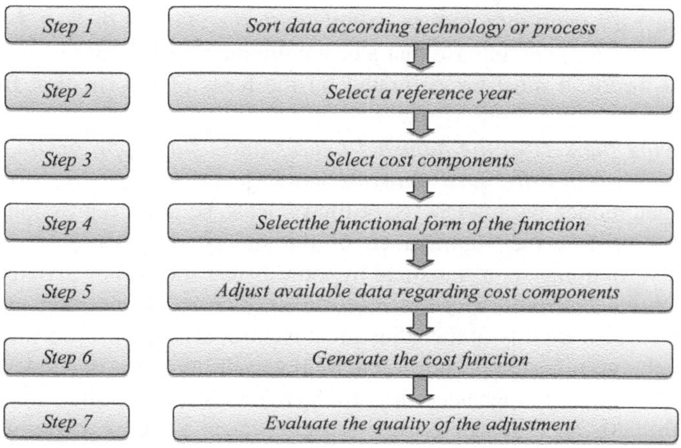

Figure 2.2 Steps for cost function modelling.

(1) *Sort through the data basis of technology.* Sorting means distinguish between the various options for saving energy or achieving other objectives previously defined.
(2) *Choose a reference year for economic valuation.* Due to the difficulty to obtain economic data in the framework of wastewater treatment, sometimes the reference year of all available information is not homogeneous. In this case, it is necessary to choose a reference year, which generally is the year of analysis. The costs for other years must be updated.
(3) *Decide on the cost components that will be included in the cost functions.* Usually, the treatment capacity of the plant is considered the most important factor to determine IC and OMC. In this sense, it is very important choose the size measure of the facilities. In WWTPs, two functional units can be used, namely population equivalent and volume of wastewater treated.
(4) *Choose the functional form of the cost function.* The formulation of IC and OMC functions is based on the assessment of the relationship between the dependent variable C (cost) and the independent variables X (volume treated or population equivalent). For this purpose, different models can be used, such as:

Inverse: $C = a + \frac{b}{X}$

Power: $C = aX^b$

Logarithmic: $C = a + b\,lnX$

Quardratic: $C = a + bX + CX^2$

where a, b, c are the parameters of the model to estimate.

(5) *Adjust all available data to comply with the choices in Step 3 regarding cost components.* In case that a cost component is missing from the report cost figure, it must be estimated on the basis of information from other sources.
(6) *Having the sets of the adjusted figures and using appropriate statistical methods, 'best-fit' cost functions are generated.* A common method to get model parameters is ordinary least squares regression analysis. Subsequently, the significance of the independent variables should be tested. In doing so, a statistical hypothesis test should be carried out.
(7) *Evaluate the quality of the adjustment.* The most common indicator to evaluate the quality of the adjustment is the coefficient of determination (R) which measures the proportion of total variability of the dependent variable relative to its average according to the regression model. Its value is ranged within [0, 1]. If the determination coefficient value is 1, the adjustment between actual and estimated data is perfect. A value of 0 indicates that there is no relationship between the variables.

2.2.2.2 Internal income

In a general study of the economic feasibility of a WWTP, internal income includes revenues obtaining from the sale of the by-products that can be recovered during the wastewater treatment process. In areas under water stress, the sale of the recycled water may play a vital role to ensure the economic feasibility of some water reuse projects. It should be taken into account that if the reclaimed water is used in agriculture, the nitrogen and phosphorus content in the water entails a saving in the fertiliser costs (Nogueira *et al.* 2013). Other incomes may be obtained from the sale of nutrients (mainly phosphorus) recovered during wastewater treatment and from the sale of stabilised sewage sludge to be used after composting.

Focusing on the implementation of technologies for improving energy efficiency, additional incomes must be quantified and incorporated in the economic feasibility study. Some technologies involve a reduction in the consumption of energy; therefore, there is an economic saving that should be taken into account. Other processes allow recovering energy from wastewater or from sewage sludge that can be used in the WWTP itself or sold, which supposes an additional income that cannot be overlooked in the economic feasibility study.

Taking into account internal cost and internal income, the internal benefit for one year is expressed as follows (Eq. 2.4). To estimate the NPV for the life-span of the project, the internal benefit must be updated using Eq. (2.2).

$$B_I = \sum_i AUM_i \cdot SPM_i + \sum_j AE_j \cdot SPE_j - (IC + OMC) \tag{2.4}$$

where B_I is the internal benefit (€/year); AUM_i is the annual unit of the material i recovered such as reclaimed water (m^3), phosphorus (kg), nitrogen (kg), composted sludge (kg), and so on; SPM_i is the selling price of recovered material, i (€/m^3 or €/kg); AE_i is the annual energy recovered through anaerobic digestion, sludge incineration, and so on, in the form, j such as heat, electricity, and so on (kWh); SPE_j is the selling price of the recovered energy, j, (€/kWh); IC are the investment costs (€/year); and OMC are the operational and maintenance costs (€/year).

2.2.3 External benefit

An externality is an effect of a purchase or use decision by one party (or group of parties) on another party who did not have a choice and whose interests were not taken into account (Hussen, 2004). In other words, an externality is generated when an economic operation between agents A and B, produces effects on a third agent C, without any monetary transaction between A and C, or between B and C. However, the absence of market does not imply the absence of value.

While any internal impact can be calculated directly in monetary units, the quantification of external impacts requires the use of economic valuation methods due to the absence of market prices. This requirement is a major difference in applying CBA rather than of financial analysis (Molinos-Senante *et al.* 2013a).

Following the same approach as for the internal benefit, the external benefit for one year is expressed as follows (Eq. (2.5)). As well as internal benefit, it should be updated for the life-span of the project.

$$B_E = P_E - N_E \tag{2.5}$$

Where B_E is the external benefit (€/year); P_E are the positive externalities such are health and environmental benefits (€/year); N_E are the negative externalities such as GHG emissions (€/year). External benefits should include the value of avoided damage costs due to the operation of the plant for example, the value of avoided damages to recreation.

2.2.3.1 External cost

The benefits of wastewater treatment are obvious, however treatment processes also result in environmental impacts (Friedrich *et al.* 2009), such as eutrophication, and contributions to climate change (Lassaux *et al.* 2007).

Due to social and political concerns about climate change, there is growing interest in minimising the consumption of energy in WWTP. Energy consumption is twofold from the perspective of assessing the economic feasibility of the wastewater treatment process. On the one hand, as it has been pointed previously, it is an internal cost. On the other hand, energy consumption is a negative externality, which should not be overlooked since WWTPs consume a significant amount of electricity which involves the indirect emission of GHG.

Although IPCC Guidelines (2007) state that CO_2 emissions have an impact factor of 0 kg CO_{2eq} when CO_2 has biogenic origins (Doorn *et al.* 2006) nowadays, there is an increasing interesting in estimating not just indirect GHG emissions from energy consumption, but also direct GHG emissions. This is because it has been verified that IPCC guidelines underestimate the values of GHG emissions regardless of its origin (biogenic or not) (Foley *et al.* 2010).

Subsequently, a methodology is described that estimates the economic value of the GHG emissions, that is, to estimate the value of the negative externalities associated to wastewater treatment (Molinos-Senante *et al.* 2013a).

Indirect GHG emissions should be estimated based on WWTP energy demands. At first, taken into account the national electrical production mix (national scheme of electrical production), each GHG emission can be estimated. Subsequently, both direct and indirect emissions should be converted to equivalent CO_2 emissions using 100-year global warming potential coefficients (IPCC, 2007).

Once total GHG emissions have been quantified in physical terms, the next step is to express them in monetary units. For this purpose, it should be noted that in the context of the Kyoto Protocol, a well-organised emissions trading has been developed. For example, in Europe the European Union's Emissions Trading System (EU ETS) was implemented in 2005, which integrates more than 11,000 power stations and industrial plants accounting for the 40% of total GHG emissions in the European Union. The price of CO_2 emissions depends on supply and demand, as well as other macroeconomic factors (Molinos-Senante *et al.* 2013a).

The average price paid through the EU ETS (or other CO_2 market) during a time period may be used as a proxy to the price of CO_{2eq} emissions. As a reference and based on SENDECO database, the average market price of CO_2 from 2009 to 2012 was 11.9 €/t (SENDECO, 2013). However, there is some concern that the European carbon market currently set lower prices for CO_2 emissions.

2.2.3.2 External benefits

In the context of wastewater treatment, the US Environmental Protection Agency (EPA) identified that wastewater regeneration and reuse provides the following environmental benefits (EPA, 1998): (i) decreased diversion of freshwater from sensitive ecosystems; (ii) decreased discharge to sensitive water bodies; (iii) recycled water may be used to create or enhance wetlands and river banks; and (iv) recycled water can reduce and prevent pollution.

Different methodologies for the quantification and internalisation of environmental externalities arising from investment projects have been developed from economic theory. Conventional valuation methods can be classified as follows (Molinos-Senante *et al.* 2012):

- *Methods not based on demand curves* such as the replacement cost method, opportunity cost method, dose-response method, among others. They use production or cost functions and provide a 'value to cost' type approach.

From a methodology point of view, these methods are not complex but they require considerable experimental information.

- *Methods based on demand curves.* They belong to the 'value to value' approach and they are used to determine the total economic value of goods and services that have no market (Hanley & Barbier, 2009). They are classified as:
  - Indirect methods such as travel cost method and hedonic price method. They rely on the use of data from actual transactions by individuals. The value of the environmental good is deducted from the complementary relationship between it and other goods with market price (Pearce & Turner, 1990).
  - Direct methods. They are known as stated preference methods since they are based on the demand approach (Hanley *et al.* 2006). This approach responds to the neoclassical view that economic value arises from the interaction between an individual and an environmental asset as an expression of individual preference, assuming that these preferences are a reflection of the maximum utility. The primary categories of stated preference methods are the contingent valuation method and choice modelling techniques (see Figure 2.3). The contingent valuation method is based on the creation of a hypothetical market through a surveying process where individuals declare their willingness to pay (WTP) (or be compensated) for an improvement (or degradation) of the quality of the environmental good being analysed (Genius *et al.* 2005). There are several ways to ask WTP questions in contingent valuations surveys, which are known as elicitation methods. As it is shown in Figure 2.3 there are four types of elicitation methods. In the open-ended format, repondents are asked to state their maximum WTP for the amenity to be valued while in dichotomous choice, respondents are asked if they are WTP single randomly assigned amount on all-or nothing basis. The iterative bidding is a series of dichotomous choices questions starting with an initial low bid that nearly all respondents who have a WTP > 0, would be willinging to pay. Finally, in the payment card format, respondents might announce their WTP to the values listed on the card. Alternatively, the choice modelling techniques are based on ranking or rating a series of "product profits" that characterise products with specific attribute levels (Pearce & Özdemiroglu, 2002). The idea of the contingent ranking method is to give a set of alternatives which consists of a given amount or a given level of a specific good and a corresponding realistic price. The alternatives specified in advance are then ranked (ranking contingent), scored (rating contingent) or selected (discrete choice experiments) (Slothuus *et al.* 2002). In most of the applications related to water resources, the quantification of these externalities has been made using the stated preference methods (Guimaraes *et al.* 2011).

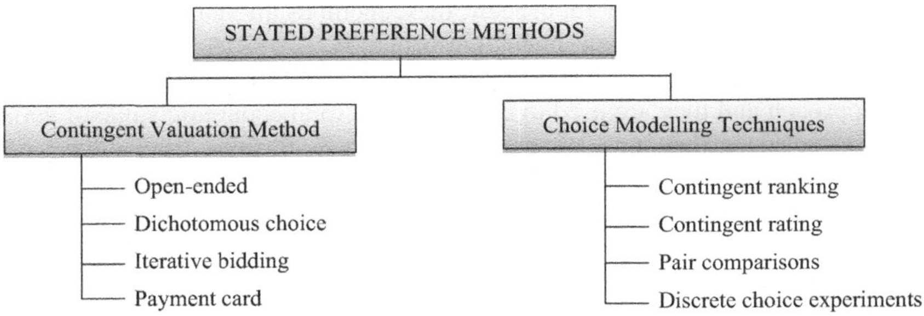

Figure 2.3 Scheme of the stated preference methods.

Alternatively to methodologies based on the demand approach and from the pioneering work by Färe *et al.* (1993) a stream of research has been produced within the framework of efficiency studies that aims to provide a valuation methodology for those undesirable outputs that have no market. Based on the cost production approach and using the concept of distance function or directional distance function, a shadow price is calculated for undesirable outputs associated to production processes. Wastewater treatment can be considered as a production process in which a desirable output (treated water) is obtained together with a series of pollutants (organic matter, suspended solids, nutrients). Contaminants extracted from wastewater are considered undesirable outputs because if they were dumped in an uncontrolled manner they would cause a negative impact on the environment (Molinos-Senante *et al.* 2010).

The shadow prices of undesirable outputs can be interpreted as an estimation of the environmental benefits gained from wastewater treatment, that is, they are a proxy to the value of the positive externalities associated with avoiding the discharge of pollution into water bodies.

The distance function provides the distance of a vector of outputs from the maximum output frontier and starts from a vector of constant inputs. Assuming that the production process uses a vector of N inputs $x \in R_+^N$ to produce a vector of M outputs $u \in R_+^M$, the distance function is defined as in (Eq. (2.6)):

$$D_0(x,u) = \text{Min}\left\{\theta : \left(\frac{u}{\theta}\right) \in P(x)\right\} \tag{2.6}$$

where $P(x)$ is a vector of outputs that are technically viable and use the vector of inputs x, (u/θ) is the outputs ratio in production frontier, while θ is a ratio between zero and one, that is, $D_0(x, u) \in [0,1]$.

The relationship of duality between the distance function and the revenue function (Shephard, 1970) is the basis to estimate shadow prices since it creates the link between relative and absolute price. The relationship between the two functions can be expressed as in (Eq. (2.7)):

$$R(x,\ u) = \text{Max}\ u\{ru:\ D_0(x,\ u) \leq 1\}$$
$$D_0(x,\ u) = \text{Max}\ r\{ru:\ R(x,\ u) \leq 1\} \tag{2.7}$$

where $R(x, u)$ is the revenue function and r represents output prices. Under the assumption that distance and revenue functions are differentiable, the Lagrange multiplier method and Shephard's dual lemma enable us to calculate shadow prices. This deduction of shadow prices for undesirable outputs means assuming that the shadow price of an absolute desirable output coincides with the market price. If m is a desirable output (treated wastewater or reclaimed water in our case) whose market price is r_m equal to its shadow price (r_m^0), and if m' is undesirable output (a pollutant removed from wastewater) and $r_{m'}$ is the shadow price of each undesirable output, for all $m' \neq m$, the absolute shadow prices are given by (Färe *et al.* 1993) (Eq. (2.8)):

$$r_{m'} = r_m^0 \frac{\partial D_0(x,\ u)/\partial u_m}{\partial D_0(x,\ u)/\partial u_{m'}} \tag{2.8}$$

In most of the applications of the Färe's methodology in the framework of wastewater treatment (Molinos-Senante, 2011) the translog function has been used as distance function due to its great flexibility. When applied to a problem with k units, n inputs and m outputs the formula is (Eq. (2.9)):

$$\begin{aligned} \ln D_0(x^k,\ u^k) = {} & \alpha_0 + \sum_{n=1}^{N} \beta_n \ln x_n^k + \sum_{m=1}^{M} \alpha_m \ln u_m^k \\ & + \frac{1}{2}\sum_{n=1}^{N}\sum_{n'=1}^{N} \beta_{nn'}(\ln x_n^k)(\ln x_{n'}^k) \\ & + \sum_{m=1}^{M}\sum_{m'=1}^{M} \alpha_{mm'}(\ln u_m^k)(\ln u_{m'}^k) + \sum_{n=1}^{N}\sum_{m=1}^{M} \gamma_{nm}(\ln x_n^k)(\ln u_m^k) \end{aligned} \tag{2.9}$$

To calculate all the parameters of the translog function linear programming should be used (Molinos-Senante, 2011).

The advantages of the cost production approach to estimate positive externalities from wastewater treatment include the following: (i) it can help society understand the benefits generated as a result of environmental improvement programs; and (ii) it offers economists a further check on estimated measures to willingness to pay that are produced by alternative models (stated preference methods) (Färe *et al.* 2011). It is worth emphasising that costs incurred to determine the environmental benefits based on the estimation of shadow prices of pollutants are much lower than in the case of the traditional methodologies (demand approach) since it is not required any surveying process.

Nevertheless, the quantification of environmental benefits using the shadow price methodology also has some limitations in relation to the stated preference methods since they may be more appropriate than the shadow price method when the aim is to estimate the total economic value. The cost production approach methodology may be useful to quantify environmental impacts derived from production processes while demand approach methodologies can be applied in a wider context.

2.3 CONCLUSIONS

Economic analysis provides tools, information, and instruments for streamlining the decision-making process. Hence, in the field of wastewater treatment, economic feasibility studies are a useful tool for selecting the most appropriate option from among a range of technological alternatives.

Within methodologies to evaluate the economic feasibility of any project, cost benefit analysis (CBA) provides a comprehensive assessment since, unlike financial analysis, CBA integrates not just the costs and income with the market value but also with the positive and negative externalities.

The current chapter presents a framework to assess the economic feasibility of any innovative technology taking into account both internal and external impacts. Regarding internal costs it has been illustrated that using cost functions is a common methodology to estimate both investment, as well as operation and maintenance costs. Since externalities are not considered by the market, their quantification requires economic valuation methods. There are two main approaches to estimate the positive externalities associated to wastewater treatment namely the demand approach and cost production approach.

As a general conclusion, we emphasise that when the economic feasibility of a wastewater treatment technology is assessed, water companies and/or water management authorities should consider impacts with and without market values. Otherwise, the quality and relevance of the results will be seriously biased and an mis-estimation of benefits will occur. Moreover, uncertainty could influence the economic feasibility of wastewater treatment technologies. To narrow uncertainty, it is essential to perform a sensitivity analysis based on statistical methodologies or follow the *'ceteris paribus' approach.*

2.4 REFERENCES

Almansa C. and Martínez-Paz J. M. (2011). What weight should be assigned to future environmental impacts? A probabilistic cost benefit analysis using recent advances on discounting. *Science of the Total Environment*, **409**, 1305–1314.

Barton D. N., Saloranta T., Bakken T. H., Solheim A. L., Moe J., Selvik J. R. and Vagstad N. (2005). Using Bayesian network models to incorporate uncertainty in the economic analysis of pollution abatement measures under the water framework directive. *Water Science and Technology: Water Supply*, **5**(6), 95–104.

Benedetti L., Bixio D. and Vanrolleghem P. A. (2006). Benchmarking of WWTP design by assessing costs, effluent quality and process variability. *Water Science and Technology*, **54**(10), 95–102.

Bonilla M., Casasús T., Medal A. and Sala R. (2004). An efficiency analysis with tolerance of the Spanish port system. *International Journal of Transport Economics*, **XXXI**(3), 379–400.

Chen R. and Wang C. (2009). Cost-benefit evaluation of a decentralized water system for wastewater reuse and environmental protection. *Water Science & Technology*, **59**(8), 1515–1522.

De Anguita P. M., Rivera S., Beneitez J. M., Cruz F. and Espinal F. M. (2011). A GIS Cost-Benefit analysis-based methodology to establish a payment for environmental services system in watersheds: application to the Calan River in Honduras. *Journal of sustainable Forestry*, **30**(1), 79–110.

Doorn M. R. J., Towprayoon S., Manso-Vieria S. M., Irving W., Palmer C., Pipatti R., and Wang C. (2006). Wastewater treatment and discharge. In: Waste. 2006 IPCC Guidelines for National Greenhouse Gas Inventories, S. Eggleston, L. Buendia, K. Miwa and K. Tanabe (eds), IGES, Japan, pp. 6.1–6.28.

EPA (US Environmental Protection Agency) (1998). Water recycling and reuse: the environmental benefits. Ed. Water Division Region IX. http://www.epa.gov/region09/water/recycling/brochure.pdf (accessed 28 October 2013).

Färe R., Grosskopf S., Lovell C. A. and Yaisawarng S. (1993). Derivation of shadow prices for undesirable outputs: a distance function approach. *Review of Economics and Statistics*, **75**(2), 374–380.

Färe R., Grosskopf S. and Weber W. L. (2001). Shadow prices of missouri public conservation land. *Public Finance Review*, **29**(6), 444–460.

Flores-Alsina X., Corominas L., Neumann M. B. and Vanrolleghem P. A. (2012). Assessing the use of activated sludge process design guidelines in wastewater treatment plant projects: a methodology based on global sensitivity analysis, *Environmental Modelling and Software*, **38**, 50–58.

Foley J., de Haas D., Yuan Z. and Lant P. (2010). Nitrous oxide generation in full-scale biological nutrient removal wastewater treatment plants. *Water Research*, **44**(3), 831–844.

Friedrich E., Pillay S. and Buckley C. A. (2009). Environmental life cycle assessments for water treatment processes – A South African case study of an urban water cycle. *Water SA*, **35**(1), 73–84.

Gavasci R., Chiavola A. and Spizzirri M. (2010). Technical-economical analysis of selected decentralized technologies for municipal wastewater treatment in the city of Rome, *Water Science and Technology*, **62**(6), 1371–1378.

Genius M., Manioudaki M., Mokas E., Pantagakis E., Tampakakis D. and Tsagarakis K. P. (2005). Estimation of willingness to pay for wastewater treatment. *Water Science and Technology: Water Supply*, **5**(6), 105–113.

Gonzalez-Serrano E., Rodriguez-Mirasol J., Cordero T., Koussis A. D. and Rodriguez J. J. (2006). Cost of reclaimed municipal wastewater for applications in seasonally stressed semi-arid regions. *Journal of Water Supply: Research and Technology- AQUA*, **54**(6), 355–369.

Guimaraes M. H., Sousa C., Garcia T., Dentinho T. and Boski T. (2011). The value of improved water quality in guadiana estuary-a transborder application of contingent valuation methodology. *Letters in Spatial and Resource Sciences*, **4**(1), 31–48.

Hanley N. and Spash C. (1993). Cost – Benefit Analysis and the Environment. Edward Elgar, Cheltenham.

Hanley N., Colombo S., Tinch D., Black A. and Aftab A. (2006). Estimating the benefits of water quality improvements under the water framework directive: are benefits transferable? *European Review of Agricultural Economics*, **33**(3), 391–413.

Hanley N. and Barbier E. B. (2009). Pricing Nature: Cost-Benefit Analysis and Environmental Policy. Edward Elgar, Cheltenham.

Hernández-Sancho F., Molinos-Senante M. and Sala-Garrido R. (2011). Cost modelling for wastewater treatment processes. *Desalination*, **268**, 1–5.

HM, Treasury. (2003). The Green Book Appraisal and Evaluation in Central Government. TSO, London.

Hussen A. M. (2004). Principles of Environmental Economics, Routledge, New York.

IPCC (Intergovernmental Panel on Climate Change) (2007). Summary for Policy Makers. In: Climate Change 2007: The Physical Science Basis. Contribution of Working Group I to the Fourth Assessment Report of the Intergovernmental Panel on Climate Change, S. Solomon, D. Qin, M. Manning, Z. Chen, M. Marquis, K. B. Averyt, M. Tignor and H. L. Miller (eds), Cambridge University Press, Cambridge and New York.

Kafetzis A., McRoberts N. and Mouratiadou I. (2010). Using fuzzy cognitive maps to support the analysis of stakeholders' views of water resource use and water quality policy. *Studies in Fuzziness and Soft Computing*, **247**, 383–402.

Lassaux S., Renzoni R. and Germain A. (2007). Life cycle assessment of water from the pumping station to the wastewater treatment plant. *International Journal of Life Cycle Assessment*, **12**, 118–126.

Le Bozec A. (2004). Methodology for the Determination of Production Costs and Full Costs of Municipal Solid Waste Treatments. Deliverable Report of the Project: Aid in the Management and European Comparison of Municipal Solid Waste Treatment methods for a Global and Sustainable Approach. Cemagert, France.

MacLeod S. P. and Filion Y. R. (2012). Issues and implications of carbon-abatement discounting and pricing for drinking water system design in Canada. *Water Resources Management*, **26**(1), 43–61.

Molinos-Senante M. (2011). Techno-economical efficiency, cost modelling and economic valuation of environmental benefits from wastewater treatment: methodology and applications. PhD dissertation, University of Valencia.

Molinos-Senante M., Hernández-Sancho F. and Sala-Garrido R. (2010). Economic feasibility study for wastewater treatment: a cost-benefit analysis. *Science of the Total Environment*, **408**, 4396–4402.

Molinos-Senante M., Hernández-Sancho F. and Sala-Garrido R. (2012). Economic feasibility study for new technological alternatives in wastewater treatment processes: a review. *Water Science and Technology*, **65**(5), 898–906.

Molinos-Senante M., Hernández-Sancho F. and Sala-Garrido R. (2013b). Cost modeling for sludge and waste management from wastewater treatment plants: an empirical approach for Spain. *Desalination and water treatment*, **51**(28–30), 5414–5420.

Molinos-Senante M., Hernández-Sancho F., Sala-Garrido R. and Cirelli G. (2013a). Economic feasibility study for intensive and extensive wastewater treatment considering greenhouse gases emissions. *Journal of Environmental Management*, **123**, 98–104.

Nogueira R., Ferreira I., Janknecht P., Rodríguez J. J., Oliveira P. and Brito A. G. (2007). Energy-saving wastewater treatment systems: formulation of costs functions. *Water Science and Technology*, **56**(3), 85–92.

Nogueira S. F., Pereira B. F. F., Gomes T. M., de Paula A. M., dos Santos J. A. and Montes C. R. (2013). Treated sewage effluent: agronomical and economical aspects on bermudagrass production. *Agricultural Water Management*, **116**, 151–159.

Panagiotakopoulos D. (2004). Sustainable Municipal Solid Waste Management. 2nd edn. Zygos. Thessaloniki, Greece.

Pearce D. W. and Nash C. A. (1981). The Social Appraisal of Projects: A Text in Cost-Benefit Analysis. Macmillan, Basingstoke.

Pearce D., Groom B., Hepburn C. and Koundouri P. (2003). Valuing the future: recent advances in social discounting. *World Economics*, **4**, 121–141.

Pearce D., Özdemiroglu E., Bateman I., Carson T., Day B., Hanemann M., Hanley N., Hett T., Jones-Lee M., Loomes G. and Mourato S. (2002). Economic valuation with stated preference techniques. Summary guide. Department for Transport, Local Governments and the Regions, London. Available at: http://www.communities.gov.uk/documents/corporate/pdf/146871.pdf (accessed 18 November 2014).

Pearce D. and Turner R. K. (1990). Economics of Natural Resources and the Environment. Baltimore: The Johns Hopkins University Press.

Prat P., Benedetti L., Corominas L., Comas J. and Poch M. (2012). Model-based knowledge acquisition in environmental decision support system for wastewater integrated management. *Water Science and Technology*, **65**(6), 1123–1129.

Segui L. (2007). Economic analysis of the reuse of water reclamation in the irrigation of vineyards. *International Journal of Agricultural Resources, Governance and Ecology*, **6**(1), 124–139.

SENDECO (Carbon Dioxide Emission Allowances Electronic Trading System) (2013). Available from: http://www.sendeco2.com/index-uk.asp (accessed 28 October 2013).

Shephard R. W. (1970). Theory of Cost and Production Functions. Princeton: Princeton University Press.

Sipala S., Mancini G. and Vagliasindi F. G. A. (2005). Development of a web-based tool for the calculation of costs of different wastewater treatment and reuse scenarios. *Water Science and Technology: Water Supply*, **3**(4), 89–96.

Slothuus U., Larsen M. L. and Junker P. (2002). The contingent ranking method – a feasible and valid method when eliciting preferences for health care? *Social Science and Medicine*, **54**(10), 1601–1609.

Sugden R. and Williams A. (1988). Principles of Practical Cost-Benefit Analysis. New York: Oxford University Press.

Termes-Rife M., Molinos-Senante M., Hernández-Sancho F. and Sala-Garrido R. (2013). Life cycle costing: a tool to manage the urban water cycle. *Journal of Water Supply: Research and Technology- AQUA*, **62**(7), 468–476.

Tsagarakis K. P., Mara D. D. and Angelakis A. N. (2003). Application of cost criteria for selection of municipal wastewater treatment systems. *Water Air Soil Pollution*, **142**, 187–210.

Woods G. J., Kang D., Quintanar D. R., Curley E. F., Davis S. E., Lansey K. E. and Arnold R. G. (2013). Centralized versus decentralized wastewater reclamation in the houghton area of Tucson, Arizona. *Journal of Water Resources Planning and Management*, **139**(3), 313–324.

Xu Y.-P. and Tung Y.-K. (2008). Decision-making in water management under uncertainty. *Water Resources Management*, **22**(5), 535–550.

Chapter 3

Introduction to energy management in wastewater treatment plants

Catarina Silva, Helena Alegre and Maria João Rosa

Urban Water Division, Hydraulics and Environment Department, LNEC – National Civil Engineering Laboratory, Av. Brasil 101, 1700-066 Lisboa, Portugal

3.1 ENERGY MANAGEMENT OF WASTEWATER TREATMENT PLANTS PUT INTO CONTEXT

Water and energy nexus within the urban water services is a key topic in the utility managers and researchers' agenda. The main drivers are of economic and of environmental nature. Energy costs typically represent a main component of wastewater utilities' operational costs (e.g., UNESCO, 2014). According to the US Environmental Agency, "for many local governments, drinking water and wastewater plants are one of the largest energy consumers, accounting for 30–40 percent of total energy consumed. Because these services are so energy intensive, they provide an excellent opportunity for efficiency, savings, and reductions in greenhouse gas (GHG) emissions"[1]. Any savings may have a significant impact on the economic efficiency of the organisations. From the environmental viewpoint, the carbon footprint of the wastewater service provision is by no means negligible and has, in general, significant room for improvement.

A part of the energy consumption refers to the drainage network, mainly due to pumping another part refers to energy required for the wastewater treatment process. The relative importance of each of these parts in the overall operating costs depends on topography, network layout, treatment processes and technologies, as well as on the operation and maintenance practices and procedures. Wastewater utilities also have energy consumption associated to

[1]www.epa.gov/statelocalclimate/local/topics/water.html.

the supporting services, such as in vehicles and office and workshop buildings, which is not addressed in this chapter.

Wastewater treatment plants (WWTPs) are pieces of a more general complex puzzle – the wastewater system as a whole. While recognizing the importance of improving the efficiency of use of energy in economic and environmental terms, most utilities and research teams tend to explore and implement sectorial approaches – equipment or process-based, thus lacking a global analysis of the whole system functioning. Instead, a strategic approach may lead to much better service at lower total asset life-cycle costs. Selection of plant location, targets of treatment process effectiveness and reliability, and control of rain water inflow are examples of aspects that should be addressed in an integrated way (i.e., understanding the behaviour of the wastewater system as a whole), and incorporating a long term analysis planning horizon of WWTPs.

Let us take a simple example to illustrate the idea. An energy manager needs to compare the behaviour and potential for improvement of two similar systems in terms of energy, in order to prioritize intervention efforts (Figure 3.1). Both systems start at a pumping station at level 0.00 m, that pumps the wastewater from a gravity network to a wastewater treatment plant. The diameter, material and length of the pressurized sewer are identical in both cases. The level of the WWTP inflow point is 65.00 m in System 1 and 45.00 m in System 2. A volume of 800 m^3/day is pumped daily during 8 h per day,in both cases. There is a flow control valve upstream the WWTP in the case of System 2.

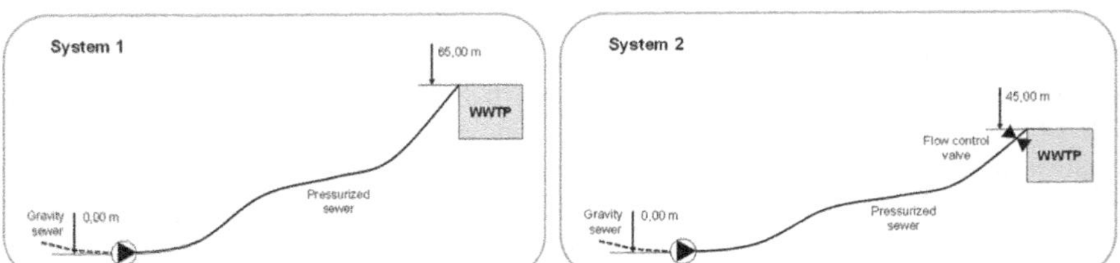

Characteristic	System 1	System 2
Pumping head, H_t (m)	80	72
Pump efficiency, η	82%	85%
Specific consumption (kWh/m^3)	0.266	0.231

Figure 3.1 Comparing energy efficiency in two pressurised wastewater systems.

The question this energy manager needs to answer is ‘which system is less efficient and, therefore, has more potential for improvement?’. The table in Figure 3.1 shows that System 1 has a lower pump efficiency, and a higher specific and total energy consumption. These are typical indicators to assess energy efficiency. When comparing the two systems using these indicators, it seems safe to reply ‘System 1’ to the question. However, a system analysis may provide a different view. Table 3.1 shows the calculations of the global energy efficiency, assessed as the ratio between the energy supplied to each system and the energy actually supplied to the users, assuming a no-losses situation. If there were losses, the downstream flow should be the authorized consumption.

Table 3.1 Total energy efficiency of Systems 1 and 2.

	System 1	System 2
Daily energy input supplied to the system ($E_{input} = \gamma \cdot Q_t \cdot H_t/\eta \times 8$) (kWh)	212.5	184.5
Daily energy supplied to the consumers ($E_{supplied} = \gamma \cdot Q_t \cdot H_s \times 8$) (kWh)	141.6	98.0
Overall energy efficiency ($E_{input}/E_{supplied}$) (–)	1.50	1.88

Table 3.1 shows that the conclusion is opposite: System 2 is globally less efficient than System 1. Figure 3.2 shows the energy line for both systems.

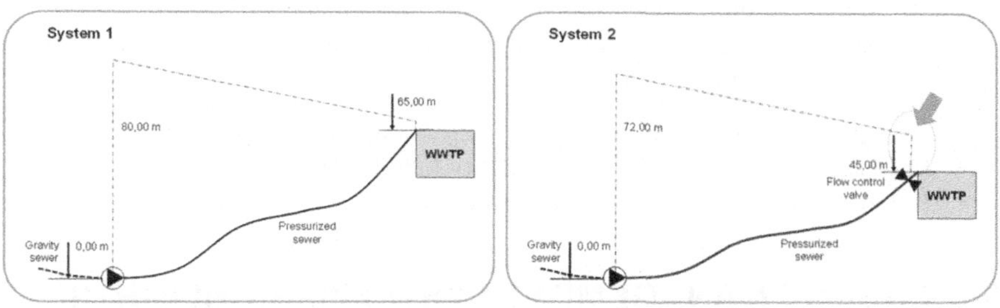

Figure 3.2 Comparison of the energy lines for Systems 1 and 2.

The head-loss caused by the flow control valve is much higher than the pump inefficiency. This example demonstrates that managing energy on an 'asset by asset' basis may easily fail to pinpoint the critical energy inefficiencies. The most cost-efficient intervention is likely to be changing the pump of System 2, in order to eliminate the need for the flow control valve.

Although not exclusively, typical sources of inefficiencies are:

- **Control valves:** as illustrated in the example, are a very typical asset where energy is wasted, particularly in the network part, in general less relevant inside the WWTP; as a result, special attention shall be paid to the valves located downstream pumps.
- **System design:** very rarely systems are solicited as planned by the time of their original planning, design and construction; as illustrated in the

example, sometimes small changes in the infrastructure may provide major positive gains.
- **Operating modes:** may affect energy consumption and intensity, as well as energy and power use costs (e.g., pump scheduling versus variable energy rates).
- **Energy and power tariffs:** often there is room for negotiation and improvement of the energy and power tariffs (e.g., in order to optimise the maximum contracted power utilities do not explore as they could the room they have for negotiation energy and power.
- **Infiltration and rainwater inflow in separated systems:** increasing the flow to be transported and treated generally impacts the energy consumption.
- **Pumping equipment:** selection, maintenance and operation modes of pumping equipment have major influence of its efficiency.
- **Other electro-mechanical equipment:** as previous, for other relevant equipment (e.g., mechanical aerators, diffusers, mixers, sludge dehydration, etc. ASK CS/MJR to correct and complete).
- **Network maintenance practices:** head-loss in pressurized sewers, influent load to the WWTPs, and treatment units and cleaning frequency of some treatment units is often affected by inadequate maintenance practices, such as cleaning.

These sources of inefficiencies are often interdependent.

In complement to fixing the inefficient uses, wastewater management also offers opportunities to explore, such as:

- Energy production from wastewater (e.g., sludge biogas production);
- Production of wind energy in the utility facility sites;
- Exploration of in-sewer treatment processes.

3.2 ENERGY MANAGEMENT SYSTEMS: HIGHLIGHTS OF THE ISO 50001

ISO 50001 (ISO, 2011) provides organizations with a structured framework to manage energy aiming at increasing energy efficiency, reducing costs and improving energy performance. This standard follows the same principles and common elements of other management systems standards, ensuring compatibility with ISO 9001 (quality management), ISO 14001 (environmental management) and ISO 55000 (asset management). Similarly to the other management systems standards, it is based on the Plan-Do-Check-Act (PDCA) cycle (Figure 3.3).

Its principles and requirements are an excellent basis for managing energy within wastewater utilities as a whole, and in wastewater treatment plants in particular.

ISO 50001 Energy Management System requires organizations to:

- Continually improve energy performance, including energy efficiency, energy use and consumption (Figure 3.4); in the case of WWTP, energy production is also an important axis of energy performance;

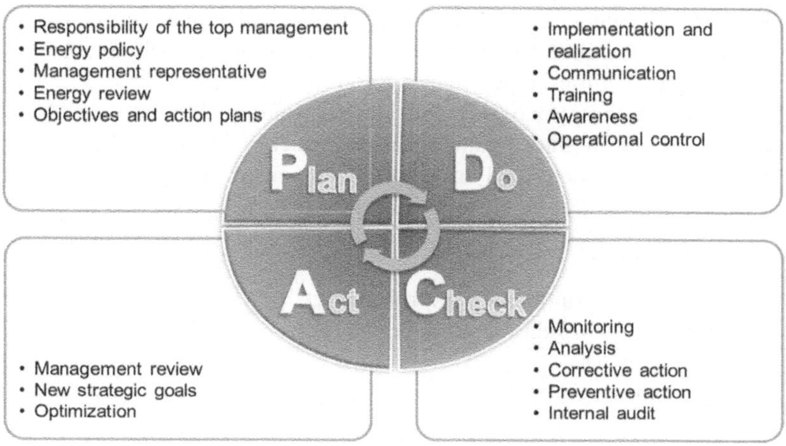

Figure 3.3 Plan-Do-Check-Act in ISO 50001: 2011.

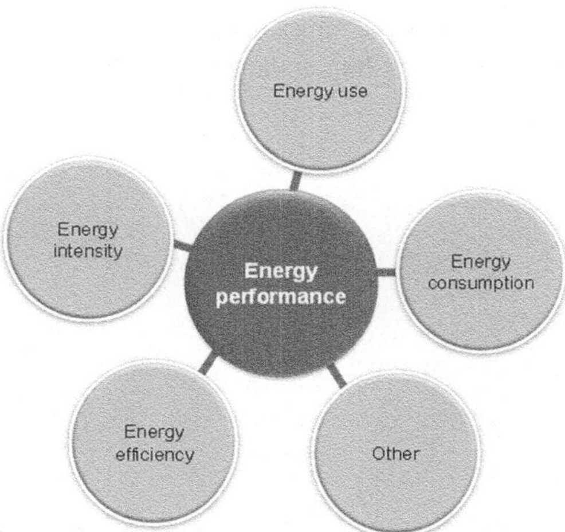

Energy use

Manner or kind of application of energy.

NOTE 1: ventilation, heating, processes, production lines

NOTE 2: The quantity is expressed as **energy consumption.**

Energy efficiency

Engineering accepted use of the term by the organization.

NOTE: conversion efficiency, energy required/energy used etc.

Energy performance

Measurable results related to energy.
Examples: energy efficiency, **energy intensity** (the inverse of energy efficiency), specific energy use, etc.

Figure 3.4 Axis of energy performance (adapted from ISO 50001: 2011).

- Review energy use, consumption and efficiency at defined intervals;
- Document the methodology and criteria used to develop the Energy Review considering facilities, equipment, systems or processes;
- Establish an energy baseline and identify Energy Performance Indicators appropriate for monitoring and measuring energy performance;
- Establish, implement and maintain documented Energy Objectives and Targets.

The benefits include:

- Identifying opportunities for improvement;
- Ensuring greater level of control;
- Enhancing image;
- Satisfying the expectations of most stakeholders;
- Reduced costs and improved business performance;
- Improving compliance with energy legislation;
- Reducing carbon emissions;
- Demonstrating transparency and commitment.

Figure 3.5 summarizes the main aims and benefits of applying ISO 50001 in the case of wastewater utilities, according to the authors' view.

Figure 3.5 Aims and benefits of applying ISO 50001 in a wastewater utility.

ISO 50001 Energy Management System is suitable for all businesses regardless of their size, geography or industry. It is particularly effective in energy intensive industries, as it is in the general situation of wastewater utilities. It is also effective if the utility faces greenhouse gas (GHG) emissions regulation or legislation, or for environmental sustainability motivations aims at decreasing its carbon foot print. ISO 50001 formalizes energy policies and objectives and embeds them into energy efficient thinking throughout the organization.

The management model advocated in ISO 50001, summarized in Figure 3.6, is very similar to the management model of the other existing management system standards.

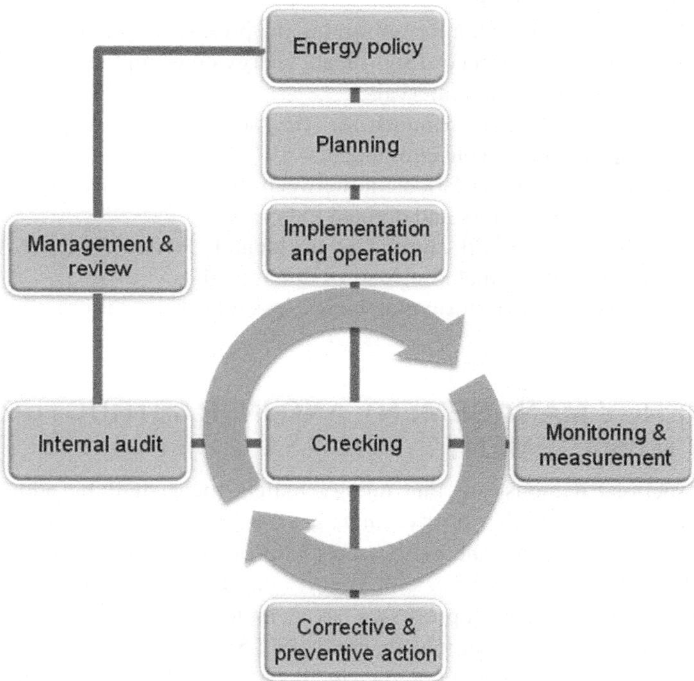

Figure 3.6 Management system model of ISO 50001 (ISO 50001: 2011).

Each box of Figure 3.6 corresponds to a set of requirements in ISO 50001: 2011. The most relevant for the establishment of a framework of energy performance assessment system for WWTPs, object of the following section, relates to the planning and checking stages of this model and of its overarching Plan-Do-Check-Act framework. For instance, under the item 'Energy objectives, targets and programmes' of the 'Plan' section, this standard states that the organization shall:

- Establish and document measurable energy objectives and targets at the relevant function and levels within the organization.
- Set specific targets for those controllable parameters that have a significant impact on energy efficiency.

- Establish and document energy management programmes which define responsibilities and the means and time frame by which individual targets are to be achieved.

With regard to 'Checking', the 'Monitoring and measuring' requirements establish that the organization shall:

- Establish the monitoring, measuring and targeting requirements of its energy management programme.
- Have a demonstrable plan for improving the provision of meters.
- Monitor, measure and record significant energy consumption and associated energy factors at defined intervals.
- Maintain records that demonstrate the accuracy and repeatability of monitoring and measuring equipment.
- Assess and review the relationship between the energy consumption and its associated energy factors and defined intervals.
- Maintain records of all significant accidental deviations from expected energy consumption, including causes and remedies.
- Compare its energy performance indicators against similar organizations or situations.

3.3 ENERGY MANAGEMENT AND INFRASTRUCTURE ASSET MANAGEMENT

Previous sections demonstrate that energy management shall be addressed in harmonization with the other utility's management processes, infrastructure asset management (IAM) being one of the most important. In fact, there is a lot in common between these two processes. By bringing the IAM principles and tools into energy management, both processes have a lot to gain.

The ISO standards on asset management (ISO 55000/55001/55002) define the basic requirements that any asset management system shall comply with. The key management principles are common and also follow a continuous improvement, plan-do-check-act (PDCA) loop.

However, the fact that wastewater systems are network-based public infrastructures demands the adoption of complementary principles, methods and tools that are not commonly needed and explored in other asset systems. The main differentiating characteristics are the indefinite life of public infrastructures, the long duration of most physical assets, and the fact that wastewater infrastructures have a system behavior: individual sewers do not have a function or performance by themselves, but only as part of the functional unit; symptoms (e.g., overflows) and causes (e.g., infiltration of insufficient capacity) typically occur in different locations. Diagnosis, prioritization, design of intervention works and selection of the best intervention alternatives need to be based on functional units, and not on an asset by asset basis.

The IAM analyses shall therefore be based on the long term behavior of functional units of assets. Inevitably, assets in different stages of their life cycle coexist. A time window of some decades needs to be used in the analysis, in order to embrace the life cycle of the key assets (Alegre & Coelho, 2012).

The IAM AWARE-P methodology, developed in Portugal, incorporate industry's best practices and is currently being promoted and adopted internationally (e.g., by the international Water Association), associating the ISO 24500 (ISO 24510/24511: 2007) and the ISO 55000 (ISO 55000/ISO 55001/ISO 55002: 2014) principles with the characteristics of the urban water systems. In the core of the methodology (Figure 3.7) are the establishment of clear corporate objectives and an educated choice of assessment criteria, supported by adequate metrics and quantifiable targets (Alegre & Covas, 2010; Almeida & Cardoso, 2010; Alegre & Coelho, 2012; Alegre *et al.* 2013).

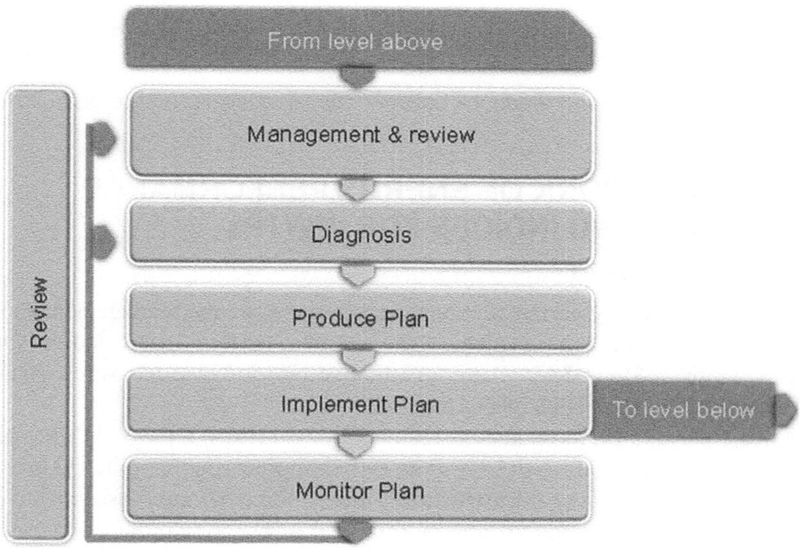

Figure 3.7 IAM planning process at each decision levels (Alegre *et al.* 2010).

The process is applied at the strategic, tactical and operational decisional levels in the utility, striving for alignment of objectives, metrics and targets between levels, as well as consistent feedback across levels (Figure 3.8).

One of the main advantages of such a structure approach is the establishment of a decision process that is transparent, defendable and yet simple, allowing to compare and prioritize intervention alternatives potentially of a very different nature.

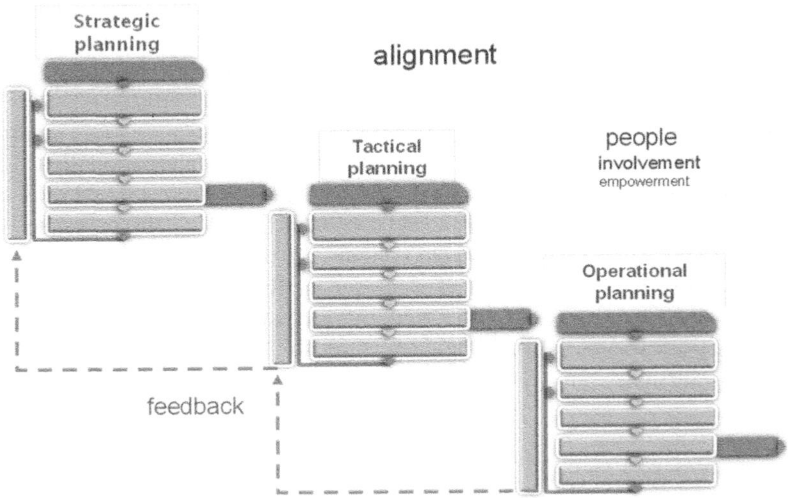

Figure 3.8 Alignment and feedback between decision levels (Alegre *et al.* 2010).

3.4 A FRAMEWORK OF ENERGY PERFORMANCE INDICATORS AND INDICES FOR WWTPs

3.4.1 Background

As demonstrated in the earlier sections, wastewater systems' management should ensure the continuous improvement of the systems' performance through PDCA cycles (Figures 3.3 and 3.6), where quantitative performance measures, such as performance indicators (PIs), play a key role.

In this context, a Performance Assessment System (PAS) for WWTPs has been developed in Portugal (at LNEC – National Civil Engineering Laboratory) to assist the benchmarking of these infrastructures, that is, the continuous assessment and improvement of performance (Cabrera *et al.* 2011). The system herein presented for WWTPs (and the analogous one developed for WTPs) is currently in its 3rd generation (Silva *et al.* 2013). The 1st generation of PASs were tested in four WTPs (Vieira *et al.* 2008; Silva, 2008; Vieira, 2009) and two WWTPs (Quadros 2010; Quadros *et al.* 2010). The 2nd generation was an output of «PASt21» – the Portuguese initiative for performance assessment of water and wastewater treatment plants (2009–2011), a project involving 18 institutions, 17 WWTPs and 10 WTPs distributed nationwide (Rosa *et al.* 2010; Silva *et al.* 2012).

Following the PDCA approach, the developed PAS is objective-driven and assumes two general objectives for any undertaking with regard to WWTP performance: (1) its effectiveness and reliability, that is, the compliance over time with the quality requirements of the treated water and (2) its efficiency (in terms of resources utilization) and sustainability (economic and environmental) (Rosa *et al.* 2010; Silva *et al.* 2012).

The system comprises (i) a PI system for the overall assessment of the plant, on an annual-basis, in terms of treated water quality, plant efficiency and reliability (ER), use of natural resources and raw materials (RU), by-products management (BP), safety, personnel, financial resources (Fi), and planning and design (Silva *et al.* 2012) and (ii) a system of performance indices (PXs) for assessing the daily performance in terms of treated water quality (Silva *et al.* 2014a), operating conditions and removal efficiencies (Silva *et al.* 2014b). The integrated analysis of PIs and PXs allows identifying improvement actions.

Energy costs may vary significantly from one utility to the next, with estimates ranging from 2%–60% of total operating costs (NREL, 2012). In WWTPs, energy represents one of the higher costs of wastewater services, and is the second largest part of the running costs of a WWTP, right after the personnel costs (CEC & AWWARF 2003; PG&E, 2003; WERF, 2010; Rodriguez-Garcia *et al.* 2011; Silva *et al.* 2012).

The energy consumption in WWTPs depends on treatment processes (Burton 1996; WEF, 2009) and plant fingerprint, mass removed (Lingsten & Lundkvist, 2008; Hernández-Sancho *et al.* 2011) and treated wastewater quality requirements (carbon or carbon and nutrients control, disinfection, etc.), treated wastewater volume (Burton, 1996; Lingsten & Lundkvist, 2008; Mizuta & Shimada, 2010; Yang *et al.* 2010; Hernández-Sancho *et al.* 2011; WERF, 2011) and the percent of facility design capacity at which a plant is operating (WERF, 2011), as well as on the operation and maintenance practices (USEPA, 2008; Guimet *et al.* 2010). Pumping requirements associated with wastewater drainage system depend on topography and network layout and not on the treatment plant itself, and are therefore out of the scope of the PAS developed for WWTPs.

Energy performance indicators (PIs) and indices (PXs) are thus core measures of the PAS for WWTPs, and were further developed in the current 3rd generation of this system (Silva *et al.* 2013), including the reference values for judging the performance, taking into account the earlier aspects affecting the energy consumption.

3.4.2 Energy performance indicators

The performance indicators were formulated according to the IWA approach (Matos *et al.* 2003; Alegre *et al.* 2006) and the principles established in ISO 24510-11-12:2007(E). The developed PIs are defined as ratios between variables (of the same or of different nature), and may be therefore dimensionless (e.g., %, –) or intensive (e.g., kWh/m^3). The numerator expresses the PI objective and, to allow for comparisons, the denominator represents one dimension of the system (Alegre *et al.* 2006). The PIs are calculated for a reference assessment period, preferably a year.

The full portfolio of PIs available in PAS for WWTPs is very broad and is not to be calculated in every application, plant and assessment period. Instead, from the

PI system one should select a set of PIs according to the assessment objectives, in this case the WWTP energy performance in a given year.

The energy performance is transversal to four assessment groups described above, ER, RU, BP and Fi. Hence, the PI set for energy includes (Silva *et al.* 2013): (i) four 1st level PIs (Table 3.2), one from each of those groups, and (ii) 14 complementary PIs (Table 3.3), related to WWTP reliability (e.g., adequacy of plant treatment and pumping capacities, recycling and aeration control, and inspection of key equipment) and renewable (wind and solar photovoltaic) energy production.

The reference assessment period is the calendar year, though PIs expressed in temporal terms are formulated to accommodate other reference assessment periods. In order to ensure unit coherence and allow for comparisons, for these PIs (e.g., wtER35.1, Table 3.3), values calculated for other reference assessment periods are converted into annual values, multiplying by '(365 days/year)/Assessment period (day)' (Silva *et al.* 2012).

The PIs are identified by a code that includes six or eight fields (the last two are optional). These fields identify the system (*t* for WTP and *wt* for WWTP), the assessment group (e.g., ER for the plant Efficiency and Reliability assessment group) and the number (e.g., 01) of the PI. Two additional fields (number and/or letter) can be used to identify an alternative processing rule (designated by a different number) or a speciation (using a different letter for each species). Examples of alternative processing rules are wtER35.1 and wtWER35.2 for Pump inspection or Inspected pumps, respectively (Table 3.3) and wtRU03.1, wtRU03.2 and wtRU03.3 for energy consumption per m^3 of treated wastewater, per kg BOD removed or kg COD removed, respectively (Table 3.2).

PIs are defined by default per m^3 of treated wastewater (e.g., wtRU03.1, Table 3.2) but they may be expressed per kg BOD or kg COD mass removed by dividing the obtained value by the explanatory factor wtEF03 for BOD or wtEF04 for COD (Silva & Rosa, 2014). The explanatory factors (EFs, Table 3.4) evaluate complementary aspects of the plant performance that assist the correct interpretation of some PIs.

Tables 3.2 and 3.3 also present, whenever applicable, the PI reference values defining three performance levels: "good" (●), "acceptable" (■) and "unsatisfactory" (♦).

The reference values are based on the PI results of 17 Portuguese WWTPs in a 5-year period (2006–2010) and on the relations derived from literature data. The reference values for the unit energy consumption reflect the inverse relations with the volume treated and are specific for activated sludge systems (conventional, with coagulation/filtration (C/F) and with nitrification and C/F) and trickling filters. The reference values for energy production were derived based on the methane generation potential and literature data; those of net use of energy were considered the difference between the references for energy consumption and energy production (Silva & Rosa, 2014).

Examples of application of the energy PIs are given in Silva and Rosa (2014).

Table 3.2 First level PIs for energy performance (adapted from Silva & Rosa, 2014).

PI code, units, processing rule and reference values for "good" (●), "acceptable" (■) and "unsatisfactory" (♦) performance			
wtRU03.1 – Energy consumption [kWh/m³]			
Energy consumption (kWh)/Treated wastewater (m³)			
Trickling filters	● ≤0.185 + 1127/TW	■]0.185 + 1127/TW; 0.231 + 1409/TW[	♦ ≥0.231 + 1409/TW
Activated sludge (AS)	● ≤0.280 + 1192/TW	■]0.280 + 1192/TW; 0.350 + 1490/TW[	♦ ≥0.350 + 1490/TW
AS + Coagulation/Filtration	● ≤0.325 + 1384/TW	■]0.325 + 1384/TW; 0.406 + 1730/TW[	♦ ≥0.406 + 1730/TW
AS with nitrification + Coagulation/Filtration	● ≤0.424 + 1362/TW	■]0.424 + 1362/TW; 0.530 + 1703/TW[	♦ ≥0.530 + 1703/TW
wtRU03.2 – Energy consumption [kWh/kg BOD removed]			
Energy consumption (kWh)/BOD mass removed (kg)			
It may be computed from wtRU03.1 and wtFE03, that is, wtRU03.2 = wtRU03.1/wtFE03			
	● ≤2	■]2; 10[	♦ ≥10
wtRU03.3 – Energy consumption [kWh/kg COD removed]			
Energy consumption (kWh)/COD mass removed (kg)			
It may be computed from wtRU03.1 and wtFE04, that is, wtRU03.2 = wtRU03.1/wtFE04			
wtBP18.1 – Production of energy from biogas [kWh/m³]			
Energy produced from biogas (kWh)/Treated wastewater (m³)			
	● ≥0.0009 BOD5	■ [0.0007 BOD5; 0.0009 BOD5[	♦ <0.0007 BOD5

(Continued)

Table 3.2 First level PIs for energy performance (adapted from Silva & Rosa, 2014) (*Continued*).

PI code, units, processing rule and reference values for "good" (●), "acceptable" (■) and "unsatisfactory" (♦) performance			
wtBP18.2 – Production of energy from biogas [%]			
Energy produced from biogas (kWh)/Energy consumption (kWh)*100			
wtER08 – Net use of energy from external sources [kWh/m^3]			
(Energy acquired to external sources – Energy sold to external users (kWh))/Treated wastewater (m^3)			
Trickling filters	● ≤0.185 + 1127/TW − 0.0009 BOD5	■]0.185 + 1127/TW – 0.0009 BOD5; 0.231 + 1409/TW – 0.0007 BOD5[	♦ ≥0.231 + 1409/TW − 0.0007 BOD5
Activated sludge (AS)	● ≤0.280 + 1192/TW − 0.0009 BOD5	■]0.280 + 1192/TW – 0.0009 BOD5; 0.350 + 1490/TW – 0.0007 BOD5[	♦ ≥0.350 + 1490/TW − 0.0007 BOD5
AS + Coagulation/Filtration	● ≤0.325 + 1384/TW − 0.0009 BOD5	■]0.325 + 1384/TW – 0.0009 BOD5; 0.406 + 1730/TW – 0.0007 BOD5[	♦ ≥0.406 + 1730/TW − 0.0007 BOD5
AS with nitrification + Coagulation/Filtration	● ≤0.424 + 1362/TW − 0.0009 BOD5	■]0.424 + 1362/TW – 0.0009 BOD5; 0.530 + 1703/TW – 0.0007 BOD5[	♦ ≥0.530 + 1703/TW − 0.0007 BOD5
wtFi05 – Electrical energy costs [€/m^3]			
Electrical energy costs (€)/Treated wastewater (m^3)			

TW = treated wastewater (m^3/d); BOD5 = influent BOD5 (mg/L); PI code, units, processing rule and reference values

Table 3.3 Complementary PIs for assessing WWTP energy performance (adapted from Silva *et al.* 2013).

PI code, units and processing rule	Ref. Values
wtER13 – Adequacy of plant capacity [%] $$\left(1-\frac{\sum_{d=1}^{n} Qt_d \times J_d + \sum_{d=1}^{n} Qt_d \times K_d}{\sum_{d=1}^{n} Qt_d}\right)\times 100$$ Qt_d = plant capacity (daily average flowrate) at day *d'* (m^3/day) Qr_d = daily average flowrate recorded at day *d'* (m^3/day) n = assessment period (day) $J_d = 1$, if $Qr_d > 0.95\ Qt_d$ in day d; 0, if $Qr_d \leq 0.95\ Qt_d$ in day d $K_d = 1$, if $Qr_d > 0.7/S\ Qt_d$ in day d; 0, if $Qr_d \leq 0.7/S\ Qt_d$ in day d S = Daily correction for flowrate seasonality (varies from 1 to 3) $S = Qr_{30\,max.}/Qr_d$ *or* $S = fs$, the lowest value; If $S < 1$, $S = 1$; If $S > 3$, $S = 3$ fs = Seasonality factor (varies from 1 to 3) $fs = Qr_{30\,max.}/P20$ $Qr_{30\,max.} = Qr_d$ maximum average for 30 consecutive days (m^3/day) ↔ high season $P20 = Qr_d$ percentile 20 (m^3/day) ↔ low season	● [80; 100] ■ [60; 80[♦ [0; 60[
wtER15 – Adequacy of BOD_5 mass capacity [%] = wtER13 with $Q = BOD_5$ mass capacity	
wtER16 – Adequacy of COD mass capacity [%] = wtER13 with Q = COD mass capacity	
wtER17 – Adequacy of TSS mass capacity [%] = wtER13 with Q = TSS mass capacity	
wtER18 – Adequacy of nutrient mass capacity [%] = wtER13 with Q = nutrient mass capacity	
wtER19 – Adequacy of aeration capacity [%] = wtER13 with Q = plant aeration capacity	

(*Continued*)

Table 3.3. Complementary PIs for assessing WWTP energy performance (adapted from Silva *et al.* 2013) (*Continued*).

PI code, units and processing rule	Ref. Values
wtER20 – Pumping utilisation upstream of the WWTP [%] = Maximum daily volume of raw wastewater (RW) pumped (m^3)/RW maximum pumping capacity (m^3)*100 wtER21 – Pumping utilisation downstream of the WWTP [%] = Maximum daily volume of treated wastewater (TW) pumped (m^3)/TW maximum pumping capacity (m^3)*100	● [80; 90] ■ [70; 80[or]90; 95] ◆ [0; 70[or >95
wtER28 – Recycling control [%] Recycling pumps with automatic control (no.)/Recycling pumps (no.)*100	
wtER29 – Aeration or mixing control [%] Aerators or mixers with automatic commands (no.)/Aerators or mixers (including reserves) (no.)*100	
wtER35.1 – Pump inspection [no./(pump.year)] (Pump inspections (no.) × 365 (day/year)/Assessment period (day))/Pumps (no.)	
wtER37.1 – Aerator or mixer inspection [no./(aerator or mixer.year] (Aerator or mixer inspections (no.) × 365 (day/year)/Assessment period (day))/Aerators or mixers (no.)	
wtER47 – Dewatering and sludge transportation equipment inspection [no./(pump.year)] (Dewatering and sludge transportation equipment inspections (no.) × 365 (day/year)/Assessment period (day))/ Dewatering and sludge transportation equipment (no.)	
wtER35.2 – Inspected pumps [%/year] Inspected pumps (no./year)/Pumps (no.) × 100 wtER37.2 – Inspected aerators or mixers [%/year] Inspected aerators or mixers (no./year)/Aerators or mixers (no.) × 100	● 100 ■]90; 100[◆ [0; 90]
wtBP19 – Production of wind and solar energy [%] Energy produced by wind power and solar photovoltaic (kWh)/Energy consumption (kWh)*100	

"good" (●), "acceptable" (■) and "unsatisfactory" (◆) performance

Table 3.4 Explanatory factors (Silva & Rosa, 2014).

FE code, units and processing rule
wtEF03 – BOD mass removed/Treated wastewater ratio [kg BOD_5/m^3] BOD mass removed (kg)/Treated wastewater (m^3)
wtEF04 – COD mass removed/Treated wastewater ratio [kg COD/m^3] COD mass removed (kg)/Treated wastewater (m^3)

3.4.3 Energy performance indices

The performance indices (PXs) are obtained by applying a processing rule (performance function, Figure 3.9) that converts state-variable data, expressing the relevant operational performance assessment aspects of the plant, into a dimensionless performance index.

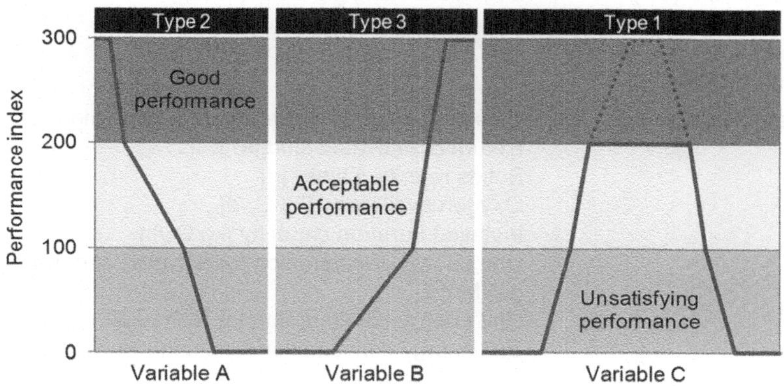

Figure 3.9 Types of performance functions to assess the operational performance of WWTPs (Silva *et al.* 2013).

The PXs range between zero and 300, in which: 300 corresponds to a situation where the performance is 'excellent'; values between 300 and 200 reflect 'good' performances; values between 200 and 100 are 'acceptable'; 100 corresponds to the 'minimum acceptable' performance; values below 100 reflect 'unsatisfactory' performance (Silva *et al.* 2013, 2014a, b).

To define the performance function, reference values are defined for each performance level based on the ranges recommended in the literature for each unit operation/process variant.

A portfolio of state-variables of energy PXs is proposed in Table 3.5. The reference values for energy PXs are under development. Whenever necessary, equations are being developed to produce the reference values as a function of other key-parameter(s).

Table 3.5 State-variables relevant for assessing the WWTP energy performance (adapted from Silva *et al.* 2013).

Treatment unit or step (type)	State-variables relevant for energy performance [Units]
Main pumping	Unit energy consumption [kWh/m^3]
Preliminary treatment	Unit energy consumption [kWh/m^3]
Chemical addition	Average velocity gradient [s^{-1}] Detention time [s] Unit energy consumption [kWh/m^3]
Flocculation	Average velocity gradient [s^{-1}] Detention time [min] Unit energy consumption [kWh/m^3]
Primary sedimentation[2]	Unit energy consumption [kWh/m^3]
Trickling filters (TF) Rock (low, intermediate and high rate) and plastic packing (high rate)	Organic loading [kg BOD/(m^3.d)] Recirculation ratio [-] Unit energy consumption [kWh/m^3]
Activated sludge (AS)[3]	Dissolved oxygen in the aeration tank [mg O_2/L] Return of activated sludge[4] [%] Solids retention time[3] [d] Oxygen availability [kg O_2/d] Installed aeration capacity [kg O_2/L] Unit energy consumption for aeration and mixing [kWh/m^3] Unit energy consumption for recirculation [kWh/m^3] Unit energy consumption for sludge wasting [kWh/m^3]
Biofilters (BF)	Air and/or water backwash flowrate [m/h] Air and/or water backwash time [min] Organic loading [kg BOD/($m^3 \cdot$ d)] Filtration rate [m/h] Unit energy consumption [kWh/m^3]
Microscreening	Unit energy consumption [kWh/m^3]
Granular filtration[5]	Air and/or water backwash flowrate [m/h] Air and/or water backwash time [min] Filtration rate [m/h] Unit energy consumption [kWh/m^3]
Membrane filtration (MF, UF)	Membrane hydraulic permeability [L/($m^2 \cdot$ h $\cdot$ bar)] Pressure increase (bar) Fouling rate (Pa/min) Net treatment capacity[6] [m^3/(m^2. day)] Unit energy consumption [kWh/m^3]

(*Continued*)

Table 3.5 State-variables relevant for assessing the WWTP energy performance (adapted from Silva *et al.* 2013) (*Continued*).

Treatment unit or step (type)	State-variables relevant for energy performance [Units]
UV disinfection	UV dose [mJ/cm²] Unit energy consumption [kWh/m³]
Thickening[7]	Daily operating hours of mechanical thickeners [hours/day] Relative centrifugal force (g-force) [× g] Unit energy consumption [kWh/m³]
Stabilization (aerobic and anaerobic digestion)	Solids retention time (without recirculation) [d] Biogas production[8] [m³/kg converted VSS] Production of energy from biogas [kWh/m³] Unit energy consumption [kWh/m³]
Dewatering[9]	Daily operating hours [hours/day] Rotational speed[10] [rpm] Relative centrifugal force (g-force)[10] [× g] Unit energy consumption [kWh/m³]

3.4.4 Methodology for PAS application

The first step of PAS application is the definition of the objectives and of the assessment criteria for a given WWTP or a group of WWTPs. The PIs must then be selected and calculated accordingly, and analysed against the references values. Further insights may require the use of complementary PIs.

The next step is the selection and calculation of PXs to complement the information provided by the homologous PIs. The latter assess the overall performance in the

[2]Applicable to conventional (with or without sludge return) and enhanced primary sedimentation.
[3]Variables applicable to several AS variants: complete mix; plug flow (conventional, extended aeration); oxidation ditch (C removal, C + N removal); A/O; MLE; Bardenpho (4-stage, 5-stage); A2/O; UCT; VIP; SBR (C removal), SBR (C + N removal) and SBR (C + N + P removal). The performance functions are AS-type specific.
[4]Except for SBR.
[5]Variables applicable to different filter packing (mono and dual media) and height; performance functions are filter-type specific.
[6]Net treatment capacity accounts for the water filtered minus the water spent for cleaning and the filtration time plus the cleaning time.
[7]Excluding the g-force, the variables listed are applicable to gravity, dissolved air flotation, gravity belt and rotary drum thickening. The performance functions are thickener-type specific.
[8]Applicable only to anaerobic digestion.
[9]Applicable to centrifuge, belt filter press, recessed-plate filter press and sludge drying beds. The performance functions are dewatering-type specific.
[10]Applicable only to centrifuge.

assessment period (usually, a calendar year) and the PXs assess 'where' (unit operations/processes and equipment) and 'when' the performance did satisfy or fail the pre-established objectives and the distance that remains to achieve these targets (Silva *et al.* 2014a), enabling the identification of improvement actions.

As introduced earlier, prior to WWTP energy efficiency one should verify the plant effectiveness and reliability, that is, the compliance over time with the quality requirements of the treated water (Silva *et al.* 2014a). One may then select the parameters and unit operations or processes for which one intends to assess and optimise the removal efficiency to enhance the treatment reliability (Silva *et al.* 2014b).

The continuous improvement of WWTP performance requires the verification and, eventually, the (re)definition of objectives and (re)selection of the corresponding PIs and PXs, which restarts the PDCA cycle of PAS application (Figure 3.10).

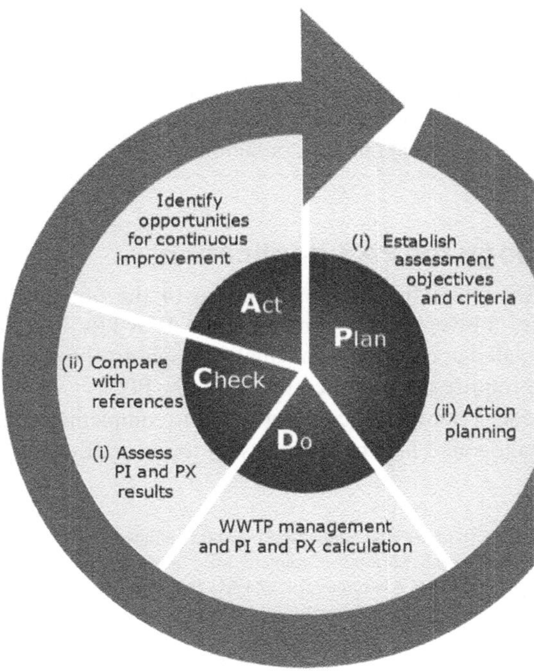

Figure 3.10 PDCA methodology for the continuous improvement of WWTP energy performance.

3.5 REFERENCES

Alegre H., Baptista J. M., Cabrera Jr. E., Cubillo F., Duarte P., Hirner W., Merkel W. and Parena R. (2006). Performance Indicators for Water Supply Services. 2nd edn, Manual of Best Practices Series, IWA Publishing, London, ISBN: 1843390515.

Alegre H. and Coelho S. T. (2012). Infrastructure Asset Management of Urban Water Systems. Chapter 3 of "Water Supply System Analysis", Avi Ostfeld (ed.), Open access at: www.intechopen.com/books/water-supply-system-analysis-selected-topics, ISBN: 978-953-51-0889-4 (accessed March 11, 2014).

Alegre H., Coelho S. T., Covas D., Almeida M. C. and Cardoso M. A. (2013). A utility-tailored methodology for integrated asset management of urban water infrastructure. *Water Science & Technology: Water Supply*, **13**(6), 1444–1451.

Alegre H. and Covas D. (2010). Infrastructure Asset Management of Water Services. An Aproach Based on Rehabilitation. Technical Guide 16. ERSAR, LNEC, IST, Lisbon, 472 pp. (in Portuguese) ISBN: 978-989-8360-04-5.

Almeida M. C. and Cardoso M. A. (2010). Infrastructure Asset Management of Wastewater and Stormwater Services. An Approach Based on Rehabilitation (in Portuguese). Technical Guide 17. ERSAR, LNEC, IST, Lisboa (in Portuguese) ISBN: 978-989-8360-05-2.

Burton F. L. (1996). Water and Wastewater Industries: Characteristics and Energy Management Opportunities. Burton Environmental Engineering, Los Altos, CA. Prepared for the Electric Power Research Institute. Palo Alto, California. Report CR106941. September.

CEC & AWWA RF (2003). Water and Wastewater Industry Energy Efficiency: A Research Roadmap. California Energy Commission and American Water Works Association Research Foundation. Prepared by Means E., McGuire Environmental Consultants, Newport Beach, CA.

Guimet V., Kelly R., Doung F., Rosina M., Audic J. M. and Terry J. M. (2010). Green energy resource: research and learning experiences from wastewater treatment plants. *Proceedings of IWA World Water Congress*, Montreal (Canada).

Hernández-Sancho F., Molinos-Senante M. and Sala-Garrido R. (2011). Energy efficiency in Spanish wastewater treatment plants: a non-radial DEA approach. *Science of the Total Environment*, **409**, 2693–2699.

ISO 24510: 2007(E) – Activities Relating to Drinking Water and Wastewater Services – Guidelines for the Assessment and for the Improvement of the Service to Users. 1st edn, International Organization for Standardization, Geneva.

ISO 24511: 2007(E) – Activities Relating to Drinking Water and Wastewater Services – Guidelines for the Management of Wastewater Utilities and for the Assessment of Wastewater Services. 1st edn, International Organization for Standardization, Geneva.

ISO 24512: 2007(E) – Activities Relating to Drinking Water and Wastewater Services – Guidelines for the Management of Drinking Water Utilities and for the Assessment of Drinking Water Services. 1st edn, International Organization for Standardization, Geneva.

ISO 50001: 2011. Energy Management Systems. Requirements with Guidance for Use. 1st edn, International Organization for Standardization, Geneva.

ISO 55000: 2014. Asset Management — Overview, Principles and Terminology. 1st edn, International Organization for Standardization, Geneva.

ISO 55001: 2014. Asset Management — Management Systems — Requirements. 1st edn, International Organization for Standardization, Geneva.

ISO 55002: 2014. Asset Management Guidelines for the Application of ISO 55001. 1st edn, International Organization for Standardization, Geneva.

Lingsten A. and Lundkvist M. (2008). Description of the Current Energy Use in Water and Wastewater Systems in Sweden (in Swedish). The Swedish Water & Wastewater Association, SWWA.

Matos R., Cardoso A., Ashley R., Duarte P., Molinari A. and Schulz A. (2003). Performance Indicators for Wastewater Services. Manual of Best Practices Series, IWA Publishing, London, ISBN: 19002229006.

Mizuta K. and Shimada M. (2010). Benchmarking energy consumption in municipal wastewater treatment plants in Japan. *Water Science and Technology*, **65**(10), 2256–2262.

NREL (2012). Energy Efficiency Strategies for Municipal Wastewater Treatment Facilities. National Renewable Energy Laboratory, Technical Report NREL/TP-7A30-53341. Colorado, January.

PG&E (2003). Municipal Wastewater Treatment Plant Energy Baseline Study. Pacific Gas and Electric Company (PG&E). San Francisco, CA, June.

Quadros S. (2010). Desenvolvimento de um sistema de avaliação de desempenho de estações de tratamento de Águas residuais urbanas. PhDThesis, Azores University, Angra do Heroísmo, p. 286 (in Portuguese).

Quadros S., Rosa M. J., Alegre H. and Silva C. (2010). A performance indicators system for urban wastewater treatment plants. *Water Science and Technology*, **62**(10), 2398–2407.

Rodriguez-Garcia G., Molinos-Senante M., Hospido A., Hernández-Sancho F., Moreira M. T. and Feijoo G. (2011). Environmental and economic profile of six typologies of wastewater treatment plants. *Water Research*, **45**, 5997–6010.

Rosa M. J., Ramalho P., Silva C., Vieira P., Quadros S. and Alegre H. (2010). *PASt21 – Iniciativa Nacional de Avaliação de Desempenho de ETA e ETAR Urbanas* ('PASt21' – the Portuguese initiative for performance assessment of water and wastewater treatment plants). In: *Proc. 10.° Congresso de Água*. Alvor (Algarve). March, p. 14 (in Portuguese).

Silva C. (2008). *Aplicação de medidas de avaliação de desempenho a estações de tratamento de Água da Águas do Algarve* (Application of performance assessment measures to Águas do Algarve water treatment plants). MSc Thesis. Algarve University, Faro, p. 286 (in Portuguese).

Silva C. and Rosa M. J. (2014). Energy Performance Indicators of Wastewater Treatment – A Field Study with 17 Portuguese Plants. IWA World Water Congress, Lisbon, September.

Silva C., Ramalho P., Quadros S., Alegre H. and Rosa M. J. (2012). Results of 'PASt21' – the portuguese initiative for performance assessment of water and wastewater treatment plants. *Water Science and Technology – Water Supply*, **12**(3), 372–386.

Silva C., Ramalho P. and Rosa M. J. (2013). Curative actions in wastewater treatment systems. Chapter 2 of PART I of report *D43.1 – Results from Overall Exploration of Potential for Sustainability Improvements*. TRUST Project.

Silva C., Quadros S., Ramalho P. and Rosa M. J. (2014a). A tool for assessing treated wastewater quality in urban WWTPs. *Journal of Environmental Management*, **146**, 400–406.

Silva C., Quadros S., Ramalho P., Alegre H. and Rosa M. J. (2014b). Translating removal efficiencies into operational performance indices of wastewater treatment plants. *Water Research*, **57**, 202–214.

UNESCO (2014). The United Nations World Water Development Report 2014: Vol. 1 – Water and energy (WWDR 2014). United Nations Educational, Scientific and Cultural Organization, UNESCO 2014. ISBN 978-92-3-104259-1.

USEPA (2008). Ensuring a Sustainable Future: An Energy Management Guidebook for Wastewater and Water Utilities. US Environmental Protection Agency, January.

Vieira P. (2009). *Avaliação de desempenho de estações de tratamento de água para consumo humano* (Performance assessment of drinking water treatment plants). PhDThesis, Algarve University. Lisboa: Teses e Programas de Investigação LNEC, p. 451, ISBN 978-972-49-2186-0 (in Portuguese).

Vieira P., Alegre H., Rosa M. J. and Lucas H. (2008). Drinking water treatment plants assessment through performance indicators. *Water Science and Technology – Water Supply*, **8**(3), 245–253.

WERF (2010). Overview of State Energy Reduction Programs and Guidelines for the Wastewater Sector. Water Environment Research Foundation. Co-published by IWA Publishing, Alexandria, VA.

WERF (2011). Energy Management. Exploratory Team Report Executive Summary. Water Environment Research Foundation, Alexandria, VA.

Yang L., Zeng S., Chen J., He M. and Yang W. (2010). Operational energy performance assessment system of municipal wastewater treatment plants. *Water Science and Technology*, **62**(6), 1361–1370.

Chapter 4

Innovative energy efficient aerobic bioreactors for sewage treatment

Eoin Syron

School of Chemical and Bioprocess Engineering, University College Dublin (UCD), Belfield, Dublin 4, Ireland; OxyMem, Blyry, Athlone, Ireland

4.1 INTRODUCTION

Sewage treatment accounts for over 1% of the total electrical requirements in the developed world and with increasingly tighter effluent discharge regulations the energy requirement for acceptable sewage treatment will increase even further. With this in mind there is a growing demand for energy efficient technologies (Caffoor, 2008). These technologies must achieve equivalent or superior levels of effluent treatment using fewer resources and therefore having less total impact on the environment. Biological treatment technologies which have been the cornerstone of Sewage Treatment Plants (STP) for the past 100 years have traditionally fallen into two categories (1) Extensive and (2) Intensive. While extensive, low or zero energy systems can be used in rural or remote areas, for urban settlements intensive treatments systems are required to reduce the space required for the STP and hence a high energy demand is associated with these technologies. In recent decades there has been significant development and improvement in intensive biological treatment processes with the development of processes such as the Membrane BioReactor (MBR) and the Moving Bed Biofilm Reactor (MBBR). While these technologies have reduced the volumetric capacity required for treatment they also have a high energy consumption rate. Anaerobic treatment processes on the other hand require little or no energy input above pumping and mixing requirements. There have been major advances in anaerobic treatment with the direct anaerobic treatment of sewage possible in warm climates, (Sghezzo *et al.* 1988) and currently there is a strong focus on the development of the anaerobic MBR for the treatment of sewage. Despite these advances aerobic treatment is still required for the removal of many different pollutants which are not broken down anaerobically for example, ammonia, and also to treat the wastewater to a sufficient quality so that it can be discharged back into the surface water. Recently there has been a move toward innovation in aerobic biological reactors, with reduced energy requirement being of key importance. Not only does this save on operating cost and carbon footprint but there is also the added incentive of achieving an energy neutral STP by maximizing the energy being produced and minimizing energy consumed. With this in mind this chapter will focus on innovative aerobic biological processes and technologies which are still in development or are at the early stages of development. The main aspect of these technologies will be their potential for significant overall energy reduction. While some of these bioreactors may or may not be utilised for the direct treatment of the sewage stream they still have the potential to be used in a STP for the treatment of one of the streams and offset the total energy demand and operational cost.

4.2 AERATION

In STPs the single largest user of energy is the aeration of the biological treatment processes. The current standard technology of using fine bubbles created via blowers and diffusers, has been studied for many years and despite many improvements being made throughout this time, the process remains highly inefficient, with a Standard Oxygen Transfer Efficiency (SOTE) of 5% per m water depth, resulting in only 30% of the oxygen supplied being transferred to the water in a 6 m deep tank. Activated Sludge aeration requires 55.6% of the total energy consumption in a treatment plant having an average energy requirement of 0.634 kWh/m^3 sewage treated (Caffoor, 2008). Therefore the direct aeration energy requirements are 0.348 kWh/m^3 of sewage treated. It is estimated that thru optimisation of operations, retrofitting of more energy efficient blowers and diffusers and innovation in the process design used in existing wastewater treatment systems a reduction of the energy consumption by 20% is quite feasible. (UKWIR Report 2010) While it is important that this is carried out on existing systems to give a 'quick win' the potential for further energy savings still exist through innovative aerobic treatment processes.

4.2.1 Innovative process design and improvement

The simplest way of reducing the aeration requirement is through effective primary treatment, increasing the percentage of biologically degradable solids removed reduces the load of BOD entering into the aeration basin and requiring oxidation. An increase of 10% in the BOD removal in the primary treatment stage can result in an energy saving of 2MWh per annum per 1000PE (@1 kgO_2/kWhr). If the removed BOD is anaerobically digested it can result in a total energy swing of 9.3MWh per annum per 1000PE (@3.5 kWh/KgBOD). To achieve this increased solids removal two potential approaches can be taken, Chemically Enhanced

Primary Sedimentation (CEPS) or Mechanically Enhanced Primary Sedimentation (MEPS). The implementation of CEPS requires very little if any additional capital expenditure and can increases the treatment capacity of a primary settler by a factor of 2, but does create an additional operational expense, as it requires the addition of a coagulant. In an economic comparison by Rashed *et al.* (2013), CEP resulted in an overall 23% reduction of sewage treatment cost. MEPS on the other hand does require a capital investment to acquire the screens or sieves and there is also an energy and maintenance requirement to operate the machinery. Ruiken *et al.* (2013), estimated the total energy saving including wastewater treatment, sludge treatment and incineration at 40% due to the implementation of 0.35 mm sieves as compared to the scenario with fine mesh sieving.

Additionally significant aeration energy savings can be made by creating pre-anoxic zones for denitrification. By creating internal recycle streams within the sewage treatment plant, nitrate produced by the oxidation of ammonia can be returned toward the inlet and produce an oxygen credit (2.86 g O_2/g NO_3-N), lowering the overall oxygen requirement for the treatment of the BOD and resulting in an energy saving of 6–8 MWh per annum per 1000PE. Additional benefits to this operational regime have been identified by Rosso and Stenstrom (2007), where the biosorption of readily biodegradable surfactants in the pre anoxic zone has a positive effect on the subsequent aerated section, increasing the α-factor by up to 2 fold. In some cases the increase in alpha factor can be the most significant energy saving mechanisim, because although the installation of a pre-anoxic zone reduces the BOD load of the wastewater treatment plant it also creates an additional pumping overhead with recirculation rates up to several multiples of the inlet flow being required. Therefore the additional savings from the increase in α-factor can result in a further halving of air supplied and the energy required. The most effective reduction of energy requirement in aerobic bioreactors for sewage treatment plants today comes at the design stage where consideration of the overall energy footprint can be made. Notwithstanding the optimisation of current aeration systems, the entire oxygen transfer process remains highly inefficient with oxygen and energy literally being blow away. These inefficiencies leave a large scope for improvement and innovation to provide the oxygen in a more efficient manner.

4.3 INCREASING OXYGEN TRANSFER FROM A BUBBLE

4.3.1 Fine bubble diffusers and oxygen transferring technologies

Haney's (1954) evaluation of bubble aeration effectively outlines the basic controlling parameters for subsurface aeration design are: (1) bubble size, (2) relative velocity and (3) residence time. By influencing these parameters it is possible to increase the OTR and the OTE from a rising bubble in a Wastewater Treatment

Tank, but in most cases the influence of the parameter requires additional energy and the modification may have an energy neutral or even energy negative impact.

E.G., By increasing the tank water depth from 3 m to 6 m a rising bubble has approximately twice the residence time in the water with which to transfer oxygen. But for air to exit a diffuser plate underneath 6 m of water it must have pressure greater than 6 m hydrostatic head. This results in an increase of 80% in the theoretical energy required to achieve a two fold increase in OTE.

4.3.1.1 Smaller bubbles

Decreasing the bubble size which results in an increased gas-liquid surface area available for mass transfer (Table 4.1) has so far achieved a relatively good success at increasing the overall energy efficiency. This has resulted in fine bubble aeration becoming the standard aeration system recommended a head of coarse and medium bubble aeration, despite fine bubble aeration having a higher maintenance requirement.

Table 4.1 Increase in bubble surface area available for mass transfer due to decreasing bubble size.

	Increase in contact area	
Bubble diameter (mm)	**Number of bubbles per m³**	**Total bubble surface area (m²)**
15	5.66×10^5	400
10	1.91×10^6	600
3(FBDA)	7.07×10^7	2000
1(UFBD)	1.91×10^9	6000

Smaller than fine bubbles are ultrafine bubbles or micro bubbles, but concerns still exist regarding these technologies for the aeration of wastewater. Microbubbles can produce a much higher gas liquid interfacial surface area, but they also have a higher energy cost than typical gas distributors, the microbubble generators can allow downsizing of the aeration tank and shorten the overall residence time of the wastewater so that a reduction in the overall cost can be achieved (Terasaka *et al.* 2011). Despite having close to 100% Oxygen Transfer Efficiency it has been observed that microbubble generation is influenced by fouling which can occur on the porous membranes (Liu *et al.* 2013) and the aeration process can have adverse effects on the overall process, causing floatation and/or the break-up of activated sludge flocs. (Liu *et al.* 2012). Disadvantages of micro bubble aeration include higher capital cost, a higher head loss across the diffuser, increased air filtration requirements, and a tendency for the microbubble generation membrane to tear when over-pressurized.

4.3.2 Increasing contact time

Along with smaller and smaller bubbles other approaches have been taken to increasing the oxygen transfer efficiency for bubble aerated systems, many of these have looked at providing more contact time between the introduced air and the liquid.

4.3.2.1 Diffusaire

With a view to achieving a higher Oxygen Transfer Efficiency with respect to the energy input Duiffusaire, an Israeli company have developed a system to increase the retention time of the bubble in an activated sludge tank. Air is added to the wastewater in a vertical tube placed in the Activated Sludge tank. The wastewater to be aerated is then pumped in such a manner as to create a downward flow in the tube and thereby increase the residence time of the bubble in the liquid, and allow more time for oxygen transfer. (Yousfan *et al.* 2011) The aerated liquid along with some entrapped air then escapes from the bottom of the tube back into the activated sludge tank. It is claimed that diffusaire proprietary technology can reduce the energy requirements for aeration by up to 50% (Diffusaire, Advanced Aeration Solutions).

4.3.2.2 Sorubin

Although somewhat similar in outer appearance to the diffusaire technology, OptusAir developed by a Sweedish company Sorub in creates a vortex in the centre of the vertically mounted tube through the use of an impellor placed at the bottom of the tube. The vortex entraps air from the surface and disperses this air liquid mixture from the bottom of the vertical tube into the wastewater treatment tank. Optusflow could enhance the effect of aeration and other treatment methods by up to 50%.

4.3.2.3 Sansox OY

The OxTubeBySansox (Finland), also increases the contact time and relatively velocity between the gas and liquid, but do so in a pipe. Using a specially developed static mixer to break up the flow pattern and create lots of eddies the OxTube minimises the boundary layer diffusional resistance between the liquid and gas bubble (Sansox, Oxytube).

4.4 BUBBLELESS AERATION–MEMBRANE AERATED BIOFILM REACTOR

With increasing oxygen transfer efficiencies coming from smaller and smaller bubbles, the next step has been to develop aeration system without bubbles altogether. This has resulted in the use of diffusive gas permeable membranes for aeration.

Although the initial goal for the development of this technology was to increase the oxygen transfer efficiency to the wastewater, the subsequent biofilm which formed on the gas permeable membrane surface, which was initially seen as barrier to oxygen transfer, was later discovered to take an active role in the wastewater treatment process and lead to the development of the Membrane Aerated Biofilm Reactor (MABR) (Timberlake *et al.* 1988). As a means to increase the oxygen transferred from the air to the wastewater, by placing the oxygen containing gas inside of amembrane, the residence time of the air in the wastewater is decoupled from the buoyancy forces which normally limit the contact time between the air and water. Therefore it is possible to dramatically increase the Oxygen Transfer Efficiency (OTE) up to a theoretical maximum of 100%. By controlling the flowrate and the partial pressure of Oxygen through the membranes the OTE and the Oxygen Transfer Rate (OTR) can be controlled. This allows for an additional level of control over the oxygen transfer process for an operator. Although a biofilm colonises the membrane surface preventing the transfer of oxygen to the wastewater the membrane supported biofilm takes an active part in the wastewater treatment and because the consumption of oxygen occurs locally to the membrane surface the active biofilm increases the rate of oxygen flux across the membrane (Shanahan & Semmens, 2006).

The unique counter diffusional concentration profile which is created through the membrane aeration, also results in the natural development of aspatially unique biofilm. This membrane aerated biofilm is a counter diffusional biofilm (Figure 4.1) as opposed to a conventional co-diffusional biofilm.

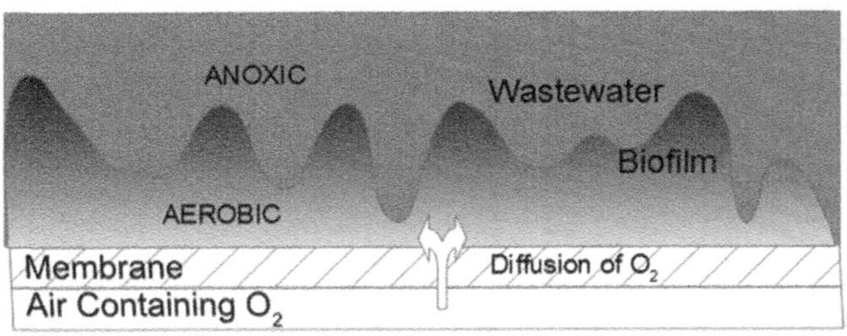

Figure 4.1 Schematic of a membrane aerated biofilm.

Along with increasing the OTE the gas pressure required to supply the oxygen to the MABR is dramatically reduced and subsequently the overall energy requirement. The structural integrity of the membranes themselves prevents the hydrostatic pressure closing off the lumen of the membrane and if non-porous membranes are used no flow of water back into the membrane lumen can occur. Therefore the supplied gas does not have to overcome the hydrostatic head and the only pressure which the supplied gas has to overcome is the pressure drop due to the air flow within the membranes. It must also be noted that because oxygen is not being transferred to the wastewater there are no wastewater surface tension forces to be overcome, therefore the alpha factor does not play a role in the oxygen transfer in a MABR.

While university based research groups have been examining this technology for some time with one of the first papers identifying the wastewater treatment potential being published by Timberlake *et al.* 1988, until recently the technology had not progressed beyond the lab scale. During this time the MABR has been tested for different applications including high rate BOD removal, tertiary nitrification and simultaneous nitrification and denitrification. These have all been identified as areas where the MABR has a significant Advantage over conventional systems and due to the increased level of process control the MABR can be tailored for each of these processes or for a multiple of different processes sequentially.

Despite this concept being around for many years 2 major stumbling blocks have prevented it from being commercialised.

(1) The availability of cheap suitable membranes to exploit the cost saving which results from the reduced energy requirement.
(2) The ability to achieve effective long term stable performance, this has typically been most difficult in wastewaters with high BOD loading rates.

In an economic evaluation of the MABR (Casey *et al.* 2008), the two major economic factors for the commercialisation of the MABR were identified as membrane replacement cost and energy cost as these directly influence the increased capital cost associated with membranes in a tank and the operational saving. Today with more and more membrane providers coming to the market and increased membrane operational and production knowledge, membrane technology has now become commoditised and the cost of membranes has reduced significantly over the past 10 years. This coupled with the increased process knowledge, higher levels of biofilm understanding and the ability to have effective biofilm control has led to a number of companies commercialising the MABR concept.

Two different approaches have been taken to the scale-up of the membrane aerated biofilm concept.

4.4.1 Submerged membrane aerated biofilm reactors

Both Oxymem and GE Water and Process technologies have taken a more conventional approach towards the MABR by placing the gas permeable membranes into a wastewater treatment tank in place of other aeration devices. This has been the approach taken by many research groups including those of Semmens (WERF, 2005) and Nerenberg (Downing, 2008) and is summarised in the reviews Syron and Casey (2008) and Martin and Nerenberg (2013). This configuration of the MABR allows the wastewater to flow around the gas permeable membranes and the oxygen/air is then supplied via a pipe network to interior of the membranes. The biofilm grows on the outside of the membrane and can be removed or have its outer layers sloughed off by a change in shear force usually carried out by through intermittent coarse bubbles scouring or a change in liquid flow direction. The mixing and oxygenation are independent in this system and can be controlled or modified according to the requirement or needs of the system. The configuration also allows for the use of oxygen enriched air or even pure oxygen to be used with up to 100% oxygen transfer efficiency (VOSS, 1994). To make the systems very compact there is a tendency to provide as much surface area per unit volume although this has led to problems with biofilm over growth.

Both Oxymem and GE have chosen to uses dense (non-Porous) membranes which although having a higher initial resistance to mass transfer than hydrophobic microporous membrane, are not effected by long term operation (Semmens, 2005). The GE MABR uses very fine bore polymethylpentane (PMP) membranes while Oxymem have developed a PolyDimethylSiloxane (PDMS) membrane for use in their MABR (Figure 4.2).

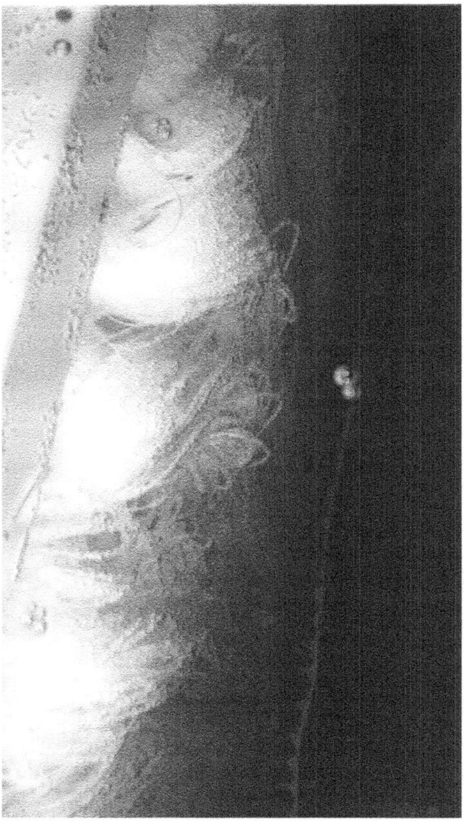

Figure 4.2 Photograph of gas permeable membranes covered with biofilm (provided courtesy of Oxymem.)

A summary of the membranes used from both systems is given in Table 4.2.

Table 4.2 Submerged MABR data taken from (1) Syron *et al.* (2014), (2) Stricker *et al.* (2011), (3) Adams *et al.* (2014).

Summary of OXYMEM and GE Zeelung results			
	Membranes		
	OxyMem	**GE (Zeelung)**	**GE membrane**
Configuration	Hollow fibre	Micro bore hollow fibre	Hollow fibres around a cord
Wall thickness	100 μm	5–20 μm	15–20 μm
Internal diameter	300 μm	30–60 μm	80 μm
Outer diameter	500 μm	50–70 μm	100 μm
Material	PDMS	PMP	PMP
Arrangement	Vertically in bunches of 400	Vertically in large bunches	Placed around a central cord kept in place by a warf
Specific surface area m^2/m^3	250–400	776–845	655
Air Pressure mBar	100 mBar	410 mBar	550 mBar
OTE	>50%	31%	>60%
SOTE kgO_2/kWh	8		6
Energy Requirement (kWh/m^3)	<0.1		

Operational data has been presented by both of these companies at international conferences and it is very likely that larger scale systems will be installed in the coming years.

4.4.2 Passively membrane aerated biofilm reactors

In a concept similar to a trickling filter or biotower two companies Emefcy and BioGill pump the wastewater to be treated to the top of their membrane aeration unit. The wastewater then flows down through the inside of the membranes with the oxygen/air surrounding the membranes and the oxygen in the air at atmospheric conditions diffusing through the membrane into the pollutant degrading biofilm inside the membrane. This approach requires no energy for an air blower or mixing, only pumping energy is required.

The Spiral Aerobic Biofilm Reactor or SABRE has been developed by Emefcy an Israli company. The technology uses a membrane envelope wound into a spiral with a defined space between each coil so that air can flow all around the membrane. The wastewater to be treated is then pumped into the top of the membrane envelope, and flows around the spiral to the end where it exits the membrane. The biofilm grows on the inside of the membrane and any sloughed or detached biomass leaves with the treated wastewater to be removed in a subsequent unit operation. Energy requirements of 0.02 kwh/m^3 of water treated have been reported for this system (Spiral Aerobic Biofilm Reactor-Emefcy).

Another similar concept which pumps the wastewater to be treated to the top of an above ground membrane unit is the Biogill developed by Biogill Operations Pty ltd (Australia). Biogill utilises folded Nano Ceramic membranes and the wastewater flows downward inside the membranes with a biofilm developing on the inside of the membrane. The nanoceramic membranes chosen by Biogill are porous to water and some of the wastewater permeates through the membranes creating a wet membrane surface and an environment suitable for the growth of fungi when contributes to nutrient removal. The wastewater is then collected in a decant tank underneath the membrane unit where the detached biomass is allowed to settle out of the wastewater. Based on the design parmeters given in the Biogill Technical System and specification guide the energy required for treating the water is 0.17kWh/m^3 at the scenario given in Table 4.3.

Table 4.3 Energy requirements for biogill, using operational data available through www.biogill.com.

Biogill energy requirements	
BOD	500 mg/l
Flow rate	8000 l/hr
Recirculation rate	15
Recirculation pump flowrate	120 m^3/hr
Height	2.5 m
Pump efficiency	0.6
Pumping energy	1.36 kw
Energy required per unit volume	0.17 kw/m^3

4.5 LOW ENERGY AMMONIA REMOVAL

4.5.1 Ammonia removal

Biological nitrogen removal is an oxygen intensive process with 4.57 g of O_2 required for every g of N-NH_4 oxidised. Traditionally the pathway for complete nitrogen removal was through the complete oxidation of ammonia all the way to nitrate (NO_3) and then denitrifying the nitrate through the addition of a carbon source and allowing heterotrophic bacteria to utilise the nitrate as a source of oxygen. The product of this two stage biological process was nitrogen gas which escaped back into the atmosphere. Depending on the exact configuration chosen this process generally required significantly more air, additional recirculation pumps, longer biomass retention and significantly larger tanks than the more traditional biological carbon removal processes. Overall the energy increase for the addition of nitrification was estimated at 69% in a case study by Menendez and Veatch, 2010. Thanks to an increased understanding of the biological nitrogen removal process, scientists and engineers have been able to provide the suitable conditions for the each of the different bacterial groups which are involved with multiple stages of the nitrogen removal process and through the control of these conditions along with reaction time, the nitrogen removal process can be split up into its component steps.

4.5.2 Shortcut nitrification

The first step of ammonia removal is carried out by Ammonia Oxidising Bacteria (AOB), these bacteria oxidise ammonia to nitrite, while the second group of bacteria Nitrite Oxidising Bacteria (NOB) oxidise the nitrite to nitrate. To achieve complete nitrogen removal it is not necessary to oxidise ammonia all the way to nitrate, by controlling the process in such a way as to prevent the complete oxidation of ammonia to nitrate, the intermediate produced nitrite can be subsequently denitrified to N_2 resulting in a saving of 1.17 gO_2 per g of N-NH_4. Shortcut nitrification, as it known, uses effective process control to reduce the growth rate of the NOB and minimise the amount of nitrite converted to nitrate through a better understanding of the growth rates of the two groups of bacteria required for the nitrification process and their kinetic parameters such as oxygen affinity (Ciudad *et al.* 2006). Trials have shown that process can be applied to wastewater with high ammonia concentrations for example, leachate (Akerman, 2005). The process has been scaled-up and is available commercially from Veolia under the name Anita-shunt.

4.5.3 Anammox

The identification of the Anaerobic Ammonia Oxidising group of bacteria (anammox) and the ability to construct full scale anaerobic ammonia oxidising reactors further reduces the energy required for nitrogen removal. This anaerobic ammonia oxidation process has been successfully scaled up by many companies including Paques and Gronmij. To date there are many different variants of the anammox technology including suspended culture (flocs) self-supporting biofilm (granules) and attached biofilm based systems (MBBR).

Although the anammox bacteria do not require Oxygen, some oxygen is required to produce the Nitrite which the anammox bacteria utilise for the ammonia oxidation. Therefore through the implementation of anammox the oxygen requirement for the aerobic treatment is significantly reduced.

4.6 OTHER AEROBIC TECHNOLOGIES

4.6.1 Aerobic granules

The development of the Aerobic Granules technology Nerada® developed by Royal Haskoning DHV is another novel aerobic biological process which similar to Anaerobic Ammonia Oxidation and Shortcut Nitrogen Removal systems encourages the growth of the desired biomass through the effective control of the environmental growth conditions. Several full scale installations currently exist and these are showing that along with the improved process outcomes there is a significant energy saving achieved through the implementation of the Aerobic Granule technology. In Epe, Netherlands, the overall energy requirement for the wastewater treatment plant was reduced by 35% after the installation of the Nerada® process and while at Frielas, Portugal, the specific air requirement (m^3 air/kg $COD_{removed}$ day) is 60% of the parallel Conventional Activated Sludge system. (Giesen & Thompson, 2013).

4.7 CONCLUSIONS

Aerobic biological treatment has formed the corner stone of sewage treatment for the past 100 years and is likely to continue to form a major part of the sewage treatment plant of the future. Aerobic biological treatment produces clean water suitable for discharge to surface waters or further processing into usable water cheaply and efficiently without the use of chemicals or the production of more concentrated streams. Despite being an economical process the delivery of oxygen to the bacteria is still energy intensive and very inefficient, thereby presenting the opportunity for innovative aerobic biological processes which treat the sewage more efficiently and economically using less energy. Over the past 10 years several processes have been developed and are beginning to be seen at full scale sewage treatment plants. While these innovative processes reduce operation cost the overall impact on the capital cost on a new build or a upgrade site has yet determined.

4.8 REFERENCES

Adams N., Hong Y., Ireland J., Koops G. H., Côté P. (2014). An Innovative Membrane-Aerated Biofilm Reactor (MABR) for Low Energy Treatment of Municipal Sewage. Conference Proceedings Singapore International Water Week, Singapore.

Akerman A. (2005). Feasibility of nitrate-shunt (nitritation) on landfill leachate. Tesis Maestría. Lunds Tekniska Universitet. Departament of Water and Environmental Engineering. Suecia. 79 pp.

Biogill Technical and Specification Guide. http://www.biogill.com/downloads.

Chun Liu, Hiroshi Tanaka, Jin Ma, Lei Zhang, Jing Zhang, Xia Huang and Yoshiaki Matsuzawa. (2012). Effect of microbubble and its generation process on mixed liquor properties of activated sludge using shirasu porous glass (SPG) membrane system. *Water Research*, **46**(18), 6051–6058.

Chun Liu, Hiroshi Tanaka, Jing Zhang, Lei Zhang, Jingliang Yang, Xia Huang and Nobuhiko Kubota. (2013). Successful application of shirasu porous glass (SPG) membrane system for microbubble aeration in a biofilm reactor treating synthetic wastewater. *Separation and Purification Technology*, **103**, 53–59.

Ciudad, G., Werner, A., Bornhardt, C., Munoz, C., Antileo, C. (2006). Differential kinetics of ammonia- and nitrite-oxidizing bacteria: a simple kinetic study based on oxygen affinity and proton release during nitrification. Process Biochemistry, **41**, 1764–1772.

Diffusair (2014). Diffusaire, Available at: www.diffusaire.com.

Emefcy-BioEnergy Systems (2014). SABRE-Spiral Aerobic Biofilm Reactor-EMEFCY Available at: http://www.emefcy.com/product.php?ID=47 (accessed 15 September 2014).

Energy Efficiency in the Water Industry: A Compendium of Best Practices and Case Studies UKWIR Report REF No 10/CL/11/13.

Eoin Casey, Eoin Syron, John Shanahan and Michael J. Semmens (2008). Comparative economic analysis of full scale MABR configurations. IWA North American Membrane Research Conference.

Giesen and Thompson (2013). Aerobic Granular Biomass for Cost-Effective, Energy Efficient and Sustainable Wastewater Treatment. 7th European Waste Water Management Conference.

Haney P. D. (1954). Theoretical principles of aeration. *Journal American Water Works Association*, **46**(4), 353–376. (accessed 15 September 2014)

Ibrahim Gar Al-Almrashed, Ahmed El-Morsy, Mohamed Ayoub (2013). A new approach for upgrading of sewage treatment plants to accommodate excess organic and hydraulic loads. *Journal of Water Sustainability*, **3**(3), 153–163, University of Technology Sydney & Xi'an University of Architecture and Technology.

Issy Caffoor (2008). Environmental Knowledge Transfer Network, Business Case 3: Energy EfficientWater and Wastewater Treatment – A Priority Technology for the UK.

Kadar Y. and Siboni G. (2006). Optimization of Energy Economy in the Design and Operation of Wastewater Treatment Plants, Mekorot Water Company, Ltd. and DHV-MED, Ltd. Tel-Aviv, Israel, 17th Congress of world energy congress, Houston, Texas.

Koichi Terasakan, Ai Hirabayashi, Takanori Nishino, Satoko Fujiokaa, Daisuke Kobayashi (2011). Development of microbubble aerator for waste water treatment using aerobic activated sludge. *Chemical Engineering Science*, **66**, 3172–3179.

Leon Downing and Robert Nerenberg (2008). Total nitrogen removal in a hybrid, membrane-aerated activated sludge process. Water Research, **42**(14), 3697–3708.

Lucas Seghezzo, Grietje Zeeman, Jules B. and van Lier H. V. M. (1998). Hamelers, GatzeLettinga, a review: the anaerobic treatment of sewage in UASB and EGSB reactors. *Bioresource Technology*, **65**(3), 175–190.

Marco R. Menendez P. E., Black and Veatch (2010). How we use Energy at Wastewater Plants...and How We Can Use Less. Annual Conference Technical Papers – NC AWWA-WEA.

Martin K. J. and Nerenberg R. (2012). The membrane biofilm reactor (MBfR) for water and wastewater treatment: principles, applications, and recent developments. Bioresource Technology, **122**, 83–94.

Ronen Itzhak Shechter, Lior Eshed, Eytan Baruch Levy, Tamar Ashlagi Amiri. Diffusion aeration for water and wastewater treatment. US 20120273414 A1, EMEFCY.

Rosso D. and Tenstrom M. K. (2007). Energy-saving benefits of denitrification. *Environmental Engineering*, **43**(3), 29–38.

Ruiken C. J., Breuer G., Klaversma E., Santiago T. and van Loosdrecht M. C. M. (2013). Sieving wastewater – cellulose recovery, economic and energy evaluation. *Water Research*, **47**(1), 43–48.

Sanox OY Ltd (2014). OxTube: Available at: http://www.sansox.fi/oxtube.html (accessed 15 September 2014).

Semmens M. J. (2005). Membrane Technology: Pilot Studies of Membrane-Aerated Bioreactors. Water Environment Research Foundation Report, Treatment Processes, Final Report, 2005.

Shanahan J. W. and Semmens M. J. (2006). Influence of a nitrifying biofilm on local oxygen fluxes across a micro-porous flat sheet membrane. *Journal Of Membrane Science*, **277**(1–2), 65–74.

Sorubin (2014). Sorubin, Available at: http://www.sorubin.se/index.php?lang=en&tab=4&document=16 (accessed 15 September 2014).

Stricker A.-E., Lossing H., Gibson J. H., Hong Y. and Urbanic J. C. (2011). Pilot scale testing of a new configuration of the membrane aerated biofilm reactor (MABR) to treat high-strength industrial sewage. *Water Environment Research*, **83**(1), 3–14.

Syron E. and Casey E. (2008). Membrane-aerated biofilms for high rate biotreatment performance appraisal, engineering principles and development requirements. *Environmental Science & Technology*, **42**(6), 1833–1844.

Syron E., Vale P. and Casey E. (2014). Where did the Bubbles go? How to Reduce the Energy Requirements for Municipal Wastewater Treatment. (Conference proceedings) IWA LET, Abu Dhabi.

Timberlake D. L., Strand S. E. and Williamson K. J. (1988). Combined aerobic heterotrophic oxidation, nitrification and denitrification in a permeable-support biofilm. *Water Research*, **22**(12), 1513–1517.

Voss M. A. (1994). Membrane Gas Transfer: Practical Applications. MS Thesis, University of Minnesota, MN.

Yousfan and Shechter (2011). Device and Method for Dissolving Gas into a Liquid. US Patent application US 2011/0309034 A1.

Chapter 5

Integration of energy efficient processes in carbon and nutrient removal from sewage

Simos Malamis, Evina Katsou and Francesco Fatone

Department of Biotechnology, University of Verona, Strada Le Grazie 15, 37134, Verona, Italy

5.1 INTRODUCTION

The anthropogenic activities of the combustion of fossil fuel, the production of nitrogen and phosphorus fertilizers, intensive cultivation as well as other actions have resulted in accelerating the global nitrogen and phosphorus cycles (Vitousek *et al.* 1997). As a result, several environmental problems have occurred to aquatic ecosystems which include groundwater pollution by nitrate, acidification and cultural eutrophication of water bodies and toxic effects caused to aquatic organisms by organic and inorganic nitrogen forms (Cervantes, 2009). Consequently, nutrient removal from sewage is important within wastewater treatment plants (WWTPs) and can help alleviate the aforementioned environmental problems related to the discharge of nutrient rich effluents.

Wastewater facilities are increasingly being required to implement treatment process improvements in order to meet stricter discharge limits with respect to nitrogen and phosphorus. Wastewater treatment is an energy intensive process, accounting for approximately 3% of electricity use in developed countries. The greenhouse gases (GHG) emissions, energy consumption and the cost of wastewater treatment increase when nutrient removal is implemented (Keller & Hartley, 2003; USEPA, 2007). Balancing between nutrient removal efficiency and the cost of treatment is critical in order to implement a solution that will meet the required limits at an acceptable cost (WERF, 2011).

This chapter reviews the conventional organic carbon and nutrient removal bioprocesses and the latest technologies for nutrient removal, which include partial nitritation coupled with the anoxic ammonium oxidation (anammox), nitritation/denitritation and via nitrite/nitrate phosphorus accumulation. The various biological nutrient removal (BNR) processes are evaluated based also on the energy aspect.

5.2 REGULATORY BACKGROUND

The urban wastewater treatment directive (91/271/EEC and its amendment 98/15/EEC) of the European Union (EU), is an emission oriented regulation which defines the required treated effluent quality of municipal wastewater discharged into water bodies in terms of chemical oxygen demand (COD), biochemical oxygen demand (BOD_5), total suspended solids (TSS), nitrogen and phosphorus concentration. The Directive imposes nutrient limits in the treated effluent that is discharged to sensitive water bodies subject to eutrophication; these limits are based on agglomeration size (i.e., population equivalent PE) (Table 5.1). The EU Nitrate Directive (91/676/EEC) imposed limits to the amount of nitrate that is applied to land within nitrate vulnerable zones, in order to protect the surface and groundwater.

Table 5.1 EU Nutrient limits imposed for the discharge of treated effluent from municipal WWTPs into sensitive water recipients (Council Directive 91/271/EEC).[a]

Parameter	Concentration[b]	Minimum decrease of load
N	15 mg/L (PE 10,000–100,000) 10 mg/L (PE > 100,000)	80
P	2 mg/L (PE 10,000–100,000) 1 mg/L (PE > 100,000)	70–80%

[a]Either the concentration or the minimum decrease conditions should be satisfied.
[b]Concentration refers to annual mean values.

The Water Framework Directive (WFD – 2000/60/EC) presented a breakthrough in European Water Policy and set the nutrients as a major element of 'good water status' to be reached by 2015 for water bodies, confirming the importance of nutrient control to prevent water pollution. As the WFD focuses on the quality of the water bodies it signals an official move from emission to imission based standards. Responding to the need for upgrading the quality of water bodies EU countries has set strict nutrient limit values to treated effluents discharged into specific water bodies and/or to nutrient limits to the actual water bodies. Such examples are the Lagoon of Venice in Italy and Asopos River in Greece (Table 5.2).

In the United States of America (USA) and Canada the nutrient effluent permits are determined at the state or provincial level based on imission criteria. In this approach, the capacity of the receiving water bodies are considered and are translated to different effluent requirements at a local level. As a result, certain water bodies such as Canada's Great Lakes region have strict limits concerning the treated effluents (0.25–0.5 mgP/L for phosphorus and lower than 6 mgN/L for nitrogen). The US Environmental Protection Agency (US EPA) has set very strict nutrient limits for 14 selected eco-regions in the USA in order to address the total maximum daily load (TMDL) standards for surface waters. In some regions the WWTPs have to meet the limit of treatment technology (LOT). The LOT is defined as the lowest economically achievable effluent quality, which for nitrogen is <1.5 to 3 mgN/L and for phosphorus 0.07 mg/L to 0.1 mgP/L (Barnard, 2006; Oleszkiewicz & Barnard, 2006).

Table 5.2 Examples of limit values for nutrients in the treated effluent discharged into water bodies.

TN (mgN/L)	NO_3–N (mgN/L)	NO_2–N (mgN/L)	NH_4–N (mgN/L)	TP (mgP/L)	PO_4–P (mgP/L)	Location	Legislative Act
10	–	0.3	2	1	0.5	Lagoon of Venice, Italy	Ministry Decree 30/07/99 – "Ronchi-Costa"
10	7	0.5	2	1		Asopos River, Greece	KYA 20488/ 2010

5.3 ENERGY CONSIDERATIONS

The potential energy available in raw wastewater entering a municipal WWTP exceeds the electricity requirements of the treatment process. Energy contained in organics entering the plant can be related to the COD load of the influent flow. This organic load is subjected to aerobic and anaerobic degradation processes, partly releasing the captured energy. Three different types of energy can be distinguished: heat, synthesis energy and electricity. Aerobic metabolism yields a high amount of energy which can hardly be put to good use. Energy is produced by catabolic biochemical reactions as the substrate is oxidised by bacteria. The energy is used for the synthesis and maintenance of the cells. Synthesis energy is thus used by bacteria during their anabolic biochemical reactions in order to develop new biomass. Synthesis energy is related to high excess sludge production. The anaerobic digestion process generates much lower synthesis energy and less heat than aerobic processes, which has much higher impact because of high concentrations in the solids train. A major part of the energy content is captured in methane. This amount of energy can be transformed by a combined heat power (CHP) unit to electrical energy and heat, which can then be used. These energy products can be recycled to the wastewater treatment line and thus provide energy for the aeration system and heat the digesters. The process should divert as much organics as possible, from the wastewater treatment line to the anaerobic solids train in order to increase as much as possible energy recovery. However, compliance with nutrient removal requirements remains the overriding objective.

5.4 CONVENTIONAL BIOLOGICAL NUTRIENT REMOVAL PROCESSES

5.4.1 Description of alternative conventional BNR processes and configurations

Biological nutrient removal (BNR) has emerged as the preferred approach for nutrient removal as it has much lower operating expenses compared to physico chemical processes. BNR is currently integrated within the biological treatment of municipal wastewater; nitrogen is removed through the biological processes of nitrification and denitrification, and is finally converted into a gaseous form and escapes into the atmosphere. In the nitrification bioreaction, ammonium (NH_4^+) is oxidized first to nitrite (NO_2^-) by the autotrophic ammonium oxidizing bacteria (*Nitrosomonas)* and then to nitrate (NO_3^-) by the autotrophic nitrite oxidizing bacteria known as *Nitrobacter* and *Nitrospira.* These reactions take place strictly under aerobic environment. During denitrification, nitrate is biologically reduced to nitric oxide (NO), nitrous oxide (N_2O), and finally to nitrogen gas (N_2) by heterotrophic bacteria in an anoxic environment (absence of oxygen, presence of nitrate/nitrite) (Metcalf & Eddy, 2003). Figure 5.1 summarizes the nitrogen related bioreactions that can take place in a WWTP. The two stage oxidation of ammonium to nitrate consumes alkalinity (7.14 g of $CaCO_3$ are required to oxidize 1 g N) and requires oxygen. According to stoichiometry, 4.57 g O_2 is required to oxidize 1 g N, if all the nitrogen was oxidized. However, the bacteria also require nitrogen for growth, thus reducing the oxygen requirements to 4.33 g O_2 per g N removed. The denitrification process requires anoxic conditions and the presence of organic carbon as electron donor. To denitrify nitrate to gaseous nitrogen the amount of organic carbon required is given by $2.86/(1 - Y_H)$ where Y_H is the heterotrophic biomass yield.

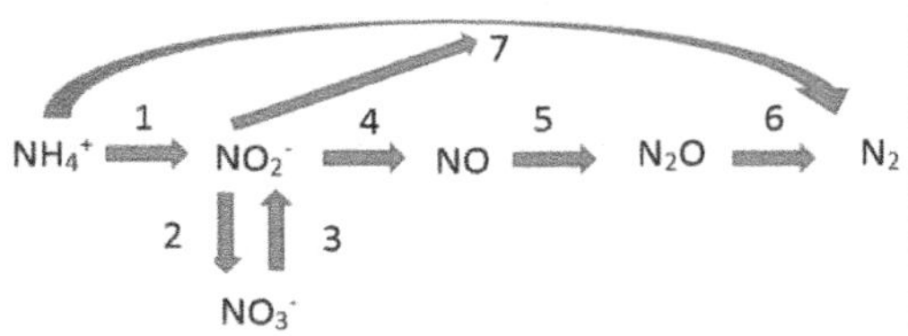

1. Aerobic ammonium oxidation to nitrite
2. Aerobic nitrite oxidation to nitrate
3. Nitrate reduction to nitrite
4. Nitrite reduction to nitric oxide
5. Nitric oxide reduction to nitrous oxide
6. Nitrous oxide reduction to N_2
7. Anoxic ammonium oxidation to N_2 (Anammox reaction).

Figure 5.1 Nitrogen bioreactions which can take place in a WWTP.

Phosphorus removal from wastewater effluents within WWTPs can be achieved in two fundamentally different ways: by chemical precipitation and by enhanced biological removal. In both ways, phosphorus is trapped in the solid matrix and is then separated from the liquid through the subsequent secondary sedimentation process. Chemical removal of phosphorus is accomplished through the addition of aluminium and iron coagulants or lime to form phosphorus precipitates. It is a flexible process for phosphorus removal which can be implemented at various stages, in the WWTP. Specifically, the chemicals can be applied (i) before primary sedimentation, (ii) directly inside the biological treatment process or (iii) at the secondary effluent, as a tertiary treatment process. In the first case, the phosphorus precipitates in the primary settling tank and is thus removed with primary sludge. In the second case phosphorus precipitates in the mixed liquor and is removed with waste activated sludge. The last case is less often practiced due to increased costs and the need for an additional separation stage (Morse *et al.* 1998). The chemical removal of phosphorus has the disadvantages of having higher operating expenses, producing more sludge and resulting in the addition of chemicals compared to biological phosphorus removal (US EPA, 2007). In the enhanced biological phosphorus removal (EBPR) process, phosphorus accumulates in activated sludge by phosphorus accumulating organisms (PAOs). These bacteria have the ability to accumulate much more phosphorus (up to 10% of their dry weight) compared to normal bacteria. This is accomplished through the implementation of a sequence of anaerobic and aerobic conditions. Under anaerobic conditions, the PAOs break the polyphosphate chains stored in their cells to generate energy; as a result phosphorus is released from the solid to the liquid phase. Then PAOs use the generated energy to convert the readily biodegradable organic matter (volatile fatty acids) that is present in the liquid into organic carbon which is stored internally in the form of polyhydroxyalkanoates (PHAs) as an energy and food reserve. Then, under aerobic conditions the PAOs utilize PHA as an energy reserve to uptake phosphorus (Metcalf & Eddy, 2003). Figure 5.2 depicts diagrammatically the EBPR process.

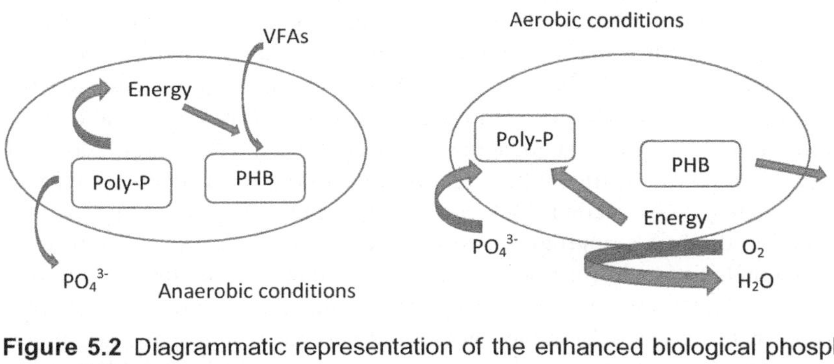

Figure 5.2 Diagrammatic representation of the enhanced biological phosphorus removal process.

Based on the above biological principles of BNR various technologies and configurations have been successfully adopted for biological nitrogen and phosphorus removal. Figure 5.3 shows the various configurations, for both suspended and attached growth processes, that can be used to remove nitrogen from wastewater. The existing conventional nitrogen removal technologies can decrease the treated effluent nitrogen concentrations to levels lower than 10 mg/L. The modified Ludzack Ettinger (MLE) is a suspended growth process, with an anoxic reactor where denitrification occurs followed by an aerobic reactor where nitrification takes place. The recirculation of mixed liquor to the anoxic reactor provides the required nitrates for denitrification. During aerobic conditions, the aerobic biodegradation of the organics takes place together with nitrification. The organic source required for denitrification of nitrate is provided by the influent sewage. The MLE can also be combined with simultaneous nitrification/denitrification (MLE/SND). In SND micro-aerobic conditions prevail, resulting in aerobic conditions at the exterior of the flocs and anoxic conditions in the interior. Thus, nitrification and denitrification takes place within the same reactor simultaneously. The four stage Bardenpho process is a suspended growth process with the sequence of anoxic/aerobic/anoxic/aerobic which could also include a SND stage. An external organic carbon source should be added in the second anoxic reactor to accomplish denitrification at a significant rate. This configuration having two aerobic and two anoxic zones results in higher nitrogen removal and in an effluent with lower nitrogen concentration than the two zone. The moving bed biofilm reactor (MBBR) is an attached growth process where many biofilm carriers are suspended inside the biological reactor. The biofilm develops on the surface of these carriers. Suitable anoxic and aerobic conditions can be maintained to favour the nitrogen removal processes. The biological aerated filter (BAF) is an attached growth process consisting of submerged media filter. Wastewater is pumped through the filter media and is treated as it comes into contact with it. The media filter provides a surface for microorganisms to grow on. Aeration is provided from the bottom part of the reactor. The BAF is a flexible system where filtration and biological treatment of wastewater are combined. The rotating biological contactor (RBC) is an attached growth process in which biofilm develops on parallel rotating disks mounted on a rotating shaft (Barnard, 2006). The sequencing batch reactor (SBR) process is a suspended growth process in which all processes take place within the same reactor in a sequencing manner. The sequence that is followed is: fill, reaction phase, sedimentation and decant. In the reaction phase aerobic and subsequently anoxic conditions can be maintained for nitrogen removal. The oxidation ditch is a continuous flow suspended growth process with uses looped channels to create time sequenced anoxic and aerobic zones to remove nitrogen. Nitrogen removal is also integrated in membrane bioreactor (MBR) processes. In these cases, suitable anoxic and aerobic reactors are integrated with the membrane modules. The latter can be placed within the aerobic tank, in a separate reactor or as an external membrane module in the external configuration.

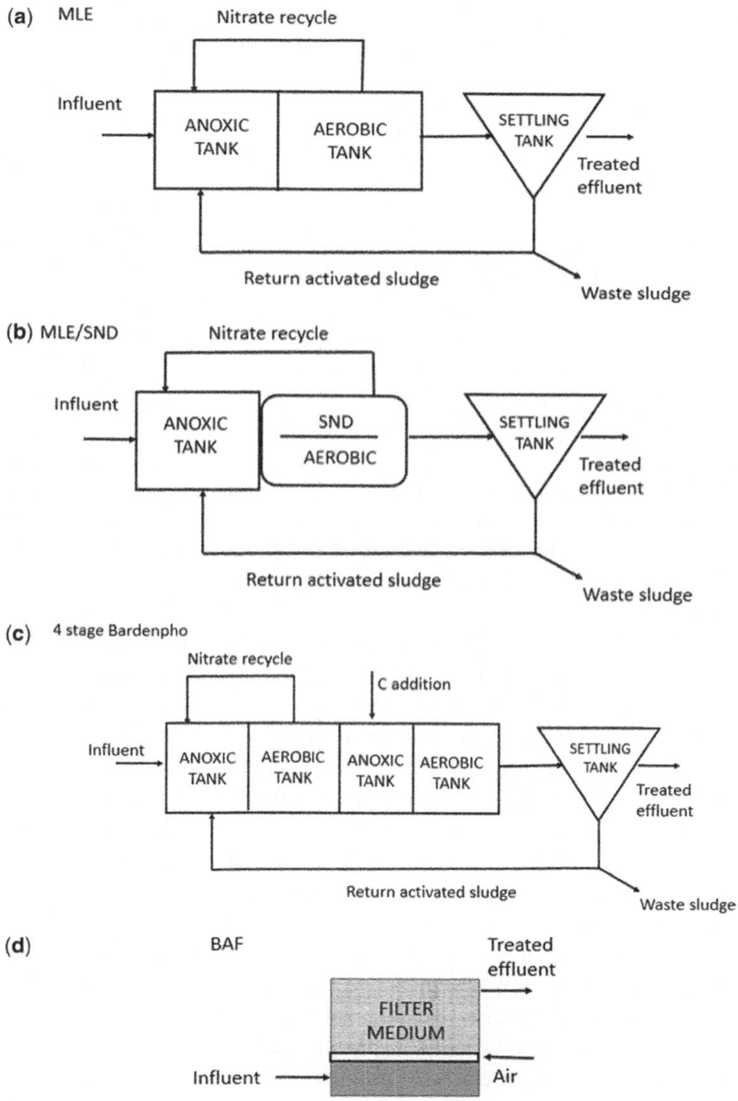

Figure 5.3 Attached and suspended biomass configurations which can be applied to remove nitrogen (SND = simultaneous nitrification/denitrification, C = dosing of external organic carbon).

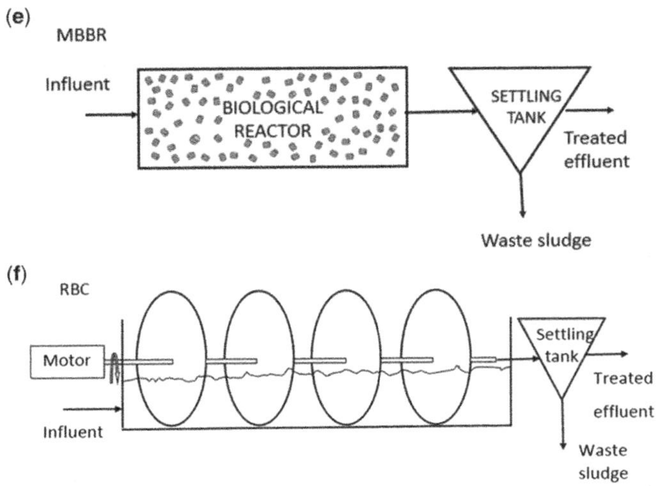

Figure 5.3 (*Continued*)

Several configurations for the EBPR have been invented, developed and implemented (Figure 5.4). In practice, the EBPR is successfully implemented when the alternation of anaerobic/aerobic is ensured. Any bioprocess in which the nitrate/nitrite is prevented from entering the anaerobic bioreactor and an adequate readily biodegradable COD is provided during the anaerobic phase can accomplish EBPR (Barnard, 2006). Therefore, EBPR can be integrated within any of the configurations described above for nitrogen removal, provided a dedicated anaerobic phase is established. The phosphorus removal configurations shown in Figure 5.4 place the anaerobic reactor upstream of the biological process. This has the advantage that volatile fatty acids present in sewage can be used by the phosphorus accumulating organisms (PAOs). The simplest configuration is the Phoredox (A/O) process with one anaerobic reactor placed before the aerobic one. In the SBR process, EBPR can be accomplished by including an anaerobic phase just after the filling and before the aerobic reaction. Thus, the sequence of anaerobic, aerobic, anoxic phase in an SBR can successfully remove phosphorus and nitrogen from wastewater. The two and four stage Bardenpho processes can be modified to remove phosphorus by inserting an anaerobic reactor upstream. These configurations are known as the three and five stage modified Bardenpho processes (Wentzel *et al.* 2008).

In the University of Cape Town (UCT) process and its modified version (which has two anoxic tanks), an internal recycle of nitrate is carried out from the aerobic tank to the anoxic one in order to achieve effective denitrification. Furthermore, to minimize the entrance of nitrates in the anaerobic zone, the mixed liquor is recycled from the anoxic tank to the anaerobic one and the settled sludge is returned to the anoxic tank. By manipulating the nitrate recycle ratio, the nitrate concentration in the anoxic tank can drop to zero. As a result, the anoxic recycle will not introduce any nitrate into the anaerobic tank. The modified version includes two anoxic tanks in order to have better control of the recycle. In the Johannesburg configuration, the elimination of nitrates in the mixed liquor recycle to the anaerobic tank is accomplished by a second anoxic tank which removes the nitrate from the recycle. The Biodenipho/Biodenitro process is a phased isolation ditch system where nitrogen removal is accomplished by the phased aerobic/anoxic reactors which are interchanged. The anaerobic conditions are accomplished upstream in a separate tank (Wentzel *et al.* 2008).

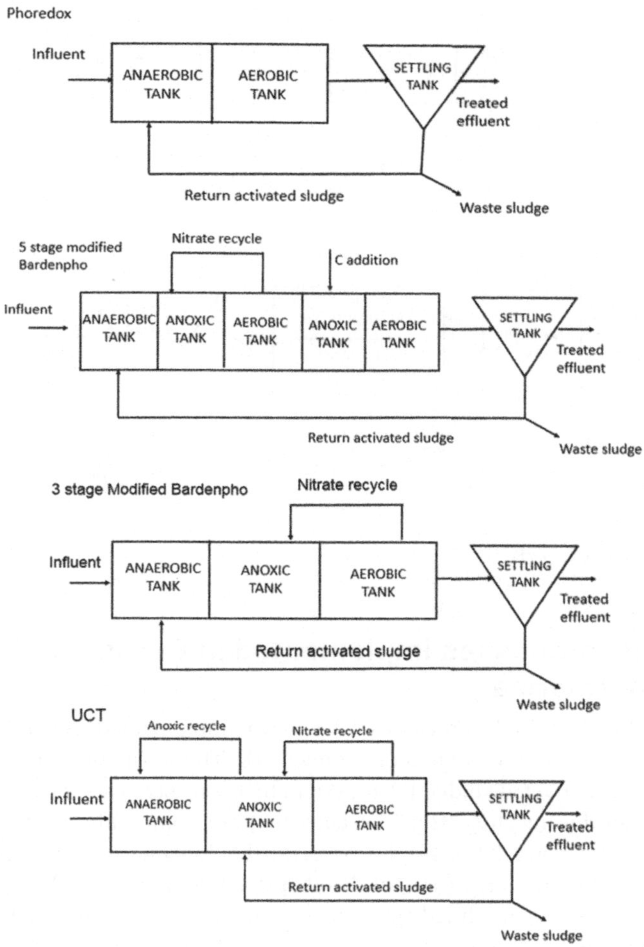

Figure 5.4 Configurations applied in WWTPs to removal phosphorus biologically (most schemes also integrating nitrogen removal).

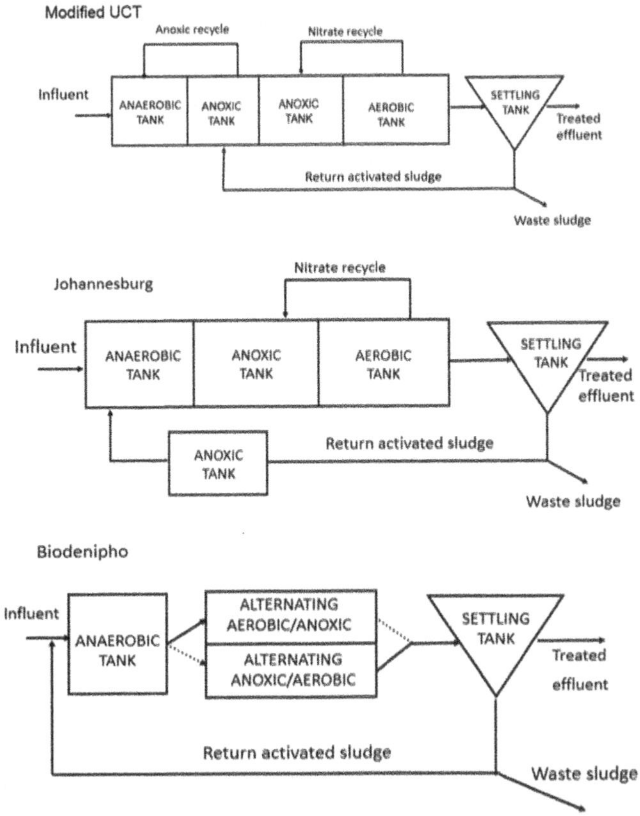

Figure 5.4 (*Continued*)

5.4.2 BNR processes implemented in Europe and Northern America

In Europe different BNR processes have been implemented depending on the financial resources and the regulatory framework. The most common process that is applied is the activated sludge BNR system that consists of a series of anaerobic, anoxic and aerobic reactors. BNR systems that are commonly used in Northern Europe are the modified Bardenpho process, the Johannesburg process and the alternating continuous processes of Biodenitro/Biodenipho. In WWTPs where EBPR is practiced, the modified Bardenpho or the Johannesburg process are often applied, having a three-stage reactor in the sequence of anaerobic, anoxic, aerobic with a pre-denitrification of the return activated sludge. Other plants employ an anaerobic zone followed by simultaneous nitrification/denitrification. The fermentation of primary sludge to produce short chain fatty acids that can be applied as carbon source to the BNR process is increasingly being applied in WWTPs in Europe. MBBR and IFAS plants are often implemented when a small footprint is required (Oleszkiewicz & Barnard, 2006). These processes use suspended carriers onto with the biofilm develops and are compact leading to significant reduction in the bioreactor volumes.

Many WWPTs in Central North America still practice only carbon removal or have recently been upgraded for BNR as a response to the imposed regulations. The tendency for upgrading is to conduct chemical precipitation of phosphorus and extend the aerobic phase in order to accomplish complete ammonium oxidation. In WWTPs where a low total nitrogen concentration limit must be met, post-denitrification using denitrifying filters is usually carried out (de Barbadillo *et al.* 2005). In several US plants the activated sludge BNR system is increasingly being adopted either as Bardenpho-like process or as MLE coupled with chemical precipitation. In western Canada the Westbank process being the most popular process. In Eastern Canada the high rate activated sludge process is typically used together with chemical precipitation. Since denitrification is not practiced, the mixed liquor contains nitrates and thus EBPR cannot be carried out. Finally, in Quebec attached growth processes (BAF) are usually employed (Oleszkiewicz & Barnard, 2006). In the last several years, nitritation/denitritation and partial nitritation with the anoxic ammonium oxidation (i.e., the completely autotrophic nitrogen removal process) are being implemented as alternative low cost treatment options for nitrogen removal. These processes are further discussed in Section 5.5.

5.4.3 Energy requirements and cost of conventional BNR processes

Figure 5.5 shows the typical energy requirements of a typical medium size WWTPs serving a population equivalent (PE) of 400,000. The majority of the energy demand (55%) is due to aeration requirements for the biological processes. The energy requirements due to pumping can vary considerably depending on the morphology and plant configuration. Given, the capital and operating expenses of the anaerobic digestion process, it is usually implemented in WWTPs having a PE higher than 40,000 in order to realize energetic benefits within a reasonable time horizon.

To evaluate the impact of the implementation of nitrogen removal on the plant efficiency, it is important to document the increase in the treatment cost that is incurred due to the adoption of nitrification/denitrification. The implementation of BNR increases the cost mainly due to the increased aeration requirements of nitrification, internal nitrate recirculation and potentially the addition of chemicals to provide the organic carbon source for denitrification and enhanced biological phosphorus removal (EPRI, 2002).

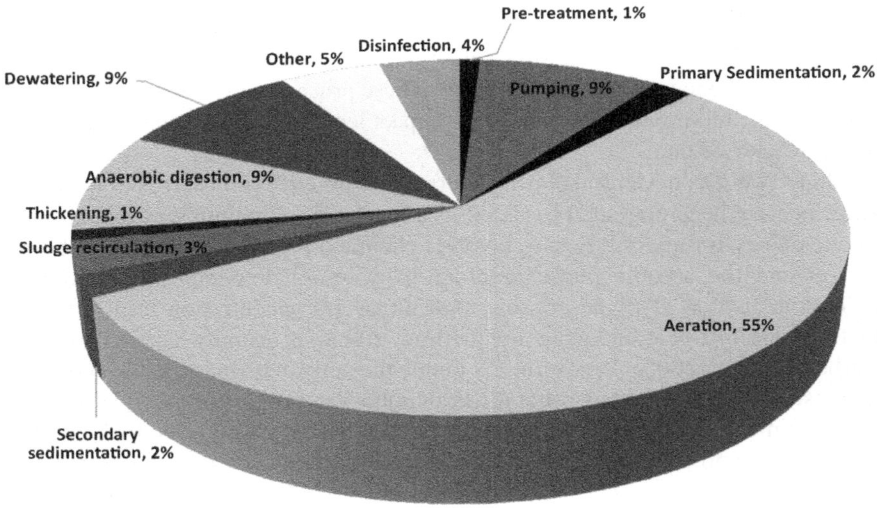

Figure 5.5 Contribution of various processes to the energy requirements of a typical medium sized WWTP having a population equivalent of 400,000.

In several cases, particularly in North America, retrofitting of existing WWTPs is required to include BNR together with organic carbon removal. In other cases new BNR WWTPs plants are developed. The BNR cost for new plants is very different from that of retrofits. The cost of new BNR plants depends on the influent loads and on the required treated effluent quality. The retrofit costs are more difficult to present and discuss as they are more site specific and vary significantly for any given plant size. They depend on the same parameters as the new BNR plants as well as on the layout and design of the existing WWTP (US EPA, 2007). Tables 5.3 and 5.4 summarize the total capital cost for the retrofitting of different BNR plants in Maryland and Connecticut respectively and Table 5.5 shows the capital and operating and maintenance (O&M) costs for new BNR plants and for retrofits. Clearly, the retrofits result in much lower construction and O&M cost compared to new BNR plants. Furthermore, the ratio of capital BNR cost/design capacity is very wide ranging from 16 $\$/(m^3/d)$ to 5234 $\$/(m^3/d)$ showing that BNR retrofit cost varies greatly depending on the specific case study. Even in WWTPs with similar design capacity, the BNR retrofit cost can vary significantly.

In terms of phosphorus removal, studies show that the cost is relatively low if the goal is to achieve total phosphorus concentration in the treated effluent up to 1 mgP/L. However, the cost increases significantly when the treated effluent should have a phosphorus concentration lower than 0.5 mgP/L.

Table 5.3 Costs for upgrading WWTPs with BNR for Maryland (US EPA, 2007).

Facilities with BNR	Design capacity (m^3/d)	Treatment process	Capital BNR cost (2006$)	Capital BNR cost/design capacity (2006 $/($m^3/d$))
Leonardtown	2461	Biolac	2,811,448	1142
Little Patuxent	68,137	A^2/O	7,263,879	107
Marlay Taylor (Pine Hill Run)	17,034	Schreiber	4,986,641	293
Maryland City	9464	Schreiber	1,375,866	145
Maryland Correctional Institute	4656	Bardenpho	2,703,932	581
Mt. Airy	2271	Activated sludge	5,235,575	2305
Northeast	7571	Activated sludge	4,225,029	558
Pakrway	28,391	Methanol	15,869,228	559
Patuxent	22,712	Oxidation ditch	2,106,763	93
Piscataway	113,562	MLE	24,778,239	218
Pocomoke City	5300	Biolac	3,924,240	740
Poolesville	2366	SBR	1,593,640	674
Princess Anne	4770	Activated sludge	4,311,742	904
Seneca	18,927	MLE	34,886,034	1843
Sod Run	45,425	MLE	21,999,198	484
Taneytown	2650	SBR	3,808,298	1437
Thurmont	3785	MLE	3,122,264	825
Western Branch	113,562	Methanol	47,132,782	415
Westminster	18,927	Activated sludge	5,274,444	279
Aberdeen	10,599	MLE	3,177,679	300

(*Continued*)

Table 5.3 Costs for upgrading WWTPswith BNRfor Maryland (US EPA, 2007) (*Continued*).

Facilities with BNR	Design capacity (m^3/d)	Treatment process	Capital BNR cost (2006$)	Capital BNR cost/design capacity (2006 $/($m^3/d$))
Annapolis	37,854	Ringlace	14,687,326	388
Back River	681,374	MLE	138,305,987	203
Ballenger	7571	Modified bardenpho	2,891,906	382
Broadneck	22,712	Oxidation Ditch	3,165,193	139
Broadwater	7571	MLE	6,892,150	910
Cambridge	30,662	Activated sludge	11,740,209	383
Celanese	4732	Sequential step feed	7,424,068	1569
Centreville	1420	SBR/Land Application	7,336,020	5166
Chesapeake beach	2839	Oxidation ditch	2,158,215	760
Conococheague	9464	Carrousel	6,620,888	700
Cox Creek	56,781	MLE	11,466,657	202
Cumberland	56,781	MLE	12,929,990	228
Denton	1703	Biolac	4,203,767	2468
Dorsey run	7571	Methanol	3,967,307	524
Emmitsburg	2839	Overland	2,562,722	903
Frederick	30,283	MLE	11,916,504	394
Freedom DISTRICT	13,249	Activated sludge	1,462,798	110
Fruitland	1893	SBR	7,546,764	3987
Hagers town	30,283	Johannesburg	11,190,344	370
Havre DeGràce	7154	MLE	7,596,882	1062
Hurlock	7571	Bardenpho	5,200,000	687
Joppatowne	3596	MLE	2,433,205	677

Table 5.4 Costs for upgrading WWTPs with BNR for Connecticut (US EPA, 2007).

Facilities with BNR	Design capacity (m^3/d)	Treatment process	Capital BNR cost (2006$)	Capital BNR Cost/design capacity (2006$/($m^3/d$))
Branford	17,034	4-stage Bardenpho	3,732,049	219
Bridgeport East Phase 1	45,425	MLE	2,323,766	51
Bridgeport West Phase 1	109,777	MLE	2,640,643	24
Bristol Phase 1	40,693	MLE	649,320	16
Derby	11,470	MLE	3,513,514	306
East Hampton	14,763	MLE	890,548	60
East Windsor	9,464	MLE	1,407,617	149
Fairfield Phase 2	34,069	4-stage Bardenpho	14,235,676	418
Greenwich	45,425	MLE	703,809	16
Ledyard	908	SBR	4,752,461	5234
Milford BB Phase 1	11,735	4-stage Bardenpho	1,407,617	120
New Canaan	5,678	MLE	1,570,463	277
New Haven Phase 1	151,416	MLE	11,134,336	74
New London	37,854	MLE	3,495,615	92
Newtown	3,528	MLE	1,436,601	407
Norwalk Phase 1	56,781	MLE	1,548,379	27
Norwalk Phase 2	56,781	MLE	7,042,287	124
Portland	3,785	MLE	1,266,843	335
Seymour	11,091	MLE	379,597	34
Stratford Phase 1	43,532	4-stage Bardenpho	1,126,094	26
Thomaston	4,543	SBR	1,451,708	320
University of Connecticut	7,495	MLE	1,489,259	199
Waterbury	94,635	4-stage Bardenpho	22,074,225	233

Table 5.5 Average BNR Costs for new small scale WWTPs (US EPA, 2007).

System	15 m³/d	40 m³/d	100 m³/d	190 m³/d	380 m³/d
MLE process					
Construction (2006$)	348,771	415,585	563,912	803,108	1,167,914
O&M (2006$)	37,263	43,515	60,553	61,636	122,699
4-Stage bardenpho process					
Construction (2006$)	448,992	491,753	634,736	889,966	1,293,524
O&M (2006$)	64,353	70,604	90,462	117,551	162,169
3-Stage process					
Construction (2006$)	388,859	444,983	589,302	837,851	1,220,029
O&M (2006$)	44,005	51,360	69,133	93,403	142,066
SBR process					
Construction (2006$)	448,992	509,125	644,090	931,391	1,290,852
O&M (2006$)	34,321	41,799	60,185	82,862	122,577
Intermittent aeration process					
Construction (2006$)	306,009	499,771	780,391	1,150,542	1,371,029
O&M (2006$)	34,321	41,799	60,185	62,862	122,577

5.5 INNOVATIVE BIOPROCESSES IN THE MAINSTREAM AND SIDESTREAM

Two innovative BNR processes are the nitritation/denitritation process and the completely autotrophic nitrogen removal process (Figure 5.6). Nitritation/denitritation is the oxidation of ammonium to nitrite and its subsequent reduction to gaseous nitrogen. The adoption of nitritation/denitritation as opposed to conventional nitrification/denitrification has significant advantages, since it theoretically reduces the oxygen demand up to 25% and requires up to 40% less external carbon. Furthermore, it decreases sludge production by 30% and carbon dioxide emissions by 20% (Gustavsson, 2010). To accomplish effective accumulation of nitrite, while arresting the formation of nitrates, the growth of the ammonium oxidizing bacteria (AOB) should be promoted, while the growth of nitrite oxidizing bacteria should be inhibited (Malamis *et al.* 2014). The NOB can be inhibited by maintaining a high free ammonia (FA) concentration (FA > 1 $mgNH_3 \cdot L^{-1}$) or high free nitrous acid (FNA > 0.02 $mgHNO_2$-N L^{-1}) concentration in the biological reactor (Anthonisen *et al.* 1976; Vadivelu *et al.* 2007; Gu *et al.* 2012). The AOB are also favoured over the NOB at low dissolved oxygen (DO) concentrations during aerobic conditions (DO = 0.4 – 1.0 mg $\cdot$ L^{-1})(Blackburne *et al.* 2008a) and high temperature (30–40°C) (Hellinga *et al.* 1998). Significant cost savings can further arise when fermented sewage sludge is supplied as carbon source to accomplish the denitrification process (Mayer *et al.* 2009; Ji & Chen, 2010; Longo *et al.* 2015). Short chain carbon sources produced from fermentation can promote nitritation/denitritation and denitrifying phosphorus removal via nitrite from the anaerobic supernatant (Frison *et al.* 2013a).

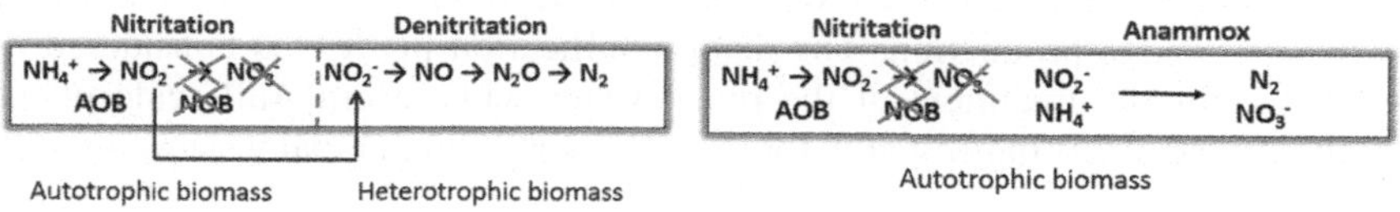

Figure 5.6 Nitritation/denitritation and nitritation/anammox processes.

The nitritation/denitritation process has been successfully implemented to treat the sludge reject water resulting from the dewatering of the anaerobically digested sewage sludge (Frison *et al.* 2013b). In this situation, the nitrite accumulation is easy to be established since the reject water is characterized by high ammonium concentration which also results in high FA (>2 $mgNH_3 \cdot L^{-1}$) for the typical pH in which it is found. The nitritation/denitritation can also be employed in the wastewater treatment line (Yang *et al.* 2007; Blackburne *et al.* 2008b). In this case the free ammonia and the temperature are lower. Therefore, effective control of the process is required to achieve the via nitrite pathway. Previous work has shown that an elevated nitrogen loading rate combined with low dissolved oxygen during the aerobic phase can accomplish complete nitrite accumulation (Katsou *et al.* 2015). According to the classification made by US EPA, the treatment of reject water through nitritation/denitritation is an innovative process with very few full scale applications worldwide (US EPA, 2013).

An alternative short-cut nitrogen removal process which is increasingly being adopted to treat nitrogenous effluents, including sludge reject water in WWTPs is the completely autotrophic nitrogen removal process or deammonification process (as it is also known) In this process ammonium is partially oxidized to nitrite (50%) and subsequently the anoxic ammonium oxidation (anammox) process is employed to remove nitrogen. In latter process the anammox bacteria convert the remaining ammonium and the produced nitrite into gaseous nitrogen under anoxic conditions (Figure 5.6). The biochemical reaction also results in the production of a small amount of nitrate. The deammonification process has been successfully implemented as a side-stream process for treating the sludge reject water produced from the dewatering of anaerobically digested sewage sludge. The relatively high temperature and high ammonia concentrations typically found in these recycle flows facilitate this process. Very few applications of the deammonification process exist for the treatment of the main wastewater treatment line. In this process, AOB should be promoted and NOB inhibited to allow nitrite accumulation. At the same time, the need for selective retention of the anammox bacteria is required. A full scale deammonification plant has been installed at the Strass WWTP in Austria where a side stream deammonification process can provide seed for bioaugmentation in the full-plant. The deammonification process has the advantages of no requirements for external carbon source (since the process is completely autotrophic), very low aeration requirements (57% reduction compared to conventional nitrification/denitrification) and very low excess sludge production. The deammonification process is considered as innovative for the treatment of the reject water with several full scale plants worldwide and emerging/research for the treatment of municipal wastewater (US EPA, 2013).

The disadvantages of the anammox process are that the anammox bacteria grow very slowly, and are very sensitive to environmental conditions. Nitrate, nitrite, dissolved oxygen and organic matter can inhibit anammox activity. Therefore, process stability can be a challenge particularly for full scale applications. Furthermore, the deammonification process does not remove phosphorus; it thus needs to be coupled with a phosphorus removal or recovery process. The DEMON®, SHARON-ANAMMOX, ANAMMOX®, Paques, ANITA-Mox, DeAmmon, CANON, OLAND are all autotrophic nitrogen removal processes. Depending on the developed process nitritation and anammox can take place in one reactor or in separate reactors. Table 5.6 summarizes the oxygen and COD requirements, the produced sludge and the treatment cost for conventional BNR, nitritation/denitritation and deammonification.

Table 5.6 Comparison of the conventional BNR with the advanced BNR processes.

Process	Oxygen requirements ($kgO_2/kgN_{rem.}$)	Organic carbon source requirements ($kgCOD/kgN_{rem.}$)	Sludge produced ($kgVSS/kgN_{rem.}$)	Total treatment cost[1] (€/kgN_{rem})
Nitrification/ denitrification	4.33	$2.86/(1-Y_H)$	0.5–1.0	3.5–5.5
Nitritation/ denitritation	3.25	$1.72/(1-Y_H)$	0.4–0.7	2.5–3.5
Deammonification	1.85	0	<0.1	1.5–2.5

[1]Total cost includes both capital and O&M cost.

Enhanced biological removal via nitrite can also be integrated with nitritation/ denitritation in order to remove phosphorus biologically. Studies have documented the occurrence of denitrifying phosphorus accumulating organisms (DPAOs) that can utilize nitrate or nitrite as electron acceptors instead of oxygen (Carvalho *et al.* 2007). The denitrifying phosphorus uptake can reduce the requirements for organic matter and sludge production through the simultaneous denitrification and phosphorus uptake. Ji and Chen (2010) investigated denitrifying phosphorus removal via nitrite in an SBR treating synthetic wastewater using sludge fermentation liquid as an external carbon source. The soluble phosphorus removal achieved was 97.6%. The main difficulty in applying denitrifying phosphorus removal via nitrite is that the DPAO activity is inhibited when these bacteria are exposed to significant nitrite concentrations (Saito *et al.* 2004).

5.6 NITROUS OXIDE EMISSIONS IN BNR

During the operation of a WWTP several off gas emissions can occur, including nitrous oxide (N_2O), carbon dioxide (CO_2), methane (CH_4) and nitric oxide (NO). N_2O is of particular environmental concern, since it has a global warming potential that is 298 times higher than that of CO_2. In terms of CO_2 equivalents (eq) nitrous oxide contributes by 7.9% to the total anthropogenic greenhouse gas (GHG) emissions. The global production of N_2O emissions from WWTPs corresponds to approximately 3.2% of the total estimated anthropogenic N_2O emissions (IPCC, 2001; Kampschreur *et al.* 2009). The N_2O emissions from wastewater management contribute by 26% to the total greenhouse gas (GHG) emissions of the water chain. The guidelines of the Intergovernmental Panel on Climate Change (IPCC, 2006) have decreased the standard N_2O emission factor from 1% to 0.5% of the influent nitrogen load of the WWTP influent (Kampschreur *et al.* 2009). It is estimated that N_2O emissions from wastewater correspond to 100 Mt CO_2 eq., while CH_4 emissions to 630 Mt CO_2 eq. for 2010 (Monni *et al.* 2006). Several practical design and operating decisions in WWTPs (including the BNR processes) have considerable impact on the overall environmental performance, including the GHG emissions (Keller & Hartley, 2003). At the level of a BNR treatment plant, the N_2O emissions can reach up to 83% of the operational CO_2 footprint (Desloover *et al.* 2011).

N_2O emissions during BNR occur during the biological processes of nitritation and denitritation (Figure 5.7). During the nitritation process, N_2O can be formed via two routes: the first pathway is as a by-product of the incomplete oxidation of hydroxylamine (NH_2OH) to nitrite. Hydroxylamine is formed through the oxidation of ammonium by ammonium oxidizing bacteria (AOB) and then NH_2OH is oxidized to nitrite a biochemical reaction which produces nitrous oxide. The second pathway of N_2O formation is attributed to the process known as nitrifier denitrification. In this biochemical process nitrite is used as electron acceptor instead of oxygen; this can occur during nitritation under limiting dissolved oxygen (DO) conditions. In this pathway, the reduction of NO_2^- to NO and to N_2O by AOB takes place. The third pathway occurs during the anoxic conditions. In this process, NO and N_2O are produced as process intermediates of nitrate reduction to N_2. This is the only stage in which N_2O is also consumed as it is reduced to N_2 (Ni *et al.* 2011).

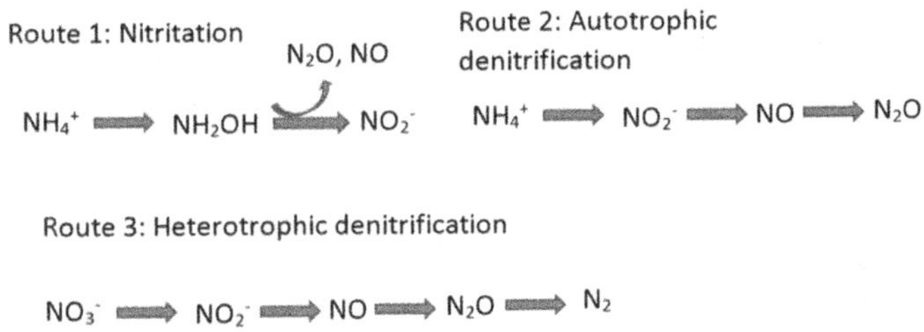

Figure 5.7 Simplified representation of biochemical processes responsible for nitrous oxide production during nitrification and denitrification: Route 1: aerobic ammonium oxidation by AOB, Route 2: nitrifier denitrification by AOB and Route 3: heterotrophic denitrification.

It is important to determine whether the implementation of the advanced bioprocesses described in the previous section, could increase the potential carbon footprint of the WWTP. Desloover *et al.* (2012) performed an overview of the quantified N_2O emissions from full scale BNR plants that apply the conventional nitrification/denitrification and advanced nitritation/anoxic ammonium oxidation (anammox) and nitritation/denitritation processes; they concluded that nitritation is the bioprocess that mainly contributes to N_2O emissions. Full-scale measurements also point to nitrite as a factor in N_2O production (Ahn *et al.* 2010). Taking into account the much higher greenhouse gas impact of N_2O compared to CO_2, it is necessary to determine whether nitrogen removal bioprocesses based on transient nitrite accumulation are systematically greater contributors of N_2O than full nitrification processes (IPCC, 2001; Ahn *et al.* 2011). N_2O emission rates can vary considerably due to the differences in the wastewater composition, the applied treatment process, the operating parameters and the environmental conditions. The most important parameters that affect the N_2O emissions include the DO concentration, the nitrite in the mixed liquor and the COD/N ratio during denitrification. Nitritation/denitritation is increasingly being applied, particularly for the treatment of nitrogenous effluents, due to the lower energy and organic carbon source requirements compared to conventional nitrification/denitrification. However, such processes can result in significant N_2O emissions (Kampschreur *et al.* 2009). Furthermore, the use of SBR technology, particularly when combined with the treatment of highly nitrogenous effluents such as sludge reject water can enhance the nitrous oxide emissions. The implementation of strategies to mitigate N_2O emissions in the BNR processes via nitrite can increase their sustainability.

5.7 CONCLUSION

Several configurations are currently applied for BNR in full scale WWTPs. Both attached and suspended growth processes are implemented. The adoption of the most appropriate BNR scheme is closely related to the treated effluent requirements for phosphorus and nitrogen. In cases where treated effluents are discharged to sensitive water bodies, the nutrient limits can be very low and the technology selection is critical. Nitritation/denitritation and the completely autotrophic nitrogen removal have emerged as alternative BNR processes compared to the conventional nitrification/denitrification. Such processes have the advantages of lower aeration requirements, resulting in lower operating expenses. They are increasingly being adopted for the treatment of the nitrogenous sludge reject water. In BNR processes it is also important to consider the GHG emissions and particularly nitrous oxide emissions. The aforementioned innovative processes should be implemented with care in order not to increase the overall carbon footprint of the WWTP.

5.8 ACKNOWLEDGEMENT

This work was supported by the Marie Curie FP7-PEOPLE-2012 project with title Low environmental footprint biological treatment processes for waste and wastewater treatment, LEF-BIOWASTE (Number 322333).

5.9 REFERENCES

Ahn J. H., Kim S., Park H., Rahm B., Pagilla K. and Chandran K. (2010). N_2O emissions from activated sludge processes, 2008–2009: results of a national monitoring survey in the United States. *Environmental Science and Technology*, **44**, 4505–4511.

Ahn J. H., Kwan T. and Chandran K. (2011). Comparison of partial and full nitrification processes applied for treating high-strength nitrogen wastewaters: microbial ecology through nitrous oxide production. *Environmental Science and Technology*, **45**, 2734–2740.

Anthonisen A. C., Loehr R. C., Prakasam T. B. and Srinath E. G. (1976). Inhibition of nitrification by ammonia and nitrous acid. *Water Pollution Control Federation*, **48**, 835–852.

Barnard J. L. (2006). Biological nutrient removal: where we have been, where we are going? Water Environment Federation's Technical Exhibition and Conference (WEFTEC), Dallas, US.

Blackburne R., Yuan Z. and Keller J. (2008a). Partial nitrification to nitrite using low dissolved oxygen concentration as the main selection factor. *Biodegradation*, **19**, 303–312.

Blackburne R., Yuan Z. Q. and Keller J. (2008b). Demonstration of nitrogen removal via nitrite in a sequencing batch reactor treating domestic wastewater. *Water Research*, **42**, 2166–2176.

Carvalho G., Lemos P. C., Oehmen A. and Reis M. A. M. (2007). Denitrifying phosphorus removal: linking the process performance with the microbial community structure. *Water Research*, **41**, 4383–4396.

Cervantes F. J. (2009). Anthropogenic Sources of N-Pollutants and their Impact on the Environment and on Public Health, Chapter 1. In: Environmental Technologies to Treat Nitrogen Pollution, F. J. Cervantes (ed.), IWA publishing, London, UK.

Commission Directive 98/15/EC of 27 February 199 Amending Council Directive 91/271/EEC with Respect to Certain Requirements Established in Annex I there of (Text with EEA relevance).

Council Directive 91/271/EEC of 21 of May 1991 Concerning Urban Wastewater Treatment. Official Journal of European Communities.

Council Directive 91/676/EEC of 12 December 1991 Concerning the Protection of Waters Against Pollution Caused by Nitrates from Agricultural Sources. Official Journal of European Communities.

de Barbadillo C., Rectanus R., Lambert S., Parker D., Wells J. and Willet R. (2005). Evaluating denitrification filters. *Water Environment & Technology*, **17**, 24–31.

Desloover J., De Clippeleir H., Boeckx P., Du Laing G., Colsen J., Verstraete W. and Vlaeminck S. E. (2011). Floc-based sequential partial nitritation and anammox at full scale with contrasting N_2O emissions. *Water Research*, **45**, 2811–2821.

Desloover J., Vlaeminck S. E., Clauwaert P., Verstraete W. and Boon N. (2012). Strategies to mitigate N_2O emissions from biological nitrogen removal systems. *Current Opinion in Biotechnology*, **23**, 474–482.

Directive 2000/60/EC of the European Parliament and of the Council of 23 October 2000 Establishing a Framework for Community Action in the Field of Water Policy, Official Journal of European Communities.

EPRI (2002). Water & Sustainability: US Electricity Consumption for Water Supply and Treatment, the Next Half Century, Palo Alto, CA:2000.

Frison N., Di Fabio S., Cavinato C., Pavan P. and Fatone F. (2013a). Best available carbon sources to enhance the via-nitrite biological nutrients removal from supernatants of anaerobic co-digestion. *Chemical Engineering Journal*, **215**, 15–22.

Frison N., Katsou E., Malamis S., Bolzonella D. and Fatone F. (2013b). Biological nutrients removal via nitrite from the supernatant of anaerobic co-digestion using a pilot-scale sequencing batch reactor operating under transient conditions. *Chemical Engineering Journal*, **230**, 595–604.

Gu S. B., Wang S. Y., Yang Q., Yang P. and Peng Y. Z. (2012). Start up partial nitrification at low temperature with a real-time control strategy based on blower frequency and pH. *Bioresource Technology*, **112**, 34–41.

Gustavsson D. J. I. (2010). Biological sludge liquor treatment at municipal wastewater treatment plants – a review. *Vatten*, **66**, 179–192.

Hellinga C., Schellen A. A. J. C., Mulder J. W., van Loosdrecht M. C. M. and Heijnen J. J. (1998). The Sharon® process: an innovative method for nitrogen removal from ammonium-rich waste water. *Water Science and Technology*, **37**, 135–142.

IPCC (2006). Guidelines for national greenhouse gas inventories. In: H. S. Eggleston , L. Buendia L, K. Miwa, T. Ngara and K. Tanabe (eds), IPCC Guidelines for National Greenhouse Gas Inventories, Vol. 5, Waste. IGES, Japan, pp. 6.24–6.26.

Ji Z. and Chen Y. (2010). Using sludge fermentation liquid to improve wastewater short-cut nitrification-denitrification and denitrifying phosphorus removal via nitrite. *Environmental Science and Technology*, **44**, 8957–8963.

Kampschreur M. J., Temmink H., Kleerebezem R., Jetten M. S. M. and van Loosdrecht M. C. M. (2009). Review: nitrous oxide emission during wastewater treatment. *Water Research*, **43**, 4093–4103.

Katsou E., Malamis S., Frison N. and Fatone F. (2015). Accepted coupling the treatment of low strength anaerobic effluent with fermented biowaste for nutrient removal via nitrite, *J. Environ. Manage.*, **149**, 108–117.

Keller J. and Hartley K. (2003). Greenhouse gas production in wastewater treatment: process selection is the major factor. *Water Science and Technology*, **47**(12), 43–48.

Longo S., Katsou E., Malamis S., Frison N., Renzi D. and Fatone F. (2015). Recovery of volatile fatty acids from fermentation of sewage sludge in municipal wastewater treatment plants. *Bioresource Technology*, **175**, 436–444.

Malamis S., Katsou E., Di Fabio S., Bolzonella D. and Fatone F. (2014). Biological nutrients removal from the supernatant originating from the anaerobic digestion of the organic fraction of municipal solid waste. *Critical Reviews in Environmental Science and Technology*, **34**, 244–257.

Mayer M., Smeets W., Braun R. and Fuchs W. (2009). Enhanced ammonium removal from liquid anaerobic digestion residuals in an advanced sequencing batch reactor system. *Water Science and Technology*, **60**, 1649–1660.

Metcalf and Eddy Inc. (2003). Wastewater Engineering, Treatment and Reuse, 4th edn, McGraw Hill, New York.

Monni S., Pipatti R., Lehtilla A., Savolainen I. and Syri S. (2006). Global Climate Change Mitigation Scenarios for Solid Waste Management. VTT Publications, Technical Research Centre of Finland, Espoo.

Morse G. K., Brett S. W., Guy J. A. and Lester J. N. (1998). Review: Phosphorus removal and recovery technologies. *Science of the Total Environment*, **212**, 69–81.

Mosier A., Kroeze C., Nevison C., Oenema O., Seitzinger S., van Cleemput O. (1999). An overview of the revised 1996 IPCC guidelines for national greenhouse gas inventory methodology for nitrous oxide from agriculture. *Environmental Science & Policy*, **2**, 325–333.

Ni B.-J., Ruscalleda M., Pellicer-Nacher C. and Smets B. F. (2011). Modeling nitrous oxide production during biological nitrogen removal via nitrification and denitrification: extensions to the general ASM models. *Environmental Science & Technology*, **45**, 7768–7776.

Oleszkiewicz J. A. and Barnard J. L. (2006). Nutrient removal technology in North America and the European union: a review. *Water Quality Research Journal of Canada*, **41**, 449–462.

Saito T., Brdjanovi D. and van Loosdrecht M. C. M. (2004). Effect of nitrite on phosphate uptake by phosphate accumulating organisms. *Water Research*, **38**, 3760–3768.

Shizas I. and Bagley D. M. (2004). Experimental determination of energy content of unknown organics in municipal wastewater streams. *Journal of Energy Engineering*, **130**, 45–53.

US EPA (2007). Biological Nutrient Removal: Processes and Costs, EPA-823-R-07-002, Washington, DC.

US EOPA (2013). Emerging Technologies for Wastewater Treatment and In-Plant Wet Weather Management, EPA-832-R-12-011, Washington, DC.

Vadivelu V. M., Keller J. and Yuan Z. (2007). Free ammonia and free nitrous acid inhibition on the anabolic and catabolic processes of Nitrosomonas and Nitrobacter. *Water Science and Technology*, **56**, 89–97.

Vitousek P. M., Aber J. D., Howarth R. W., Likens G. E., Matson P. A., Schindler D. W., Schlesinger W. H. and Tilman D. G. (1997). Human alternation of the global nitrogen cycle: sources and consequences. *Ecological Applications*, **7**, 737–750.

Water Environment Federation (WEF) and American Society of Civil Engineers (ASCE)/ Environmental and Water Resources Institute (EWRI) 2006. Biological Nutrient Removal (BNR) Operation in Wastewater Treatment Plants. McGraw Hill, New York.

Wentzel M. C., Comeau Y., Ekama G. A., van Loosdrecht M. C. M. and Brdjanovic D. (2008). Chapter 7: Phoshporus removal. In: Biological Wastewater Treatment: Principles Modelling and Design, M. Henze, M. van Loosdrecht, G. Ekama and D. Brdjanovic (eds), IWA, UK.

WERF (2011). When do the costs of wastewater treatment out weigh the benefits of nutrient removal? Striking the Balance Between Nutrient Removal in Wastewater Treatment and Sustainability (NUTR1R06n). www.werf.org/a/ka/Search/ResearchProfile.aspx?ReportId=NUTR1R06n (accessed 10 March 2014).

Wett B., Buchauer K. and Fimml C. (2007). Energy Self-Sufficiency as a Feasible Concept for Wastewater Treatment Systems. Proceeding of IWA Leading Edge Technology Conference, Singapore, Asian Water, 21–24 September. http://www.araconsult.at/site_04en.html (accessed 5 May 2014).

Yang Q., Peng Y. Z., Liu X. H., Zeng W., Mino T. and Satoh H. (2007). Nitrogen removal via nitrite from municipal wastewater at low temperatures using real-time control to optimize nitrifying communities. *Environmental Science and Technology*, **41**, 8159–8164.

Chapter 6

The aerobic granulation as an alternative to conventional activated sludge process

Anuska Mosquera-Corral[1], Ángeles Val del Río[1], Helena Moralejo-Gárate[2], Alberto Sánchez[2], Ramón Méndez[1] and José Luis Campos[1]

[1]Department of Chemical Engineering, Institute of Technology, University of Santiago de Compostela, Lope Gomez de Marzoa, s/n, 15705, Santiago de Compostela, Spain
E-mail Address: anuska.mosquera@usc.es, mangeles.val@usc.es, ramon.mendez.pampin@usc.es, joseluis.campos@usc.es

[2]Galician Water Research Center Foundation (CETaqua), Emprendia Building, Campus Vida, E-15782, Santiago de Compostela, Spain
E-mail Address: hmoralejo@cetaqua.com, asanchez@cetaqua.com

6.1 INTRODUCTION

It has already been more than 20 years since the first studies were performed in the field of aerobic granular biomass (Mishima & Nakamura, 1991). By that time, aerobic granules were developed in a UASB type reactor using pure oxygen. However, this work was not really appreciated at that moment. It was only in the late 90's when the interest on the aerobic granular systems appeared again. Initially only organic matter was removed (Morgenroth *et al.* 1997; Dangcong *et al.* 1999; Beun *et al.* 1999, 2002) and later research works indicated that the simultaneous removal of organic matter, nitrogen and eventually phosphorus compounds was feasible in the same unit (Arrojo *et al.* 2004; de Kreuk *et al.* 2005a; Bassin *et al.* 2012; Isanta *et al.* 2012). The interest for this kind of technology increased when its advantages compared to the conventional activated sludge systems were identified: excellent biomass settling properties, large biomass retention, the ability to withstand shock and toxic loadings, the presence of aerobic and anoxic zones inside the granules to perform different biological processes, and so on.

(Morgenroth *et al.* 1997; Dangcong *et al.* 1999; Beun *et al.* 1999; Peng *et al.* 1999; Tay *et al.* 2001a; Lin *et al.* 2003; Liu *et al.* 2003; Yang *et al.* 2003; Campos *et al.* 2009; Khan *et al.* 2013).

Due to this great interest the '1st IWA-Workshop Aerobic Granular Sludge' took place in Munich in 2004, where a definition for aerobic granules was stated as (de Kreuk *et al.* 2005b):

> *"Granules making up aerobic granular activated sludge are to be understood as aggregates of microbial origin, which do not coagulate under reduced hydrodynamic shear, and which settle significantly faster than activated sludge flocks."*

Furthermore, aerobic granules should fulfil the following requirements: (i) the values of the sludge volume index after 10 and 30 minutes of settling (SVI_{10} and SVI_{30}, respectively) do not differ more than a 10% (Schwarzenbeck *et al.* 2004a); (ii) the position of microorganisms inside the granules is fixed and it does not change due to the existence of a matrix of biomass and exopolymeric substances (EPS); (iii) no carrier material is intentionally involved or added; (iv) the minimum size of the granules is considered to be around 0.2 mm (de Kreuk *et al.* 2005b); (v) it is a general convention to consider those bacterial aggregates with an SVI_{10} value equal or lower than 60–70 mL/g TSS as to be aerobic granules.

Nowadays, much research work has been performed producing significant knowledge. From this knowledge, some patents belonging to different countries about aerobic granular systems have been elaborated and some full-scale plants are running in different countries. However, to the authors' knowledge, regarding this full-scale application, only one of all the technologies is a reality at the moment, the Nereda® one. In the following sections the gathered knowledge up to date in the field of aerobic granulation has been summarised. Information related to the full-scale implementation of the aerobic granular systems is also provided, along with information regarding the advantages of this technology compared to the conventional ones.

6.2 BASICS OF AEROBIC GRANULATION

The formation of aerobic granules occurs gradually during start-up period of the reactor systems in such a way that the seed sludge evolves into compact aggregates, further to granular sludge and finally to mature granules (Tay *et al.* 2001a). For the production of aerobic granular biomass, the type of reactor used, the imposed settling times, the applied organic load, and so on, are important parameters to take into account.

In principle, there are two possible strategies of operation to obtain aerobic granules. One is based on the operation of bioreactors under aerobic conditions with very short feeding periods (COD and nitrogen are simultaneously removed). The alternative strategy is based on feeding under anaerobic conditions (based on COD, nitrogen and phosphorous simultaneous removal). Parameters affecting the production of granular biomass in both alternatives are reviewed in the following sections.

6.2.1 Conditions for aerobic granular biomass formation

In order to produce aerobic granules, two operational conditions are crucial: the existence of a feast-famine regime and the application of short settling times. The appropriate feast-famine regime will contribute into selecting the organisms capable of accumulating polyhydroxyalkanoates (PHA) and form aggregates; the application of short settling times will contribute into removing the biomass fraction with low settling velocities and to concentrate the fast settling biomass, which corresponds to the granules, inside the reactor.

6.2.1.1 Feast–famine regime

Conventional activated sludge systems are facilities operated in continuous mode with biomass grown in suspension. The concentration of organic matter in the liquid media of these systems is always low. When high loaded wastewater enters the system, problems related to the dissolved oxygen concentration and, consequently, with biomass floatation frequently appear. Under these conditions, to produce aggregated biomass like aerobic granules is not possible. In order to produce the granular biomass in aerobic conditions, the heterotrophic biomass must be cyclically subjected to periods of availability (feast phase) and absence (famine phase) of organic substrate in the liquid phase. This is achieved if the fed organic matter is supplied to the system, operated in sequencing mode, within a very short time period (Figure 6.1a). In this way, the organic matter concentration, after feeding addition, is increased, leading to high concentration gradients. The dependence of the biomass aggregation capacity on the organic matter availability in the liquid is explained by Chudoba's theory of the kinetic selection (Chudoba *et al.* 1994).

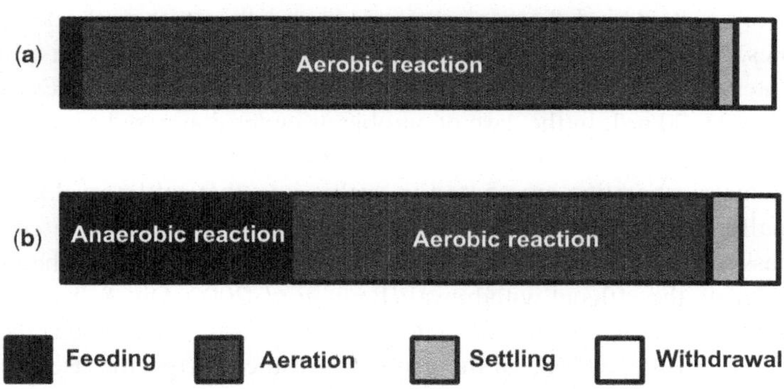

Figure 6.1 Distribution of the phases within a cycle of operation of aerobic granular systems in the case of simultaneous (a) organic matter and nitrogen removal and (b) organic matter, nitrogen and phosphorous removal.

This theory states that, due to the higher maximum specific growth rate (μ_{max}) and half saturation constant (Ks) values (according to Monod kinetics), the growth of the granule forming microorganisms is favoured at high substrate concentrations. When the substrate concentration is low, the filamentous microorganisms, with low half saturation constant, grow at higher rates than the granule forming organisms and the settling properties of the sludge are deteriorated.

Furthermore, it has been observed that, during the feast phase, the organic biodegradable substrates present in the wastewater are stored inside the biomass in the form of PHA, while, during the famine phase, the PHA are used for biomass growth (Beun *et al.* 2002; Val del Río *et al.* 2013). Under these conditions, when stable granules are formed in the system, organic matter and nitrogen can be removed simultaneously.

The aforementioned procedure can also be applied to produce granular biomass when the system is operated under anaerobic-aerobic alternating conditions (Figure 6.1b). In this alternative option, the granulation occurs due to the development of slow growing organisms like the phosphorous removing bacteria (de Kreuk *et al.* 2004, 2005a). In this case, the feast phase occurs under anaerobic conditions, when the phosphorous is released from the cells while PHA is accumulated, and the famine period takes place under aerobic conditions, where the phosphate is uptaken at the expense of the stored PHA. As in the first option, the feast-famine alternating conditions function as the selective pressure to favour the development of phosphorous removing bacteria, which are slow growing organisms known to form aggregates like granular biomass (de Kreuk & van Loosdrecht, 2004). In this case organic matter, nitrogen and phosphorous are removed in the unit.

6.2.1.2 Short settling times

To select for a certain kind of organisms is not enough to guarantee the development of granules. To remove the possible competition with other bacteria with faster growth rates is also needed. The settling velocity of biomass reaches values in the range of 25–70 m/h in the case of aerobic granules compared to the 7–10 m/h in the case of activated sludge. For this purpose, the imposed settling times during biomass separation are chosen in such a way that settling velocity values are normally larger than 10 m/h (Winkler *et al.* 2013). Under these conditions, the biomass which does not settle fast enough to be retained inside the reactor is removed during the effluent withdrawal (Beun *et al.* 2002; Qin & Tay, 2004; Liu *et al.* 2005b; Val del Río *et al.* 2012b; Morales *et al.* 2013). To achieve these high settling rates, and depending on the geometry of the reactor, the use of reactors characterized by large H/D ratios (column height/column diameter), normally around 5–20 m/m, is necessary.

With regard to the biomass settling properties, the SVI is also an important parameter to take into account. The typical values of SVI are between 30 and

75 mL/g TSS in the case of aerobic granules compared to the 100–200 mL/g TSS obtained in the case of biomass from conventional activated sludge systems. These low values allow the achievement of high concentrations of compact biomass inside the reactors and facilitate the separation of the effluent after settling.

6.2.2 Sequencing batch reactors

Although some works are available where aerobic granules are produced in continuous mode (Mishima *et al.* 1991; Dangcong *et al.* 1999; Morales *et al.* 2012; Zhou *et al.* 2013), sequencing batch reactors (SBR) have been mainly applied (Morgenroth *et al.* 1997; Beun *et al.* 1999; Schwarzenbeck *et al.* 2004a; Liu *et al.* 2005b; Zima *et al.* 2007; Wang *et al.* 2009; Morales *et al.* 2013). These systems operate in cycles broken down in phases of different duration, namely: filling, reaction, settling, effluent withdrawal and idle phases (Wilderer *et al.* 2001). This operation in cycles provides SBR systems with a great flexibility to fulfil the required conditions for aerobic granular biomass production, that is existence of feast-famine regime and short settling periods (Figure 6.1). Furthermore, these systems must guarantee that the reaction phase occurs either under complete mixing conditions or in a plug flow regime.

In previous works, the air is supplied to achieve the required level of mixing, necessary to favour the mass transfer of the substrates and oxygen, necessary to carry out the aerobic reactions. However, aerobic granules can be formed and grow in mechanically stirred SBRs (Mosquera-Corral *et al.* 2011). When phosphorous removal occurs, the system operates in plug flow from the bottom. In this way, the organic matter present in the wastewater is accumulated inside the biomass in the form of PHA while phosphate is released (feast phase). When the feeding period ends and substrate is no longer present in the liquid media, aeration is switched on. At this point the accumulated PHA is used to uptake the phosphate released in the previous period (famine phase).

The feeding can be performed within very short periods (Figure 6.1a) when the reactor operates under aerobic conditions or in plug flow under anaerobic conditions (Figure 6.1b). The establishment of the feast-famine regime or periodic starvation (Wilderer *et al.* 2001), where alternating presence-absence of organic carbon compounds occurs in the liquid, is possible due to the feeding strategy applied. Furthermore, the possibility to achieve the separation of the biomass inside the reactor instead of in a secondary separate settler is an important advantage of the SBRs.

The idle phase connects the successive cycles and does not always exist. A general scheme of the SBR operation is shown in Figure 6.2. The duration of each phase within a cycle is variable and influenced by the level and variability of the hydraulic load as well as the organic load, the oxygen requirements for removal, the necessity of low energy consumption, and so on.

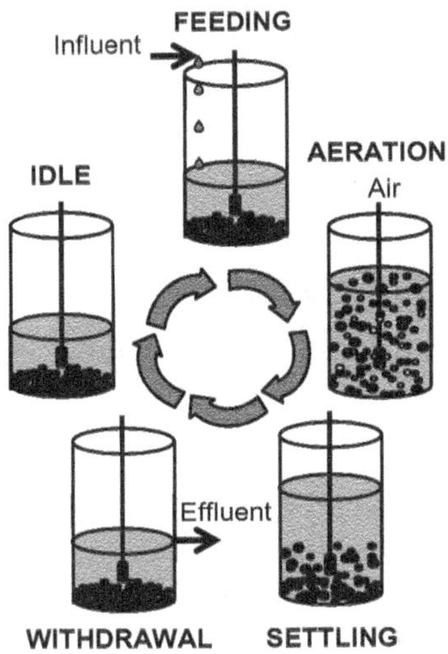

Figure 6.2 Operational phases of a cycle applied in a sequencing batch reactor.

However, to take benefit from the advantages of a SBR system, some special features should be considered in the design stage of the full scale plants such as the fact that SBR implies higher costs in control, data acquisition and instrumentation in comparison to continuous processes due to its high level of sophistication. Furthermore, the use of several SBRs in parallel is necessary to cope with high flow rates. A previous tank is usually installed to homogenise the peaks of flow to be treated along the working day. Although this homogenisation tank presents the advantage to avoid high fluctuations in the influent composition of the SBR, it also means an increase in the installation and control system costs. Another point to have in mind is the quality of the produced effluent which depends on the settling properties of the granular sludge. The detriment of the settling properties can occur if the denitrification process takes place during the sedimentation phase, with the consequent production of nitrogen gas which causes the sludge bed flotation. To finalize, an equalization step after the SBR might be a potential requirement depending on the downstream processes.

6.2.3 Factors affecting aerobic granule characteristics and stability

Once the operational conditions are determined to produce aerobic granular biomass, it is still important to take into account the potential effect of other parameters on the properties and stability of the aerobic granules. These parameters include the substrate composition, the organic loading rate and concentration, EPS formation, presence of divalent cations, dissolved oxygen (DO) concentration and hydrodynamic shear forces.

6.2.3.1 Substrate composition

Aerobic granules can be produced in SBR systems when the readily biodegradable organic matter content of the treated wastewater is larger than 75 mg COD/L (Mosquera-Corral *et al.* 2011). Lower concentrations do not provide the minimum concentration gradient conditions for the establishment of the required feast-famine regime. Taking into account that the aerobic granules are a special kind of 'biofilm', the knowledge obtained regarding the effect of the organic matter composition on the biofilm formation and characteristics could be applied in the case of the aerobic granule formation. In this sense, previous works on biofilms have demonstrated the relationship between the degree of reduction of the substrate and the biofilm density (Mosquera-Corral *et al.* 2003). For example, acetate results in high maximum biomass growth rates ($\mu_{max} = 9.6\ d^{-1}$) and produce less dense biofilms (18 g $VSS/L_{biofilm}$) than methanol resulting in slow maximum biomass growth rates ($\mu_{max} = 5.26\ d^{-1}$) and high biofilm density (80 g $VSS/L_{biofilm}$). In the case of aerobic granules it is necessary to take into account that the organic compounds present in the wastewater are first transformed into storage compounds (PHA), during the feast period, and these are latter used for growth during the famine period. In these conditions, slow growing microorganisms are developed because the biomass growth rate using an intracellular polymer (PHA) is slower than using a simple extracellular compound. Furthermore, some authors have proposed to operate SBR systems to select for Poly-P accumulating bacteria (PAO) and glycogen accumulation organisms (GAO) which are known to grow at slow rates (de Kreuk & van Loosdrecht, 2004).

From a microscopic point of view, previous works have also indicated that substrates containing carbohydrates favour the uncontrolled growth of filamentous bacteria (Schwarzenbeck *et al.* 2005), while acetate is more suitable to produce aerobic granules. In general, simple substrates are better than complex ones because are easy to store by the biomass. Aerobic granules have been successfully cultivated on a wide variety of substrates (Figure 6.3) including glucose, acetate, ethanol, peptone, phenol and industrial wastewater (Morgenroth *et al.* 1997; Beun *et al.* 1999; Yi *et al.* 2003; Arrojo *et al.* 2004; Schwarzenbeck *et al.* 2004b; de Kreuk & van Loosdrecht, 2006; Sun *et al.* 2006; Figueroa *et al.* 2008; Ho *et al.* 2010; Val del Río *et al.* 2012a).

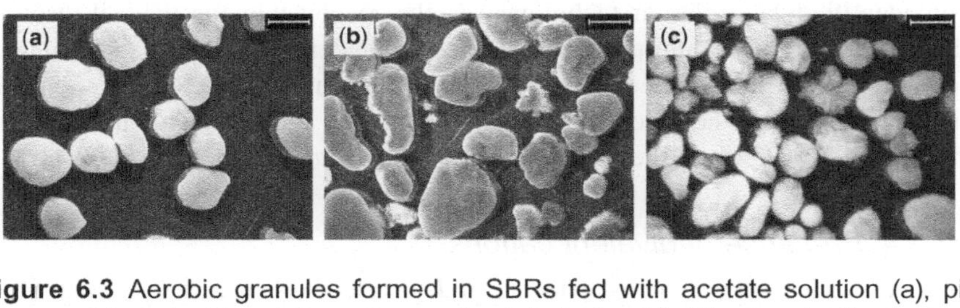

Figure 6.3 Aerobic granules formed in SBRs fed with acetate solution (a), pig manure (b) and effluent from the seafood industry (c). The bar in the images corresponds to 2 mm.

6.2.3.2 Organic loading rate

The properties of the aerobic granules depend on the organic loading rate (Tay *et al.* 2004; Adav *et al.* 2010). High organic loading rates result in the production of less dense and resistant granular biomass (Moy *et al.* 2002). This can be related to the fact that there is a limit to the storage rate of the biomass. Furthermore, high organic loading rates are related to limitations of dissolved oxygen concentration which are known to promote the development of filamentous organisms and the deterioration of the physical properties of the aggregates (Mosquera-Corral *et al.* 2005a).

The organic load treated is directly related to the biomass concentration in the reactor. In aerobic granular systems the biomass concentration ranges normally from 4 to 15 g VSS/L (Beun *et al.* 1999; Di Iaconi *et al.* 2004; Val del Río *et al.* 2012a). These values are higher than those conventionally achieved in activated sludge reactors of 1–2 g VSS/L. The systems can operate under an organic loading rate of up to 19 g COD/(L · d) on such a high level of biomass concentration (Adav *et al.* 2010).

6.2.3.3 Exopolymeric substances

EPS produced in the granules include different kinds of compounds in variable amounts comprising proteins, polysaccharides, lipids, glycoproteins, humic-like substances, nucleic acids, among others. These compounds are produced under different conditions such as substrate or oxygen limiting conditions, ionic strength, temperatures of operation and so on. There are conflicting results reported in literature regarding the function of the EPS on the formation of aerobic granules (de Kreuk *et al.* 2005b). It is believed that one of the EPS functions is to act as a 'glue' among the microorganisms present in an aggregate due to its physical properties (Tay *et al.* 2001a). Moreover, according to previous research, the EPS content increases with granulation, there are differences in loosely bound and tightly bound EPS and, within the granule structure, insoluble versus soluble polysaccharide gradients occur.

The polysaccharides have been identified as the major compounds forming the granular EPS and the gel-like structures responsible for bacterial granule formation (Lin *et al.* 2010). Furthermore, two different exoplysaccharides have been identified by isolation techniques as the responsible for the gel-forming matrix of the granules: alginate-like polysaccharide (Lin *et al.* 2010) and granulan (Seviour *et al.* 2011). At the moment it is still not clear which EPS is responsible for the formation of aerobic granules and, even new compounds could arise as important components of these structures (Seviour *et al.* 2012).

6.2.3.4 Presence of divalent cations

From research works there is evidence that divalent cations as Ca^{2+} and Mg^{2+} attach to negatively charged groups present on the bacteria surfaces and to the EPS molecules promoting the aerobic granulation process. Jiang *et al.* (2003) reported that the addition of Ca^{2+} accelerated the aerobic granulation process. In the case of other cations, such as Mg^{2+}, it was demonstrated that the presence of this ion in the feeding media enhanced the sludge granulation process in sequencing batch reactors (Li *et al.* 2009). The external addition of these ions should not be an option, since they are often already present in the wastewater and may have a positive role during the granulation process.

6.2.3.5 Dissolved oxygen concentration

In aerobic reactors the dissolved oxygen (DO) concentration is an important parameter and must be kept at sufficient high levels to allow the different biological processes to occur (Mosquera-Corral *et al.* 2005a), and simultaneously, avert the growth of filamentous which decrease the granule density.

In systems where the mixing is performed pneumatically, the DO supply is also directly linked to the degree of mixing. For this reason, modification of the hydrodynamic conditions of the system by reducing the supplied air flow will affect the quality of the formed granules (Tay *et al.* 2001b). However, this fact is not always taken into account.

6.2.3.6 Hydrodynamic shear forces

Evidence shows that application of high shear forces favour the formation of more compact and dense aerobic granules (Shin *et al.* 1992; Tay *et al.* 2001b; Liu & Tay, 2006; Di Iaconi *et al.* 2006). Tay *et al.* (2001a, c) reported that the production of extracellular polysaccharides, and, consequently, the stability of aerobic granules is closely associated to the applied shear forces. The collisions between the granules provoke the detachment of weakly attached materials from the surface of the aggregates helping to maintain their high densities and smooth surfaces. The detached solid parts are removed through the effluent of the reactor.

6.2.4 Biological processes inside the aerobic granules

The bacterial populations form layers in different depths inside the granules during granulation. Their position will depend on the required conditions for their growth. As a result, since oxygen is difficult to penetrate the granule, external layers will be comprised by mainly aerobic organisms (heterotrophic and nitrifying bacteria), while anoxic (denitrifying bacteria) and anaerobic populations will develop in the inner zones (Figure 6.4).

Aerobic granules were produced in SBR systems initially to oxidise the organic matter contained in the treated wastewaters (Beun *et al.* 1999). Later, the operational conditions allowed the development of aerobic granules capable to remove nutrients (nitrogen and phosphorus) simultaneously (Beun *et al.* 2001; Lin *et al.* 2003; Yang *et al.* 2003; de Kreuk & van Loosdrecht, 2004; de Kreuk *et al.* 2005a; Bassin *et al.* 2012; Isanta *et al.* 2012).

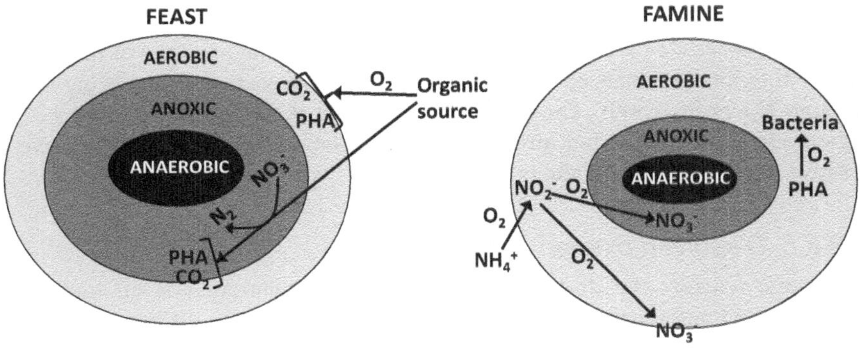

Figure 6.4 Transformation of the organic matter and nitrogen inside a granule during the feast and famine periods.

However, depending on the composition of the wastewater treated, the share of the different populations inside the granules changes and one of them can become predominant. In some extreme cases, the aerobic granules can only perform a single process, such as nitrification as has occurred in either continuous (Campos *et al.* 2000) or sequencing conditions (Mosquera-Corral *et al.* 2005b).

When the reactor is intermittently fed, a feast-famine regime is established. This alternating conditions impose different environmental conditions for the biomass inside the granules in both feast and famine periods. During the feast period the concentration of external organic carbon (e.g., acetate) is high in the liquid medium. Therefore, this substrate completely diffuses into the granules and it is stored as PHA in aerobic (where DO is present) and anoxic conditions (using NO_3^- in the inner layers of the granule). A minor transformation of the organic carbon to CO_2 also occurs. Dissolved oxygen (DO) has a small penetration depth because it is very rapidly consumed by the heterotrophic bacteria, and eventually by the autotrophic microorganisms (nitrifying bacteria), in the outer layers of the granules. Although the nitrification process can occur during this period, usually, the heterotrophic bacteria consume the most part of the oxygen for organic carbon removal. In this situation the available oxygen for nitrification is not enough. The NO_3^- formed, mainly during the famine period of the previous cycle, is used to store PHA in anoxic conditions in the inner zones of the granule.

During the famine period the DO penetration depth is larger than during the feast period, since the consumption due to the activity of the microorganisms gets lower. In the centre of the granules, the NO_3^- is present and part of it can be used as electron acceptor to degrade the stored PHA. Aerobic conversion of PHA for biomass growth and nitrification, as long as there is NH_4^+ present, occur in the aerobic layer. The autotrophic organisms oxidize NH_4^+ into NO_3^- and this compound diffuses both towards the centre of the granule and towards the liquid phase surrounding the granules. The nitrate produced is removed during feast phase of the next operational cycle.

In those cases where phosphate is simultaneously removed, the distribution of the different processes is shown in Figure 6.5. During the feast phase, anaerobic conditions are imposed to the granules, and organic matter (volatile fatty acids, VFA) is uptaken to accumulate PHA, while the Poly-P is hydrolysed as phosphate to the liquid medium to provide the energy needed for PHA accumulation (de Kreuk *et al.* 2005a). If the organic matter is in excess, glycogen accumulating organisms (GAO) will use the VFA left to produce PHA consuming the already accumulated glycogen produced for energy production (not shown in the Figure 6.5 to simplify the process). During the famine phase, the reverse process takes place: the stored PHA is consumed for biomass growth (or for eventual glycogen production) and the released energy is stored in the phosphate bonds during the poly-P formation. If nitrification also takes place during this period, the stored PHA is used for denitrification.

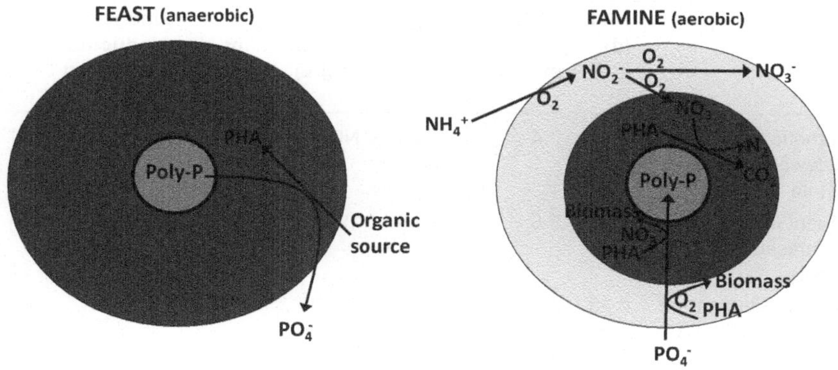

Figure 6.5 Transformation of the organic matter, nitrogen and phosphorous inside a granule during the feast and famine periods.

6.3 COMPARISON WITH ACTIVATED SLUDGE SYSTEMS

The potential of the aerobic granular technology is very promising since complete granulation with municipal wastewater as substrate has been shown and extensive nutrient removal seems well feasible (de Bruin *et al.* 2005). Besides urban, industrial wastewater can also be treated in these systems.

In addition to the advantages of this process mentioned, a comparison between the conventional activated sludge systems and the aerobic granular technology for sewage treatment indicates that the latter presents several improvements regarding costs and quality of the produced effluent. These systems perform similar or better than activated sludge in terms of process stability, sludge production and effluent quality, while required surface for implantation decreases to 25% and energy consumption to 65–75% (de Bruin *et al.* 2004). These preliminary calculations indicated that the annual costs of the granular SBR including primary and post treatment are 17% lower while mechanical/electrical works account for 40–45% of the capital costs.

Further studies in pilot plant confirmed some of the previous estimations. Di Iaconi *et al.* (2010a, b) determined a biomass productivity of 0.04–0.14 kg TSS/kg $COD_{removed}$ which represent 74% less sludge production than conventional activated sludge systems (0.4–0.6 kg TSS/kg $COD_{removed}$).

Regarding energy associated to aeration, data collected from pilot-plants operation do not provide reliable information due to the highly oversized air flows as shown in Table 6.1. Information gathered indicates that pilot plants operate at shorter HRT, around 6 hours, and higher applied organic loading rates (OLR) as high as 9 kg COD/$m^3 \cdot$ d much higher than activated sludge systems.

Table 6.1 Operational parameters of pilot-scale plants based aerobic granular biomass.

Type of feeding	HRT (d)	OLR (kg COD/ $m^3 \cdot$ d)	$V_{reactor}$ (L)	VSS effluent (mg/L)	Air ($m^3 \cdot$ d)	Sludge production (kg TSS/kg $COD_{removed}$)	Ref.
Synthetic wastewater + sodium acetate	0.33	2.4	34	ND	54	ND	[1]
Effluent from pharmaceutical industry	0.21	5.5	40	194	20	ND	[2]
Primary effluent from a municipal WWTP	0.33	1.0	1000	70–20	ND	ND	[3]
Synthetic wastewater + sodium acetate	0.25	2.4–9.7	100	<200	121	ND	[4]
Synthetic low-strength wastewater	0.23	1.74	100	ND	46	ND	[5]
Primary effluent from a municipal WWTP	0.31	2.5	800	ND	230	0.12–0.14	[6]
Primary effluent from a tannery WWTP	2	2.9	800	35–50	110	0.04–0.08	[7]
Pig slurry	0.25	2–4	100	<250	137	ND	[8]

[1] Tay *et al.* (2005); [2] Inizan *et al.* (2005); [3] Ni *et al.* (2009); [4] Jungles *et al.* (2011); [5] Isanta *et al.* (2012); [6] Di Iaconi *et al.* (2010a); [7] Di Iaconi *et al.* (2010b); [8] Morales *et al.* (2013). OLR: Organic loading rate. ND: No available data.

With regards to CAPEX and OPEX of industrial scale plants not much information is available as in the case of activated sludge due to the novel application of this technology (Table 6.2).

Table 6.2 Comparison of aerobic granular pilot and industrial scale plants to activated sludge systems.

Location Type of wastewater	$V_{reactor}$ (m^3)	CAPEX % reduction	OPEX % reduction	Sludge production % reduction	Ref.
Pilot plants					
Standard Dutch wastewater composition	–	–	7–17	–	[1]
Primary effluent from a municipal WWTP	0.8	–	–	74	[2]
Primary effluent from a tannery WWTP	0.8	–	60[a]	87[a]	[3]
Indrustrial installations					
Gansbaai Municipal + Industrial	3 × 1600	20	50	–	[4]
Epe Municipal + Industrial	3 × 4500	25	25	–	
Frielas Municipal + Industrial	1000	–	30[b]/50[c]	–	
Wemmershoek Municipal	2 × 1800	–	50/75[c]	–	

[a]After ozonation and compared to Activated Sludge+ Fenton process. [b]Electricity for aeration. [c]Total electricity consumption. References: [1] De Bruin *et al.* (2004); [2] Di Iaconi *et al.* (2010a); [3] Di Iaconi *et al.* (2010b); [4] Giesen and Thompson (2013).

Regarding the performance some aspects have to be taken into account different from activated sludge systems. During the start-up, large quantities of the added inoculum are washed out from the reactor, temporarily decreasing the quality of the produced effluent. This can be avoided by inoculating the reactor with previously developed granular biomass (Liu *et al.* 2005a) or with a mixture of crushed aerobic granules and floccular sludge (Pijuan *et al.* 2011).

The solid content of the effluent has to be reduced previously to its discharge to natural water bodies by means of filtration systems (membrane systems, settlers, sand filters, etc.).

Another aspect to take into account during operation is the aeration cost. This is relatively high due to the need of high air flows to keep the required DO concentration and the mixing. The air requirements can be reduced by the use of slow-growing microorganisms, such as the nitrifiers or phosphorous removing bacteria (de Kreuk & van Loosdrecht, 2004).

6.4 FULL SCALE APPLICATIONS OF THE AEROBIC GRANULAR TECHNOLOGIES

The research on aerobic granulation was initially focused on the use of lab-scale reactors fed with synthetic wastewater. Later, several studies with industrial wastewater have been performed. Nowadays, the technology is in the stage of demonstration and full-scale plants for the treatment of urban wastewater operate. Some applications in the industrial wastewater sector are also available.

Although several patents belonging to different companies of the water sector have been filed, only the technology Nereda® has realised at full scale. Information regarding the operation of these plants is provided in Table 6.3 and discussed in this section.

Table 6.3 Information of operational conditions of NEREDA® existing plants.

WWTP	Type of wastewater	Reactor volume (m^3)	Inhabitant equivalent	HRT (day)	WW concentration (mg COD/L)	OLR (kg COD/ $m^3 \cdot d$)
Ede The Netherlands (2005)	Cheese Industry	100	1500–5000	0.4–2	4000	2–10
Gansbaai South Africa (2006)	Municipal + Industrial	3 × 1600	63,000	0.96	1200	1.25
Rotterdam The Netherlands (2006)	Edible oil industry	2 × 1600	10,000–30,000	4.6	5000–10,000	1.1–2.2
Oosterwolde The Netherlands (2009)	Food industry	300	5000	0.6	3600	6
Epe The Netherlands (2012)	Municipal + Industrial	3 × 4500	59,000	0.375	680	1.8
Frielas Portugal (2012)	Municipal + Industrial	1000	10,000	3.3	500	0.15
Wemmershoek South Africa (2013)	Municipal	2 × 1800	40,000	7.2	870	0.121
Dinxperlo The Netherlands (2013)	Municipal	3 × 1250	15,730	0.27	570	2.1
Garmerwolde The Netherlands (2013)	Municipal	2 × 9500	140,000	0.19	525	2.76
Vroomshoop The Netherlands (2013)	Municipal	2400	25,000	0.08	800	10

The first pilot plant based on aerobic granular biomass was started up in the Netherlands in October 2003 to treat urban wastewater (de Bruin *et al.* 2004, 2005), this was the origin of the Nereda® process (Nereda® 2013). Since 2005, over 10 full-scale aerobic granular sludge technology (AGS) systems had been implemented in the Netherlands, Portugal and South Africa, for the treatment of both industrial and municipal wastewater based on the Nereda® process.

In Gansbaai (South Africa) a demonstration plant, comprising three SBRs of 1600 m^3, was constructed in 2006, to minimize the risks of implementation of the technology, for the treatment of sewage and industrial wastewater to handle capacities up to 5000 m^3/day.

The Nereda® process was also implemented as a demonstration plant in a WWTP in Portugal in 2012 (Frielas WWTP). The plant has a capacity of 70,000 m^3/day for the treatment of domestic wastewater from 250,000 inhabitants. It consists of a conventional activated sludge system with six complete mix biological reactors and 12 settlers. One of them was retrofitted into a Nereda® pilot reactor with a volume around 1000 m^3, that is working in parallel with the remaining five activated sludge reactors. The granular excess sludge from the Nereda® reactor is pumped to the activated sludge lines, improving the sludge characteristics and settling performances of the existing activated sludge plant.

After the success of the Gansbaai WWTP, the first municipal full-scale installation in the Netherlands was constructed in Epe to treat a flow rate of 1500 m^3/h. This plant consisting of three SBRs of 4500 m^3 was inaugurated in May 2012. Nowadays, three new plants are being started up in the Netherlands with this technology (Garmerwolde, Vroomshoop and Dinxperlo WWTPs). Also a second Nereda® plant with a capacity of 5000 m^3/day is under construction in Wemmershoek (South Africa). The design equivalent inhabitants of these plants ranged from 15,700 and 140,000 and the COD concentrations from 500 to 875 mg COD/L.

Furthermore, there are about 20 new plants scheduled to be built in different countries including Australia, China, Brazil, India, the Middle-East, Belgium, UK, Poland, Ireland and the USA.

Besides the application to treat municipal wastewater, a number of Nereda® plants had been constructed in the Netherlands to treat industrial wastewater from: cheese speciality industry (Ede), edible oil company (Rotterdam), food industry (Oosterwolde) and pharmaceutical industry. In this case the organic matter concentration ranged from 1200 to 10,000 mg COD/L.

With respect to the operational conditions these plants can cope with organic loading rates as high as 10 kg COD/$m^3 \cdot$ d similar to those obtained from pilot scale experiments (Table 6.1) at HRTs as short as 0.4 days. Shorter HRTs of 0.08 days are applied at Vroomshoop WWTP treating municipal wastewater with a COD concentration of 800 mg COD/L.

Although Nereda® process is the clearest example of AGS technology application, other works were also performed at pilot scale. Tay *et al.* (2005) operated a pilot plant for the development of aerobic granular biomass with a height of 1.6 m and a diameter of 0.19 m (working volume of 34 L). These authors treated a synthetic effluent. Inizan *et al.* (2005) performed experiments at pilot scale with a reactor of 1.8 m of height and 0.2 m of diameter (working volume of 40 L) and using either a synthetic medium or the effluent of a pharmaceutical company; the aerobic granulation successfully occurred. Later, Ni *et al.* (2009) used a pilot reactor of 1 m^3 to treat urban wastewater reaching COD and nitrogen removal percentages of 90–95%.

From the basis of the AGS but using a contention system for the granules, a Sequencing Batch Biofilter Granular Reactor (SBBGR) with a volume of 3.1 m^3 was developed by IRSA (Istituto di Ricerca Sulle Acque, Italy). Different studies were carried out in this plant treating sewage from an Italian WWTP (Di Iaconi *et al.* 2008; Di Iaconi *et al.* 2009; Di Iaconi *et al.* 2010).

Further research focused on the development of new technologies based on aerobic granular biomass is under development nowadays at pilot scale (Isanta *et al.* 2012; Morales *et al.* 2013), and, it is expected that in the future, new technologies will be available as an alternative to the already existing Nereda® process.

6.5 ACKNOWLEDGEMENTS

This work was funded by the Spanish Government through the NOVEDAR_Consolider (CSD2007-00055) and Plasticwater (CTQ2011-22675) projects. Anuska Mosquera-Corral, Ángeles Val del Río, Ramón Méndez and José Luis Campos belong to the Competitive Group of Reference GRC2013/032, Xunta de Galicia.

6.6 REFERENCES

Adav S., Lee D. J. and Lai J.-Y. (2010). Potential cause of aerobic granular sludge breakdown at high organic loading rates. *Applied Microbiology and Biotechnology*, **85**(5), 1601–1610.

Arrojo B., Mosquera-Corral A., Garrido J. M. and Méndez R. (2004). Aerobic granulation with industrial wastewater in sequencing batch reactors. *Water Research*, **38**(14–15), 3389–3399.

Bassin J. P., Winkler M. K. H., Kleerebezem R., Dezotti M. and van Loosdrecht M. C. M. (2012). Improved phosphate removal by selective sludge discharge in aerobic granular sludge reactors. *Biotechnology and Bioengineering*, **109**(8), 1919–1928.

Beun J. J., Hendriks A., van Loosdrecht M. C. M., Morgenroth E., Wilderer P. A. and Heijnen J. J. (1999). Aerobic granulation in a sequencing batch reactor. *Water Research*, **33**(10), 2283–2290.

Beun J. J., van Loosdrecht M. C. M. and Heijnen J. J. (2001). N-removal in a granular sludge sequencing batch airlift reactor. *Biotechnology and Bioengineering*, **75**(1), 82–92.

Beun J. J., van Loosdrecht M. C. M. and Heijnen J. J. (2002). Aerobic granulation in a sequencing batch airlift reactor. *Water Research*, **36**(3), 702–712.

Campos J. L., Mendez R. and Lema J. M. (2000). Operation of a nitrifying activated sludge airlift (NASA) reactor without biomass carrier. *Water Science and Technology*, **41**(4–5), 113–120.

Campos J. L., Figueroa M., Mosquera-Corral A. and Mendez R. (2009). Aerobic sludge granulation: state-of-the-art. *International Journal of Environmental Engineering*, **1**(2), 136–151.

Chudoba J. and Pujol R. (1994). Kinetic selection of microorganisms by means of a selector – twenty years of progress: history, practice and problems. *Water Science and Technology*, **29**(7), 177–180.

Dangcong P., Bernet N., Delgenes J.-P. and Moletta R. (1999). Aerobic granular sludge-a case report. *Water Research*, **33**(3), 890–893.

de Bruin L. M. M., de Kreuk M. K., van der Roest H. F. R., Uijterlinde C. and van Loosdrecht M. C. M. (2004). Aerobic granular sludge technology: an alternative to activated sludge? *Water Science and Technology*, **49**(11–12), 1–7.

de Bruin L. M. M., van der Roest H. F., de Kreuk M. K. and van Loosdrecht M. C. M. (2005). Promising results pilot research aerobic granular sludge technology at WWTP ede. In: Aerobic Granular Sludge. Water and Environmental Management Series, IWA Publishing, Munich, pp. 135–142.

de Kreuk M. K. and van Loosdrecht M. C. M. (2004). Selection of slow growing organisms as a means for improving aerobic granular sludge stability. *Water Science and Technology*, **49**(11–12), 9–17.

de Kreuk M. K. and van Loosdrecht M. C. M. (2006). Formation of aerobic granules with domestic sewage. *Journal of Environment Engineering*, **132**(6), 694–697.

de Kreuk M., Heijnen J. J., and van Loosdrecht M. C. M. (2005a). Simultaneous COD, nitrogen, and phosphate removal by aerobic granular sludge. *Biotechnology and Bioengineering*, **90**(6), 761–769.

de Kreuk M. K., McSwain B. S., Bathe S., Tay S. T. L., Schwarzenbeck N. and Wilderer P. A. (2005b). Discussion outcomes. In: Aerobic Granular Sludge, Water and Environmental Management Series, S. Bathe, M. K. de Kreuk, B. McSwain and N. Schwarzenbeck (eds), IWA Publishing, Munich, pp. 165–169.

Di Iaconi C., Ramadori R., Lopez A. and Passino R. (2004). Hydraulic shear stress calculation in a sequencing batch biofilm reactor with granular biomass. *Environmental Science and Technology*, **39**(3), 889–894.

Di Iaconi C., Ramadori R., López A. and Passino R. (2006). Influence of hydrodynamic shear force on properties of granular biomass in a sequencing batch biofilter reactor. *Biochemical Engineering Journal*, **30**(2), 152–157.

Di Iaconi C., De Sanctis M., Rossetti S. and Ramadori R. (2008). Technological transfer to demonstrative scale of sequencing batch biofilter granular reactor (SBBGR) technology for municipal and industrial wastewater treatment. *Water Science and Technology*, **58**(2), 367–372.

Di Iaconi C., Moro G. D., Pagano M. and Ramadori R. (2009). Municipal landfill leachate treatment by SBBGR technology. *International Journal of Environment and Waste Management*, **4**(3–4), 422–432.

Di Iaconi C., De Sanctis M., Rossetti S. and Ramadori R. (2010a). SBBGR technology for minimising excess sludge production in biological processes. *Water Research*, **44**(6), 1825–1832.

Di Iaconi C., Del Moro G., De Sanctis M. and Rossetti S. (2010b). A chemically enhanced biological process for lowering operative costs and solid residues of industrial recalcitrant wastewater treatment. *Water Research*, **44**(12), 3635–3644.

Figueroa M., Mosquera-Corral A., Campos J. L. and Mendez R. (2008). Treatment of saline wastewater in SBR aerobic granular reactors. *Water Science and Technology*, **58**(2), 479–485.

Giesen A., van Loosdrecht M. C. M., de Bruin B., van der Roest H. and Pronk M. (2013). Full-scale experiences with aerobic granular biomass technology for treatment of urban and industrial wastewater. International Water Week Amsterdam, November.

Ho K. L., Chen Y. Y., Lin B. and Lee D. J. (2010). Degrading high-strength phenol using aerobic granular sludge. *Applied Microbiology and Biotechnology*, **85**(6), 2009–2015.

Inizan M., Freval A., Cigana J. and Meinhold J. (2005). Aerobic granulation in a sequencing batch reactor (SBR) for industrial wastewater treatment. *Water Science and Technology*, **52**(10–11), 335–343.

Isanta E., Suárez-Ojeda M. E., Val del Río Á., Morales N., Pérez J. and Carrera J. (2012). Long term operation of a granular sequencing batch reactor at pilot scale treating a low-strength wastewater. *Chemical Engineering Journal*, **198–199**(0), 163–170.

Jiang H. L., Tay J. H., Liu Y. and Tay S. T. L. (2003). Ca^{2+} augmentation for enhancement of aerobically grown microbial granules in sludge blanket reactors. *Biotechnology Letters*, **25**(2), 95–99.

Jungles M. K., Figueroa M., Morales N., Val del Río Á., da Costa R. H. R., Campos J. L., Mosquera-Corral A. and Méndez R. (2011). Start up of a pilot scale aerobic granular reactor for organic matter and nitrogen removal. *Journal of Chemical Technology and Biotechnology*, **86**(5), 763–768.

Khan M. Z., Mondal P. K. and Sabir S. (2013). Aerobic granulation for wastewater bioremediation: a review. *Canadian Journal of Chemical Engineering*, **91**(6), 1045–1058.

Li X.-M., Liu Q.-Q., Yang Q., Guo L., Zeng G.-M., Hu J.-M. and Zheng W. (2009). Enhanced aerobic sludge granulation in sequencing batch reactor by Mg^{2+} augmentation. *Bioresource Technology*, **100**(1), 64–67.

Lin Y. M., Liu Y. and Tay J. H. (2003). Development and characteristics of phosphorous-accumulating granules in sequencing batch reactor. *Applied Microbiology and Biotechnology*, **62**(4), 430–435.

Lin Y., de Kreuk M., van Loosdrecht M. C. M. and Adin A. (2010). Characterization of alginate-like exopolysaccharides isolated from aerobic granular sludge in pilot-plant. *Water Research*, **44**(11), 3355–3364.

Liu Y., Xu H. L., Yang S. F. and Tay J. H. (2003). Mechanisms and models for anaerobic in upflow anaerobic sludge blanket reactor. *Water Research*, **37**(3), 661–673.

Liu Q. S., Liu Y., Tay S. T. L., Show K. Y., Ivanov V., Benjamin M. and Tay J. H. (2005a). Start-up of pilot-scale aerobic granular sludge reactor by stored granules. *Environmental Technology*, **26**(12), 1363–1369.

Liu Y., Wang Z. W. and Tay J. H. (2005b). A unified theory for upscaling aerobic granular sludge sequencing batch reactors. *Biotechnology Advances*, **23**(5), 335–344.

Liu Y-Q. and Tay J-H. (2006). Variable aeration in sequencing batch reactor with aerobic granular sludge. *Journal of Biotechnology*, **124**(2), 338–346.

Mishima K. and Nakamura M. (1991). Self-immobilization of aerobic activated sludge – a pilot study of the aerobic upflow sludge blanket process in municipal sewage treatment. *Water Science and Technology*, **23**(4–6), 981–990.

Morales N., Figueroa M., Mosquera-Corral A., Campos J. L. and Méndez R. (2012). Aerobic granular-type biomass development in a continuous stirred tank reactor. *Separation and Purification Technology*, **89**(0), 199–205.

Morales N., Figueroa M., Fra-Vázquez A., Val del Río A., Campos J. L., Mosquera-Corral A. and Méndez R. (2013). Operation of an aerobic granular pilot scale SBR plant to treat swine slurry. *Process Biochemistry*, **48**(8), 1216–1221.

Morgenroth E., Sherden T., van Loosdrecht M. C. M., Heijnen J. J. and Wilderer P. A. (1997). Aerobic granular sludge in a sequencing batch reactor. *Water Research*, **31**(12), 3191–3194.

Mosquera-Corral A., Montras A., Heijnen J. J. and van Loosdrecht M. C. M. (2003). Degradation of polymers in a biofilm airlift suspension reactor. *Water Research*, **37**(3), 485–492.

Mosquera-Corral A., de Kreuk M. K., Heijnen J. J. and van Loosdrecht M. C. M. (2005a). Effects of oxygen concentration on N-removal in an aerobic granular sludge reactor. *Water Research*, **39**(12), 2676–2686.

Mosquera-Corral A., Vázquez-Padín J. R., Arrojo B., Campos J. L. and Méndez R. (2005b). Nitrifying granular sludge in a sequencing batch reactor. In: Aerobic Granular Sludge. Water and Environmental Management Series, S. Bathe, M. K. de Kreuk, B. McSwain and N. Schwarzenbeck (eds), IWA Publishing, Munich, pp. 63–70.

Mosquera-Corral A., Arrojo B., Figueroa M., Campos J. L. and Méndez R. (2011). Aerobic granulation in a mechanical stirred SBR: treatment of low organic loads. *Water Science and Technology*, **64**(1), 155–161.

Moy B. Y. P., Tay J. H., Toh S. K. Liu Y. and Tay S. T. L. (2002). High organic loading influences the physical characteristics of aerobic sludge granules. *Letters in Applied Microbiology*, **34**(6), 407–412.

Nereda® http://www.royalhaskoningdhv.com/nereda (accesed 25 March 2014)

Ni B.-J., Xie W.-M., Liu S.-G., Yu H.-Q., Wang Y.-Z., Wang G. and Dai X.-L. (2009). Granulation of activated sludge in a pilot-scale sequencing batch reactor for the treatment of low-strength municipal wastewater. *Water Research*, **43**(3), 751–761.

Peng D., Bernet N., Delgenes J. P. and Moleta R. (1999). Aerobic granular sludge – a case report. *Water Research*, **33**(3), 890–893.

Pijuan M., Werner U. and Yuan Z. (2011). Reducing the startup time of aerobic granular sludge reactors through seeding floccular sludge with crushed aerobic granules. *Water Research*, **45**(16), 5075–5083.

Qin L. Liu Y. and Tay J-H. (2004). Effect of settling time on aerobic granulation in sequencing batch reactor. *Biochemical Engineering Journal*, **21**(1), 47–52.

Schwarzenbeck N. and Wilderer P. A. (2005). Treatment of food industry effluents in a granular sludge SBR. In: Aerobic Granular Sludge. Water and Environmental Management Series, IWA Publishing, Munich, pp. 95–102.

Schwarzenbeck N., Erley R. and Wilderer P. A. (2004a). Aerobic granular sludge in an SBR-system treating wastewater rich in particulate matter. *Water Science and Technology*, **49**(11–12), 41–46.

Schwarzenbeck N., Erley R., Mc Swain B. S., Wilderer P. A. and Irvine R. L. (2004b). Treatment of malting wastewater in a granular sludge sequencing batch reactor (SBR). *Acta Hydrochimica et Hydrobiologica*, **32**(1), 16–24.

Seviour T., Lambert L., Pijuan M. and Yuan Z. (2011). Selectively inducing the synthesis of a key structural exopolysaccharide in aerobic granules by enriching for *candidatus "competibacter phosphatis". Applied Microbiology and Biotechnology*, **92**(6), 1297–1305.

Seviour T., Yuan Z., van Loosdrecht M. C. M. and Lin Y. (2012). Aerobic sludge granulation: a tale of two polysaccharides? *Water Research*, **46**(15), 4803–4813.

Shin H. S., Lim K. H. and Park H. S. (1992). Effect of shear stress on granulation in oxygen aerobic upflow sludge reactors. *Water Science and Technology*, **26**(3–4), 601–605.

Sun F. Y., Yang C. Y., Li J. Y. and Yang Y. J. (2006). Influence of different substrates on the formation and characteristics of aerobic granules in sequencing batch reactors. *Journal of Environmental Sciences*, **18**(5), 864–871.

Tay J.-H., Liu Q.-S. and Liu Y. (2001a). The role of cellular polysaccharides in the formation and stability of aerobic granules. *Letters in Applied Microbiology*, **33**(3), 222–226.

Tay J.-H., Liu Q.-S. and Liu Y. (2001b). The effects of shear force on the formation, structure and metabolism of aerobic granules. *Applied Microbiology and Biotechnology*, **57**(1–2), 227–233.

Tay J-H, Liu Q-S. and Liu Y. (2001c). Microscopic observation of aerobic granulation in sequential aerobic sludge blanket reactor. *Journal of Applied Microbiology*, **91**(1), 168–175.

Tay J. H., Pan S., He Y. X. and Tay S. T. L. (2004). Effect of organic loading rate on aerobic granulation. II: characteristics of aerobic granules. *Journal of Environment Engineering-ASCE*, **130**(10), 1102–1109.

Tay J. H., Liu Q.-S., Liu Y., Show K.-Y., Ivanov V. and Tay S. T.-L. (2005). A comparative study of aerobic granulation in pilot- and labratory scale SBRs. In: Aerobic Granular Sludge. Water and Environmental Management Series, IWA Publishing, Munich, pp. 125–133.

Val del Río A., Figueroa M., Arrojo B., Mosquera-Corral A., Campos J. L., García-Torriello G. and Méndez R. (2012a). Aerobic granular SBR systems applied to the treatment of industrial effluents. *Journal of Environmental Management*, **95**(Supplement), S88–S92.

Val del Río A., Morales N., Figueroa M., Mosquera-Corral A., Campos J. L. and Mendez R. (2012b). Effect of coagulant-flocculant reagents on aerobic granular biomass. *Journal of Chemical Technology and Biotechnology*, **87**(7), 908–913.

Val del Río A., Morales N., Figueroa M., Mosquera-Corral A., Campos J. L. and Méndez R. (2013). Effects of the cycle distribution on the performance of SBRs with aerobic granular biomass. *Environmental Technology*, **34**(11), 1463–1472.

Wang F., Lu S., Wei Y. and Ji M. (2009). Characteristics of aerobic granule and nitrogen and phosphorus removal in a SBR. *Journal of Hazardous materials*, **164**(2–3), 1223–1227.

Wilderer P. A., Irvine R. L. and Goronszy M. C. (2001). Sequencing Batch Reactor Technology. IWA publishing, London, UK.

Winkler M. K., Kleerebezem R., Strous M., Chandran K. and van Loosdrecht M. C. M. (2013). Factors influencing the density of aerobic granular sludge. *Applied Microbiology and Biotechnology*, **97**(16), 7459–7468.

Yang S. F., Tay J-H. and Liu Y. (2003). Effect of substrate N/COD ratio on the formation of aerobic granules. *Journal of Environment Engineering*, **131**(1), 86–92.

Yi S., Tay J. H., Maszenan A. M. and Tay S. T. L. (2003). A culture-independent approach for studying microbial diversity in aerobic granules. *Water Science and Technology*, **47**(1), 283–290.

Zhou D., Liu M., Gao L., Shao C. and Yu J. (2013). Calcium accumulation characterization in the aerobic granules cultivated in a continuous-flow airlift bioreactor. *Biotechnology Letters*, **35**(6), 871–877.

Zima B. E., Diez L., Kowalczyk W. and Delgado A. (2007). Sequencing batch reactor (SBR) as optimal method for production of granular activated sludge (GAS) – fluid dynamic investigations. *Water Science and Technology*, **55**(8–9), 151–158.

Chapter 7

Anaerobic digestion of sewage wastewater and sludge

Katerina Stamatelatou

Department of Environmental Engineering, Democritus University of Thrace, Vas. Sofias 12, 67100, Xanthi, Greece

7.1 INTRODUCTION

Anaerobic digestion has been closely related to energy production since the major final product of the process is a gaseous mixture of carbon dioxide and methane (biogas). From this aspect, anaerobic digestion is by default an energy producing technology and, depending on the substrate and the operating conditions, the energy produced may exceed any energy required so that the net energy balance may be positive.

Anaerobic digestion is practically an 'energy transfer' process. The energy captured in the chemical bonds of the organic compounds of solid or liquid media is transferred mainly to the C-H bonds of the gaseous methane where the carbon atom is at its utmost reduced state, meaning that methane is the highest energy density organic compound. This energy transfer is carried out by microorganisms growing in the absence of oxygen by degrading a great range of organic substrates or feedstocks. The microorganisms belong to groups of different physiology and substrate affinity; they co-operate to break complex compounds into successively smaller ones in a balanced anaerobic environment of natural ecosystems or robust bioreactors.

The common application of the anaerobic digestion is the treatment of sewage sludge and slurries in digesters. Landfills are also gigantic bioreactors where biogas emissions indicate the anaerobic biological activity upon the organic fraction of the municipal solid wastes. Other feedstocks considered to be ideal substrates for anaerobic microorganisms are the high organic loaded liquid or solid wastes of industries such as food, pulp, petrochemical and so on industries, livestock establishments, and so on. Recently, dedicated crops or crop residues after harvesting and several waste streams of biorefineries may be directed to centralized biogas units for producing biogas, fertilizers and a nutrient rich liquor which can be sold or become a part of the crop nutrient management plan. The anaerobic digestion is a core technology role in the integrated resource recovery systems (Batstone & Virdis, 2014).

Traditionally, municipal wastewater treatment plants are based on aerobic processes as the core biochemical conversion process and anaerobic digestion was restricted to sewage sludge treatment. Anaerobic digestion of municipal wastewater has been carried in warm climate countries but stability and other problems have arisen and there are doubts about the suitability of this technology on this type of wastewater.

However, several alternatives have been developed with the aim to replacing the energy consuming steps of a typical sewage treatment plant. The pillars for improving the performance of the anaerobic digestion in the sewage management sector are the bioreactor technology, the control and automation and the pretreatment methods. Moreover, a redesign of the sewage treatment plant configuration based on the anaerobic digestion, would involve the improvement of many more steps preceding or following the anaerobic digester(s), so that the whole plant is optimized. In what follows, the main issues of anaerobic digestion of sewage and sewage sludge will be discussed and an economical assessment of this technology will be addressed.

7.2 THE PROCESS

Anaerobic digestion is a bioprocess consisting of a complex network of individual steps catalysed by different groups of microorganisms. The microorganism groups generally grow at different rates and become sensitive at a different degree when exposed to certain environmental conditions (pH, ammonia, concentration of metabolites such as volatile fatty acids, hydrogen etc.). The basic steps of the process and the breakdown of the organic content of the initial feedstock (expressed in Chemical Oxygen Demand, COD, units) are depicted in Figure 7.1. A brief description of the steps follows:

- *Disintegration*. This step lumps processes such as lysis, non enzymatic decay, phase separation and physical breakdown (e.g., shearing) (Batstone *et al.* 2002) and is responsible for the initial separation of the organic complex (e.g., sludge particles) into carbohydrates, proteins and lipids as well as inert material. It is regarded as the slowest step when the feedstock consists of solid particles.
- *Hydrolysis*. After disintegration, hydrolytic enzymes are excreted by microorganisms to breakdown the organic polymers (carbohydrates, proteins and fats) into their respective monomers (sugars, amino acids, lipids), so that they can be taken up by the microorganisms for further degradation.

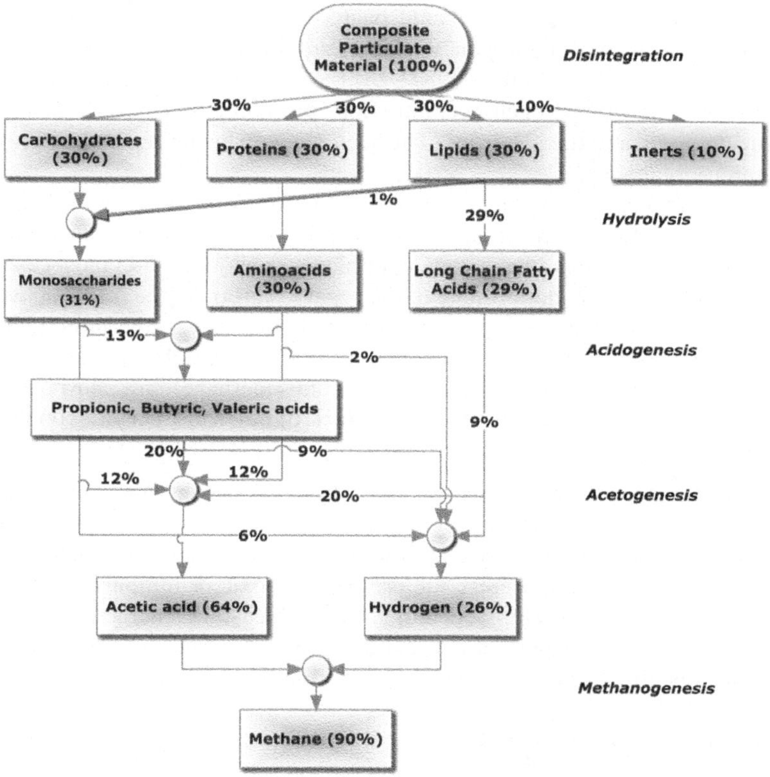

Figure 7.1 COD flux for a particulate waste consisting of 10% inerts and 30% each of the main organic polymers (in terms of COD) (Batstone *et al.* 2002).

- *Acidogenesis.* Following hydrolysis, the monomers can be converted to a mixture of volatile fatty acids, alcohols and other simpler organic compounds by a versatile group of microorganisms called acidogens. The electron donors and acceptors come from the organic compounds degraded. This 'internal' electron transfer is a characteristic of fermentation processes and often acidogenesis is regarded as identical to fermentation. The acidogens grow rather fast and are pH resistant (5–6) giving them the advantage of prevailing in the anaerobic consortium at adverse conditions. In the case of disturbances, the rapid acid formation may not be followed by degradation at the same rate, resulting in acid accumulation (and concomitant pH drop). This often brings an unbalance with an imminent potential of the process failure.
- *Acetogenesis.* What produced during acidogenesis (propionate, butyrate, valerate, lactate, alcohols etc.) are transformed to acetic acid, carbon dioxide

and hydrogen by the acetogenic bacteria. This is a thermodynamically infeasible step if hydrogen is not maintained at low levels (10^{-4}–10^{-6} atm; Harper & Pohland, 1987) by the hydrogen consuming bacteria. In the case of hydrogen accumulation, propionic and butyric acids also accumulate and the pH drops. The acetogenenic bacteria are slow growing microorganisms.
- *Methanogenesis*. Methane can be produced from acetate or hydrogen utilisation by the acetoclastic and the hydrogen utilizing methanogens respectively. The methane content of biogas is about 60% in most cases but depends on the oxidation state of the organic carbon in the initial substrate; the more reduced the carbon in the initial substrate is, the more methane will be produced (Gujer & Zehnder, 1983). Acetoclastic methanogens, which produce almost 70% of the total methane) are slow growing microorganisms and sensitive to pH, lack of nutrients, certain compounds and so on.

Any unbalance among the anaerobic digestion steps may influence the methanogenesis adversely. It has been recognised that the most important factors that may cause such an unbalance are the pH, the temperature, the nature of the feedstock (composition, nutrients), the presence of toxic or inhibitory substances and the organic loading rate. Depending on the feedstock characteristics (solid content, carbohydrate or protein concentration etc.), the rate limiting step of the process may be the disintegration/hydrolysis or methanogenesis.

7.3 THE TECHNOLOGY

Due to the variety of factors influencing the anaerobic process efficiency, many aspects of technology have been developed to improve and make the process cost-effective: reactor engineering, operation practices – integration with other processes, control and automation. All these aspects are based on the deep comprehension of the biochemical processes of anaerobic digestion (Angelidaki *et al.* 2012).

Reactor engineering in anaerobic digestion aims at (a) maintaining the microorganisms inside the reactor and sustaining sufficient contact between them and the substrate (b) increasing the reaction rates and eliminating the limiting transport phenomena, and (c) providing the microorganisms with a suitable environment to adapt and coexist under the operating conditions (Lettinga, 1995). Based on these principles, high-rate configurations were developed based on the contact process and the ability of the microorganisms to aggregate into granules or form biofilms (filters, fluidised beds, sludge blanket reactors).

The innovation connected with the boost of anaerobic digestion technology came with the development of the upflow anaerobic sludge blanket reactors (UASBR) by Lettinga *et al.* (1980). They found that, under certain conditions, the microorganisms tend to aggregate with a dense structure forming the granules.

Due to the high settling velocity, the granules are maintained in the interior of the reactor in the lower part. UASB is a vertical column filled partially with granules which are kept in suspension by introducing the wastewater from the bottom at an appropriate upward velocity (to keep suspension but not break or wash out the granules from the reactor). On the top of the reactor there is the three phase separator device, the design of which is very important. When the granules hit on the separator as they rise up, the biogas bubbles attached get separated from the granule and as a result, they fall back on the bottom while the effluent is removed from the top of the reactor. According to Tiwari *et al.* 2006) there are more than 1000 full scale applications in the world based on UASB technology. Since then, there are a lot of modifications developed combining the features of UASBR, filters etc.).

The anaerobic membrane bioreactors (AnMBR) is another type of bioreactors focusing on the complete retention of microorganisms by combinations of microfiltration (MF) or ultrafiltration (UF) modules. High conversion rates are achieved but the main disadvantages (membrane fouling and the concomitant operating cost as well as the cost due to energy required for the pressure-driven membranes to function) still remain and hinder the widespread application of this technology (Lin *et al.* 2013; Stuckey, 2012).

Besides the basic reactor configuration, other important details of an anaerobic digestion plant are the necessary number of reactors-units in series, the organic loading of each reactor, the pH and the temperature, the addition of trace metals and nutrients and so on (Stamatelatou *et al.* 2010; Takashima & Speece, 1989; Zandvoort *et al.* 2006; Lv *et al.* 2010; Demirel & Yenigün, 2002). Other tools such as molecular techniques, developments on analytical chemistry, modeling and simulation have contributed to a deeper insight of the process itself (Vanwonterghem *et al.* 2014; Batstone *et al.* 2002; Pavlostathis & Giraldo-Gomez, 1991), while developments on control and automation make the anaerobic systems more robust (Pind *et al.* 2001).

7.4 ANAEROBIC DIGESTION OF SEWAGE SLUDGE

Sewage sludge is a byproduct from biological sewage treatment. It consists mostly of particulate organics, initially contained in the sewage and removed in the primary sedimentation tanks, and microorganisms, called as 'waste activated sludge, WAS' and removed in the secondary sedimentation tanks. Sludge is produced mainly during primary and secondary treatment (Figure 7.2). The concentration of solids in primary sludge varies in the range of 2–7% total solids), while in the case of WAS is much less ranging from 0.5 to 1.5% (Turovskiy & Mathai, 2006). The composition of the mixture of the two sludge types depends on the sewage treatment plant configuration and operation and, in general determines the performance of the subsequent anaerobic digestion step. The reason for this is that the WAS is harder to biodegrade than the primary sludge. Moreover, it is often the major component of the total sludge produced, since some sewage treatment plants may not use primary treatment.

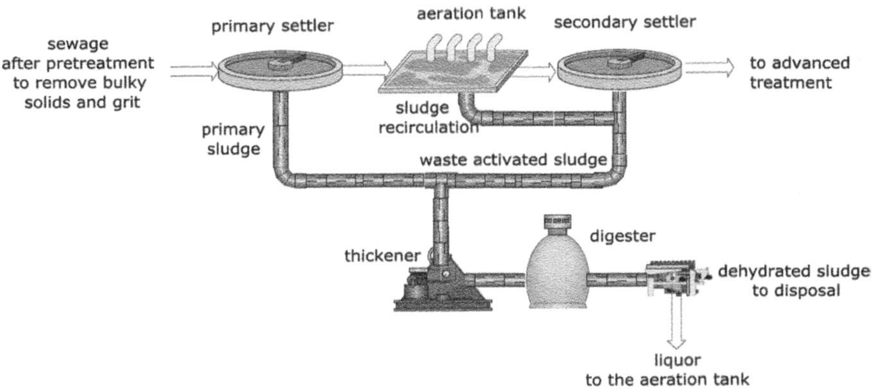

Figure 7.2 Sewage sludge route in a typical layout of a sewage treatment plant.

In general, sewage sludge management has been a serious and difficult to solve issue, because of its high moisture, its slow biodegradation rate, its poor dewaterability, its instability and odor problems induced, as well as the large quantities produced (a typical value is 20 kg/capita/y according to Bundgaard & Saabye, 1992). On the other hand, there is a great potential of material and energy recovery from sewage sludge (see Chapter 8).

Anaerobic digestion has been traditionally applied as the core technology for sewage sludge management. The energy potential of sewage sludge and its recovery in the form of biogas is an attractive option to reduce the high energy requirements of the other processes entailed in a typical sewage treatment scheme. The anaerobic digestion subunit in a sewage treatment plant consists of a thickening step to increase the solid concentration in the sludge from 0.5–1.5% to 4–6% (in the case of a mixture of primary and secondary sludge, Tchobanoglous *et al.* 2003), a heating step to increase the temperature of the sludge to mesophilic (preferably) or thermophilic temperature levels, the main anaerobic digester operating under a solid retention time (SRT) of 15–30 d and a post treatment step aiming to prepare the digested sludge (drying, composting, conversion to liquid or solid fuel) for its final use or disposal (landfill, agricultural use, biofuel). The volatile solid reduction can be correlated to the SRT (valid for a range from 15 to 20 d) according to:

$$VS_{des} = 13.7\ln(\text{SRT}) + 18.9 \tag{7.1}$$

where the VS_{des} is the destruction of the volatile solids (%) and SRT the solid retention time.

The production of biogas V_{QH4} (in m^3/d) can be estimated according to:

$$V_{QH_4} = 0.35 \cdot \mathrm{COD}_{\mathrm{des}} \cdot Q - 1.42 P_X \qquad (7.2)$$

where 0.35 is the stoichiometric coefficient of the production of methane (m^3) from 1 kg COD converted at 0°C, $\mathrm{COD}_{\mathrm{des}}$ is the reduction (destruction) in COD ($\mathrm{COD}_{\mathrm{influnet}}$-$\mathrm{COD}_{\mathrm{effluent}}$) in kg/m^3, Q is the flowrate of sludge (in m^3/d), P_X is the net biomass produced during anaerobic digestion (in kg/d). P_X is calculated according to:

$$P_X = \frac{Y \cdot \mathrm{COD}_{\mathrm{des}} \cdot Q}{1 + k_d \cdot \mathrm{SRT}} \qquad (7.3)$$

where Y is the yield coefficient (0.05–0.1 kg VSS/kg COD destroyed and k_d is the biomass decay constant (0.02–0.04 d^{-1})

The biogas can be transformed either to thermal energy in boilers or both thermal and electrical energy in combined heat units (CHP). Other alternatives are upgrading it to biomethane to resemble the natural gas.

The main barrier for the anaerobic digestion of sewage sludge is the slow hydrolysis step of the particulate matter; especially in the case of WAS, the extracellular polymeric substances (EPS) contribute into forming a complex of a slurry mass consisting of organics, inorganics and microorganisms. As a result, the enhancement of the anaerobic digestion process has been often correlated with the development of cost efficient pretreatment steps that would increase the accessibility of the hydrolytic enzymes to the extracellular complex network (formed by EPS) as well as the intracellular compounds.

Besides the energy recovery potential, anaerobic digestion of WAS results in the improvement of sludge dewaterability, especially when combined with pretreatment methods (Xu *et al.* 2011). EPS have been shown to influence the dewaterability of sludge; although they favor the flocculation of sludge, their high water affinity promote the hydration and decrease the dewaterability (Neyens & Baeyens, 2003).

The pretreatment methods considered to be suitable for enhancing the anaerobic biodegradability of sewage sludge are ultrasonic treatment, mechanical disintegration, chemical oxidation, treatment with alkali, thermal hydrolysis and biological treatment (Stamatelatou *et al.* 2012). Combinations of the above methods have also been studied (Dhar *et al.* 2012). However, with the introduction of a new technology, new problems may arise; the energy consumption is usually increased, the cost (operation and capital) is higher, environmental or other economical problems result because of the harsh conditions applied. Consequently the development of a pretreatment method is intriguing and the scientific interest on this is growing. Some of these scientific developments have been proved in numerous sewage treatment plants in the world (Table 7.1) according to a study of Jolly and Gillard (2009). In the sequel, the principles of the most common studied and evaluated economically methods are presented.

Table 7.1 Experience on sewage sludge technologies at full scale (Jolly & Gillard, 2009).

Technology	First installation of a full scale plant	Number of full scale plants	Year of record
Thermal hydrolysis	1996	24	2004
Biological – enzymatic hydrolysis	2002	11	2008
Thermophilic anaerobic digestion	1954	>20	2000
Sonication	2000	>10	2004

7.4.1 Sonication

During sonication, the ultrasound waves (sound waves at a frequency higher than 20 kHz) are emitted and transmitted into the slurry medium and create alternating regions of positive and negative pressure (compression and rarefraction regions respectively). When negative pressure is developed in a liquid medium, the gas dissolved forms bubbles (cavitation bubbles) which collide due to the vibrations of the ultrasound and become large up to a size that become unstable and collapse. This phenomenon is known as cavitation. The bubble collapse is followed by sever conditions of temperature and pressure (5000°C and pressure of 500 atmospheres) which last a few microseconds (Pilli *et al.* 2011). These harsh conditions cause high shear forces on the cells and disrupt them. Another effect of cavitation is the generation of oxygen hydroxide radicals (•OH) which also contribute into the disintegration of sludge through oxidation. However, Wang *et al.* (2005) concluded that the main mechanism for disintegration by ultrasonication is the shear forces induced by cavitation.

The ultrasonic pretreatment of sludge has been reviewed and evaluated with respect to the effect of frequency, duration, the sonication density (power supplied per volume of sludge treated), specific sonication energy (energy supplied per kg of sludge solids) on the particle size reduction, the dewaterability and settlability of sludge, the COD and nitrogen solubilization and the biogas production when anaerobic digestion is applied on the sonicated sludge (Carrere *et al.* 2010; Pilli *et al.* 2011). It would be expected that the positive effect of sonication would be enhanced as the sonication density increases. However there seems to be a certain limit beyond which, the intracellular polymers released will cause the flocculation of the small particles through hydroxyl and negatively charged carboxyl groups. Another negative effect of breaking the particles into smaller parts is that their specific surface is increased and so is the water bounded per surface unit. In any case, sonication favors COD solubilisation and, therefore, digestability. The increase of biogas up to 50% has been observed after sludge pretreatment with sonication in lab and full scale digesters (Carrere *et al.* 2010; Barber, 2005; Hogan *et al.* 2004).

7.4.2 Microwave

The microwave pretreatment is based on the electromagnetic radiation emitted at a frequency of 0.3 to 200 GHz into a polar medium. The dipoles of the polar medium (water molecules in the case of sludge) rotate to align with the alternating electromagnetic field. The molecule rotation causes friction which produced thermal energy. Due to reported results on the cell lysis and the change in the structure of the proteins of the microorganisms, microwave has been considered to be a potential method for sludge pretreatment. Compared to ultrasonic pretreatment, it was found by Park *et al.* (2004) that the same level of COD solubilisation could be achieved by providing a 3 time less energy per mass of sludge solids than in the case of ultrasonic treatment.

7.4.3 Thermal hydrolysis

During thermal hydrolysis the sludge is subjected to elevated temperature levels (150–200°C) under high pressure (600–2500 kPa). These severe conditions disrupt the microbial cell membrane and the structure of other solids present in the sludge mixture. As a result the soluble COD, which is more accessible to microorganisms, is increased. The effectiveness of thermal hydrolysis depends on the combination of temperature and pressure levels applied as well as the duration of pretreatment. Generally, thermal pretreatment is considered suitable to be implemented before mesophilic digestion and not thermophilic digestion because in the latter case, the digestion process is much more efficient than the mesophilic digestion and the thermal pretreatment does not result in substantial higher methane yields (Appels *et al.* 2008). There are several full scale applications of thermal hydrolysis as a pretreatment step of sewage sludge, developed by companies such as Cambi and Kruger Inc (a subsidiary of Veolia Water).

7.4.4 Autohydrolyis – Enzymatic hydrolysis

Enzymatic hydrolysis is based on methods that would enhance the ability of hydrolysing enzymes to promote bacterial wall lysis and breakdown of other cellular macromolecules. There is evidence that these enzymes are adsorbed on the EPS and therefore, their bioavailability is limited. The enzymatic hydrolysis methods aim to make these enzymes bioavailable through the application of mild conditions such as heat treatment at low temperatures (below 100°C), sonication, radiation, addition of surfactants and so on (Guo & Xu, 2011). Microaerophilic conditions may also activate the hydrolysis process prior to digestion (Carrere *et al.* 2010).

7.4.5 Other methods

The application of pressure has been used to enhance the cell disruption of the WAS. Specifically in the MicroSludge process, alkaline pre-treatment and milling precede the application of pressure in homogenisers to reduce the viscosity of the sludge streams. The sludge is pressurised up to 12,000 psi (82,700 kPa) and pressure is released accelerating the sludge up to sound velocity. Under this velocity, the shear forces disrupt the cells and almost liquify them. On a similar concept of pressurising and releasing, the thickened sludge can be compressed to a pressure between 3 and 6 bar (less than the MicroSludge). Under these conditions, sludge is introduced into the sludge stream and diffuses easily into the microbial cells though the cell wall to equilibrate the inner and outer cellular pressure. Upon pressure release, the inner pressure is significantly higher than the outer and the cell is burst out.

Cell disruption takes place during the OpenCel process too. The principle of this method is based on the charged and polar nature of the molecules of the cell membranes and wall. Under the influence of imposed high voltage electrical micro pulses, the cell walls are destroyed and the membranes become porous.

7.4.6 Economic analysis of the pretreatment methods

The operating cost of the pretreatment methods is mainly correlated with the energy consumption during pretreatment. However, energy consumption cannot be the only parameter to be considered since the potential for energy recovery (Dhar *et al.* 2012) as in the case of thermal pretreatment should be taken into account too. Moreover, the cost of energy required by the different methods may be different depending on the form of energy provided; for example thermal energy is required by thermal hydrolysis method and can be partially provided by the biogas transformation on-site. On the other hand, sonicators' operation is based on electricity which is more expensive than thermal energy (Carrere *et al.* 2010).

In the case of sonication, the index used for estimation of the energy required is the specific energy input (SE) expressed in kJ/kg TSS and is defined as the ratio energy to solid mass according to eq. (7.4)

$$SE = \frac{P \cdot \Delta t}{V \cdot \mathrm{TSS}} \tag{7.4}$$

where P is the power (kW), Δt is the duration of sonication (s), V is the volume of the sludge under sonication (L) and TSS is the concentration of the total suspended solids of sludge (kg/L).

The energy imparted to the sludge volume, however, is lower, due to losses from the electrical energy of the ultrasonic generator to the acoustical energy transmitted in the medium. The series of energy transformations during ultrasonic treatment is (Kobus & Kusinska, 2008): electrical → mechanical → acoustical → cavitation → thermal. The imparted energy is expressed as the acoustical energy and can be estimated from the thermal energy, assuming that the acoustical energy will finally result in heat when cavitation and bubble collapse occur. The thermal energy is the most used method for estimating the acoustic energy in the medium and is based on the temperature change of the medium with time (Kobus & Kusinska, 2008).

On the other hand, the thermal energy imparted in the medium during thermal hydrolysis can be calculated directly based on the energy (Q_s, kJ) needed to elevate the temperature of the sludge (T_o, °C) to the temperature of the pretreatment (T, °C) according to eq. (7.5):

$$Q_s = \rho_{sl} \cdot V_{sl} \cdot C_p \cdot (T - T_o) \quad (7.5)$$

where ρ_{sl} is the sludge density (kg/m^3), V_{sl} is the volume of the sludge under thermal hydrolysis and C_p is the specific thermal capacity of the sludge (4.18 kJ/kg °C). The actual thermal energy consumption is calculated based on the heat losses during heating. Moreover the recovery of thermal energy from the heated sludge should be taken into account when estimating the cost of the process.

Dhar *et al.* (2012) studied the effect of sonication (from 1000 to 10000 kJ/kg), thermal hydrolysis (from 50 to 90°C), and sonication at elevated temperature (combined ultrasonic and thermal hydrolysis treatment) on sludge and found that the increase of the ratio of soluble COD to total COD (Y, %) correlated to the imparted energy to the sludge (X, kJ): $Y = 0.247X + 7.056$, $R^2 = 0.0801$ regardless the pretreatment method used. However, the increase in the biogas production did not follow a linear correlation with the soluble to total COD ratio; Ultrasonic treatment was more effective than thermal pretreatement under the conditions tested in terms of biogas production, but the opposite was observed in terms of COD solubilisation. Thermal pretreatment results in agglomeration and increase in particle size and this could have influenced the methane yield (Bougrier *et al.* 2005). The increase in temperature and the SE input did not seem to affect the methane yield either, but the combination of both methods resulted in higher yields. Similar results were obtained with volatile sulfur compounds generated. Sludge dewaterability was not improved by temperature raise and was rather decreased at the highest SE input.

The assumptions for the cost estimation of the pretreatment technologies (ultrasonic and thermal hydrolysis) were (a) the sludge temperature was 25°C, the heat recovery from the thermally pretreated sludge was 80%. The cost for dewatering, transportation and landfilling was \$250/ton TSS, while for electricity and natural gas was \$0.07/kWh and \$0,28/m^3 respectively. The cost for biogas purification and specifically H2S removal through non regenerable KOH-AC bed was \$0.0005/m^3 biogas (cost per unit of biogas purification) and \$12/kg H2S (cost absorbent per unit of H2S removal). The results of this economic assessment showed that (Dhar *et al.* 2012):

(a) ultrasound pretreatment yielded a net saving of \$54/ton TSS at a moderate SE input (1000 kJ/kg TSS), while the net savings were negative in the case of the low and high SE inputs.
(b) the thermal pretreatment, at all temperatures tested (50–90°C), yielded a net saving from \$45/ton TSS to \$78/ton TSS.

(c) the thermal hydrolysis (at 50–90°C) combined with the untrasound pretreatment (at 1000 kJ/kg TSS yielded a net saving from \$44/ton TSS to \$66/ton TSS.

Other aspects which have an economic impact in the long term, but were not included in the assessment of Dhar *et al.* (2012), are the prevention of erosion of the equipment from sulfur compounds, the enhanced dewaterability of sludge after pretreatment and digestion as well as the optimization of the polymer dose and, finally, the investment cost for purchasing and installing the pretreatment equipment. Jolly and Gillard (2009), on the other hand, retrieved data from full scale applications of various technologies and estimated that pretreatment technologies as well as thermophilic anaerobic digestion enhanced dewaterability allowing the production of a sludge cake containing 25–32% Total Solids (TS). The cost of the polymer dose (decreasing as the volatile solid content decreases) as well as the cost of the anaerobic digestion liquor treatment (higher in ammonia concentration with increasing the process efficiency) were taken into account.

An important factor for the economic evaluation is the energy balance and the breakdown of the energy into thermal and electrical needed in each technology. The energy required depends on the performance of each method which vary according to the conditions (type of sludge – primary or secondary, solid content, temperature conditions and duration of treatment, etc.). Table 7.2 summarises the estimation of energy required in the case of some common pretreatment technologies and the biogas yields obtained. The main assumptions are also included where available.

In some cases, the energy estimates are obscure to decipher, because it is not clear if they concern the individual steps of the pretreatment or they refer to the whole anaerobic digestion unit. For example the electrical energy required for a mesophilic digester is 0.04 kWh kg^{-1}VS or 0.032 kWh kg^{-1} TS according to Carrere *et al.* (2010) and 0.150 kWh kg^{-1} TS according to Jolly and Gillard (2009). These values are not comparable probably because Jolly and Gillard (2009) have considered the electrical consumption of the whole plant (consisting of the following stages: pre-digestion thickening, pre-treatment, anaerobic digestion, CPH plant, post digestion storage, post digestion dewatering, and liquor treatment). Comparison between the different technologies with respect to the mesophilic digestion (as the control case) reveal that sonication requires much higher electrical energy than the others and all three studies agree that the energy balance is negative.

In the case of thermal hydrolysis, the assumption of Pérez-Elvira (2011) that the electrical demand is zero may be optimistic since electricity is indeed required to drive the various units of the process. In any case the electrical energy required is low compared to the thermal energy. All three studies concluded that the high thermal energy demand of thermal hydrolysis has not a negative impact on the economics since this form of energy can be recovered though the heat generated by the CPH units, the hot streams of the process itself, and if more thermal energy is needed, a part of biogas can be used in boilers (reducing the biogas available

Table 7.2 Electrical and Thermal Energy required in sewage sludge treatment and biogas yields.

Technology	Jolly and Gillard (2009)		Pérez-Elvira (2011)			Carrere *et al.* (2010)		
	Assumptions	**Electrical/ Thermal energy (kWh kg^{-1} TS)**	**Assumptions**	**Electrical/ Thermal energy (kWh m^{-3})**	**Biogas yield (L kg^{-1}VS)**	**Assumptions**	**Electrical/ Thermal Energy (kWh kg^{-1} VS)**	**Biogas yield (kWh kg^{-1} TS)**
None (only mesophilic digestion applied)	Input: 6% TS HRT: 18d TS reduction: 45%	0.150/NR	Input: 8% TS 35°C, HRT: 17d TS reduction: 42%	NR	488	Input: 6% TS, 80% VS 35°C, HRT: 20d VS reduction: 40%	0.04/0.5	1.9
None (only thermophilic)	Input: 6% TS HRT: 22d VS reduction: 56%	0.178/NR				Input: 6% TS, 80% VS 55°C, HRT: 15d VS reduction: 50%	0.03/1.0	2.4
Thermal hydrolysis	Input: 11% TS 165°C, 30 min VS reduction: 60%	0.310/ NR 11 bar steam from CHP/boiler	Input: 8% TS 170°C, 30 min TS reduction: 55%	0/NR	652	Input: 9% TS, 80% VS 170°C, 15–30 min VS reduction: 60%	0.04/2.0	2.9
Biological – Enzymatic hydrolysis	Input: 7.5% TS 42°C, 15 h VS reduction: 52%	0.304/NR				Input: 6% TS, 80% VS 70°C, 9–48 h VS reduction: 50%	0.03/1.0	2.4
Sonication	Input: 6% TS VS reduction: 55.5%	0.675/NR	nput: 8% TS 100 kWh m^{-3}	70/NR	NR	Input: 6% TS, 80% VS 100W, 16s 30 kWh m^{-3} VS reduction: 50%	0.37/0.5	2.4

NR: Not Reported.

to the CHP and thus 'consuming' the potential electrical energy that could be produced. Pérez-Elvira (2011) showed that the process can be self sustained in the case of a 13% TS, that is, the thermal energy for the pretreatment was obtained by recovering the heat from the process itself, increasing the profitability of the combined process. The author estimated that the economic benefit of treating the sewage sludge from a population of 100,000 is €132,373/yr (8% TS inlet) and €223,867/yr (13%TS). The biogas productivity reached 1.4 L L^{-1} d^{-1} compared to 0.26 observed in conventional AD systems (Pérez-Elvira *et al.* 2011).

The savings in energy is not the only criterion for selecting a treatment scheme. Mills *et al.* (2014) studied five scenarios for conversion of sewage sludge to energy. The core technology in all five scenarios was the anaerobic digestion process. Thermal hydrolysis was selected as a pretreatment step in most of the scenarios. Post treatment steps for biogas exploitation (in CHP or as biomethane) and digested sludge disposal (land application, solid fuel production or conversion to pyrolysis gas to be used in CHP) were considered as alternatives and evaluated against the conventional scenario of an anaerobic digestion unit coupled with CHP and the digested sludge be utilized in land applications. They performed a life cycle analysis (LCA) to include both inflow and outflow materials and energy of each scenario as well as the emissions to the environment on the assumption of treating 100 total dry solids per day. They also estimated the Capital Expenditure (CapEx) based on the simplified equation (eq. (7.6)) as well as the annual operating expenses (OpEx). Based on CApEx and OpEx and assumptions made for the discount rate (8%), they calculated the internal rate of return (IRR) of the investment for each scenario and considered both cases of the offer or absence of incentives by the UK state.

$$\text{CapEx} = k \cdot S^{0.6} \tag{7.6}$$

where k is the cost of the asset and S is the scale assumed.

The combination of the environmental and economical impacts for each scenario indicated the thermal hydrolysis pretreatment is preferable to the conventional anaerobic digestion scenario. The IRR was estimated to be 10.6% with incentives or 4.05 without incentives in the case of the conventional case, while the application of the thermal hydrolysis pretreatment step increases these values to 12.75% and 5.98% respectively. The environmental impact was also positive in the case of thermal hydrolysis pretreatment compared to the conventional case.

The final use of biogas and the digested sludge alter both the economics and the environmental impact of the processes. In the case of biogas upgrade to biomethane (for injecting it to the grid) and use the digested sludge to land, the IRR raised up to 18.92% (with incentives), but was negative without incentives (the investor would not see a return on the investment within the operational life of the plant) and the environmental impact was the most negative of all scenarios. The reasons for this is due to the particular incentives policy of UK, as Mills *et al.* (2014) state, which is rather high in this case, and poses high risks for the investment if this policy is adjusted or cancelled. The use of biomethane as vehicle fuel is a better option since the higher prices of biomethane as vehicle fuel would make the investment less independent on the incentives. The scenarios of using the digested sludge for solid fuel or in CHP after pyrolysis are comparable with high IRR (14.39% and 17.46% with incentives and 8.48% and 7.64%) and the most positive environmental impacts.

In the same line, Jolly and Gillard (2009) have concluded that the choice for the final disposal of sludge determines the economics and the payback period. Incineration of digested sludge, despite the high capital cost, results in more positive cost balance and shorter payback period than land application. In their study, they concluded that the thermal hydrolysis and thermophilic digestion are the best choices in this respect. They admit that their conclusion on the efficiency of thermophilic digestion should be verified by other works too, since they relied their estimations on data taken from one single plant.

Co digestion of sludge with other feedstocks such as organic residues as glycerol (Athanasoulia *et al.* 2014), landfill leachates (Pastor *et al.* 2013) agricultural wastes or energy crops (Hidaka *et al.* 2013; Galitskaya *et al.* 2014) or food wastes (Serrano *et al.* 2014; Belhadj *et al.* 2014; Powell *et al.* 2013; Dai *et al.* 2013) would increase the efficiency of the process as well as the quality of the digested liquor and sludge. The economic evaluation of developing a biogas unit within a sewage treatment plant accepting more inflows than sewage to increase the methane potential could be based on the work of Karellas *et al.* (2010). They developed an investment decision tool for biogas production from biomass feedstocks. This tool requires inputs such as the feedstock characteristics, availability and their gate fees (in the case of outputs of industrial activities) or cost (in the case of biomass), the market prices for the end products (electricity, heat, deigested sludge, liquor) and additional revenues, the total capital and annual operating costs and any economic incentives (loans, existing subsidies and grants). The output of this tool is the economic evaluation of the investment in terms of essential economic indicators such as the internal rate of return (IRR) and net present value (NPV) and so on.

7.5 ANAEROBIC DIGESTION OF SEWAGE

The anaerobic digestion process has not been limited to sludge treatment in a sewage treatment plant. WAS biodegradability energy content may be limited if the activated sludge process has been operated at high solid retention time (Bolzonella *et al.* 2005), which is a common practice leading to sludge minimisation but at the expense of less methane potential of the excess sludge and higher aeration cost. Even at typical solid retention times, the activated sludge process converts almost half of the organic matter into microbial cells that are harder to biodegrade and yield biogas. Based on this fact, an alternative option to trying to increase sludge biodegradability though energy intensive pretreatment methods is to apply anaerobic digestion upstream; (a) as a pretreatment method of sewage and (b) as a treatment method of sewage after preconcentrating it.

7.5.1 Pretreatment of sewage via anaerobic digestion

Anaerobic digestion and, particularly, UASB technology has been applied for sewage treatment in tropical and subtropical and low income countries such as Brazil (Vieira & Garcia, 1992; Vieira *et al.* 1994), Colombia (Schellinkhout & Collazos, 1992), India (Draaijer *et al.* 1992; Khalil *et al.* 2008) at ambient temperatures. In many cases, where there was not any treatment facility, anaerobic digestion was only applied to remove BOD up to 80% (Vieira *et al.* 1994). The concentration level of COD, BOD and TSS (in mg L^{-1}) in the effluent have been reported to be 140, 75 and 30 (Lew *et al.* 2004). Khalil *et al.* (2008) estimated that 80% of total UASB reactors worldwide for sewage treatment operate in India.

Campos *et al.* (2009) made an economic evaluation of a sewage treatment plant installed in a 200,000 inhabitant city in Brazil. At the time of the study, the first part of the plant had been constructed including a UASB reactor followed by an activated sludge system with air flotation. Based on the first year of operation, the first stage seemed to yield good performance (BOD removal at 20 C: 72% in UASB and 91.4% in total; TSS concentration in the effluent: 16 ± 8 mg L^{-1}). The economics and energy consumption also seemed attractive: USD$ 219.05 and 0.5233 kWh per 1000 m^3 respectively. For the second step, denitrification, coagulation applying ferric chloride and UV disinfection would be included.

Low temperatures deteriorate the UASB performance. Hybrid UASB reactors, having substituted the three phase separator on the top of the UASB with a filter was tested but only marginal improvement in COD removal but a better colloidal fraction removal was noticed (Elmitwalli *et al.* 1999; Lew *et al.* 2004). An option to increase the efficiency of UASB at low temperatures is to increase the influent concentration through co-digestion as suggested by Zhang *et al.* (2013). The increased growth of methanogens will increase their population to compensate for the negative effect of the low temperature on the growth rate.

Problems of stability and low performance also arise when the sewage is stronger. This is the case of arid areas, such as Jordan and Palestine, where the consumption of water is limited resulting in the production of sewage with a higher COD concentration than usual (COD > 1000 mg L^{-1}, the 70% of which is particulate). Moreover, the temperature fluctuations are wider (15–25°C) than other areas. Mahmoud (2008) applied a modification of the UASB reactor, called UASB-digester, in Palestine. In the UASB-digester, a parallel digester unit is added for enhanced sludge stabilisation and generation of active methanogenic sludge which are recirculated to the UASB reactor (Mahmoud *et al.* 2004). The results showed that removal of total COD increased from 54% (one stage UASB) to 72% and this system seems promising for further study.

Although the methane solubility in water is low, the huge quantities of wastewater treated per time result in an appreciable total methane mass dissolved and removed though the treated effluent. Seghezzo *et al.* (1998) estimated that the biogas recovered though digestion of sewage in a UASB was 60–80% of the anticipated amount based on a COD balance. Moreover, sulfide formation due to sulfate reduction results in mal odors; this is more intense for wastewaters with low ratio of COD per sulfate as in the case of sewage (Subtil *et al.* 2012). As a result, the pretreatment of sewage in UASB reactors may be restrained, especially near dense populated areas. Another problem with anaerobic digestion of sewage is that, due to the low growth rate of methanogens (a) more time is required for the start-up if the inoculum is not appropriately adapted, and (b) the process is more susceptible to toxic or inhibitory compounds.

The need for subsequent treatment steps to remove nutrients, pathogens and solids was addressed in all cases UASB technology was implemented. As a result, integrated processes combining anaerobic and aerobic steps have been extensively studied at lab and pilot scale, as well as at full scale. The application of UASB technology for the sewage treatment has been studied as a pretreatment method followed by an aerobic treatment step requiring less energy (even by 90%) than the conventional activated sludge process (Khan *et al.* 2011). The main combinations of anaerobic/aerobic systems are shown in Figure 7.3 and reviewed by Chan *et al.* (2009).

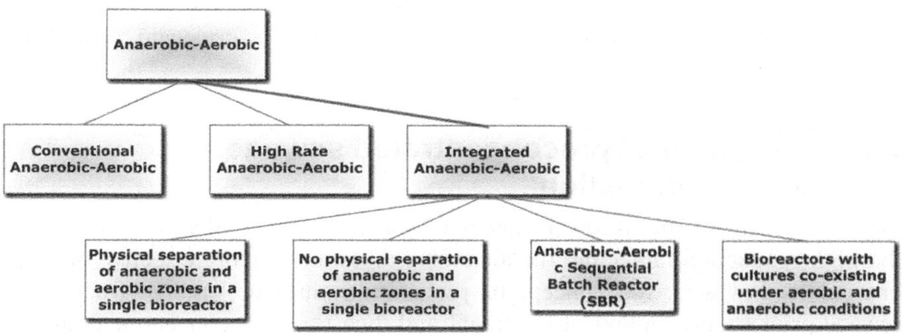

Figure 7.3 Types of combined anaerobic–aerobic systems

Khalil *et al.* (2008) based on the large experience gained in India calculated the net present value of sewage treatment plants adopting various technologies: activated sludge process, trickling filters, stabilisation ponds, sequential batch reactors, membrane bioreactors and UASB combined with a final polishing step (ponds) or extended aeration systems. A 30 year life time of the investment with a 12% interest was assumed and the year served as the basis for all calculations was 2010. The net present values were in the national currency (rupee) and the currency rate was 40 rupees per US dollar. The net present value for a plant of a 50 MLD total capacity (in rupee) was 8100 (activated sludge process), 4800 (trickling filters), 1000 (stabilisation ponds), 7600 (sequential batch reactors), 13100 (membrane bioreactors), 2700 (UASB and aeration pond), 3300 (UASB and extended aeration system). The stabilisation ponds is the cheapest technology but requires large areas which may not be available in all cases. The next cost-effective technologies are those with the UASB used as a pretreatment step, while the other aerobic technologies were far more expensive.

The anaerobic membrane bioreactors (AnMBR) have also been studied. They combine the features of anaerobic bioreactors and membranes which are able to achieve separation of solids from the mixed liquor to a high degree. As a result, the COD conversion (>85%) is similar to the one achieved in conventional aerobic MBR without the cost of aeration, the total suspended solid removal is more than 99% (Lin *et al.* 2013). This is not the case for nutrient removal though; further treatment is needed unless the treated effluent is used as a fertiliser. The two common configurations with membranes located out of the bioreactor (external) or inside the bioreactor (submerged) present advantages and disadvantages; the more direct control on fouling and replacement of membranes, the higher fluxes but more frequent cleaning, the negative effect on microbial activity due to the higher fluxes and the high energy consumption (approximately 10 kWh/m^3 product) in the case of external An MBR, and the lower energy consumption, the more simple and less frequent membrane cleaning and the milder operating conditions in the case of submerged AnMBR. Lin *et al.* (2011) concluded that the decisive factors for the life cycle cost of submerged AnMBR are the flux, the price and the lifetime of membranes.

7.5.2 Treatment of preconcentrated sewage via anaerobic digestion

Since the organic matter is low in concentration but high in total mass (considering the high production rates) and the anaerobic digestion is more effective on high organic load wastes, the concept of preconcentrating the organic fraction of sewage prior to any biological treatment and digesting this concentrated stream arose as an alternative to the disadvantages of digesting the sewage or the sludge. The preconcentration step will result in a high organic load stream suitable for digestion and a low organic load stream biodegradable further through less energy intensive processes.

A modification of the activated sludge process has been developed by Boehnke *et al.* (1997). It is based on the ability of the organics to interact with the microorganisms of sewage to form flocs with excellent adsorption ability and settling properties. This modified activated sludge process is known as AB process and is carried out under high F/M ratios (3–6 kg BOD kg^{-1} MLSS d^{-1}) in the first (A) stage, while normal loading conditions prevail in the second (B) stage. The settled sludge from the A stage is very rich in organics and, in this respect, the A stage could be applied as a preconcentration step (Verstraete *et al.* 2009). Other preconcentration techniques include membrane filtration, dynamic filter filtration, dissolved air flotation and on coagulation/flocculation by metal salt or polyelctrolyte addition. The latter method is the chemically enhanced primary treatment (CEPT) and has been studied with respect to the HRT, the types of suitable coaggulants, the dose of coaggulants etc (Tchobanoglous *et al.* 2003; Libhaber & Jaramillo, 2012; Harleman & Murcott, 1999).

Diamantis *et al.* (2013) made an economic evaluation of the CEPT process followed by anaerobic digestion for a 2000 PE scenario, 15 years lifetime and a 6% interest. They estimated the cost of the combined process as 0.2 €/m^3 (0.1 €/m^3 for the CEPT and 0.1 €/m^3 for the anaerobic digestion). The biogas produced suffices for supplying the required energy to the digester and as a result the process can be regarded as zero energy.

Moreover it seems that different strategies should be followed depending on the scale of STP. For small scale STPs where anaerobic digestion of sludge is not a feasible option, medium term storage of sludge is favored if biodegradation is hindered. This is achieved by increasing the coagulant dose which ends up in the concentrated stream inhibiting its biodegradation. Moreover, a high quality supernatant is produced, simplifying the post treatment steps and reducing the cost. On the other hand, for medium to large scale facilities, lower doses of coagulant would not cause any problem in the digestability of the energy rich concentrated stream.

Verstraete *et al.* (2009) introduced the term 'used water' for sewage regarding sewage as a resource of energy and matter and not as something useless that can be wasted. If energy, water and nutrients are recovered from the 'used water', then the whole process for treating sewage can be economically viable with no waste streams generated (Zero WasteWater). They estimated that the order of the total cost for the combined preconcentration (dissolved air flotation or dynamic sand filtration followed by ultrafiltration and reverse osmosis) and anaerobic digestion steps vary between €0.66–0.95/m^3. On the other hand, Verstraete and Vlaeminck (2011) estimated that almost €1/m^3 can be gained as profit from (a) recovering water, heat, nitrogen and phoshorous and (b) producing energy from biogas and biochar from digested sludge. This means that the zero wastewater approach can be economically viable.

7.6 CONCLUSIONS

The evaluation of a technology developed or improved to yield high efficiency and productivity should not be based solely on energy saving criteria or economic indices. It is evident that a holistic approach is imposed in the case of sewage treatment. There is a growing number of researchers that considered the sewage a resource and not as waste. Based on this concept, there is an attempt to benefit from sewage as much as possible with the least cost and minimum environmental impacts. As a result, concepts such as the 'zero wastewater' (Verstraete & Vlaeminck, 2011), the energy self-sufficient sewage treatment plants (Jenicek *et al.* 2013; Frijns *et al.* 2013), zero carbon footprint (Novotny, 2011, 2012) and so on, indicate the desired targets for the future sewage treatment plant: recover anything recoverable from a STP, no energy consumption, no environmental impacts. However, a sewage treatment plant based on such integrated concepts may be a good option for new installations, otherwise retrofitting existing conventional facilities to novel, anaerobic based facilities seems to be costly (McCarty *et al.* 2011).

The anaerobic digestion process has a key role in all these schemes since it has been related with energy and matter recovery as well as economic profit. Cost efficient technologies or practices that improve the efficiency and biogas productivity of the anaerobic digestion process are in the core of such schemes. This justifies the continuously growing effort on anaerobic digestion which, although regarded as mature technology, still remains on the top of scientific interest. It should be noted however, that the conclusion on the economic sustainability of the sewage treatment plants of the future is based on assumptions for some units of the concept. Whatever the risk of false estimation is, it is evident that a vision leading to a less energy intensive and costly sewage management is under shape and becomes inspiring. Following this vision, the modifications on existing conventional sewage plants are carried out and the positive results are demonstrated in numerous case studies (see second part of the present book as well as Dewettinck *et al.* 2001; Zeeman *et al.* 2008; Jenicek *et al.* 2013 and many others).

7.7 REFERENCES

Angelidaki I., Karakashev D., Batstone D. J., Plugge C. M., Alfons J. M. and Stams A. J. M. (2011). Biomethanation and its potential. *Methods in Enzymology*, **494**, 327–351.

Appels L., Baeyens J., Jan Degreve J. and Dewil R. (2008). Principles and potential of the anaerobic digestion of waste activated sludge. *Progress in Energy and Combustion Science*, **34**(6), 755–781.

Athanasoulia E., Melidis P. and Aivasidis A. (2014). Co-digestion of sewage sludge and crude glycerol from biodiesel production. *Renewable Energy*, **62**, 73–78.

Barber W. P. (2005). The effects of ultrasound on sludge digestion. *Chartered Institution of Water and Environmental Management*, **19**, 2–7.

Batstone D. J. and Virdis B. (2014). The role of anaerobic digestion in the emerging energy economy. *Current Opinion in Biotechnology*, **27**, 142–149.

Batstone D. J., Keller J., Angelidaki I., Kalyuzhnyi S. V., Pavlostathis S. G., Rozzi A., Sanders W. T. M., Siegrist H. and Vavilin V. A. (2002). Anaerobic Digestion Model No. 1 (ADM1), IWA, Task Group for Mathematical Modelling of Anaerobic Digestion Processes. IWA Publishing, London.

Belhadj S., Joute Y., El Bari H., Serrano A., Gil A., Siles J. Á., Chica A. F. and Martín M. Á. (2014). Evaluation of the anaerobic co-digestion of sewage sludge and tomato waste at mesophilic temperature. *Applied Biochemistry and Biotechnology*, **172**(8), 3862–3874.

Boehnke B., Diering B. and Zuckut S. W. (1997). Cost-effective wastewater treatment process for removal of organics and nutrients. *Water-Engineering and Management*, **144**(7), 18–21.

Bolzonella D., Pavan P., Battistoni P. and Cecchi F. (2005). Mesophilic anaerobic digestion of waste activated sludge: influence of the solid retention time in the wastewater treatment process. *Process Biochemistry*, **40**, 1453–1460.

Bougrier C., Carrere H. and Delgenes J. P. (2005). Solubilization of waste-activated sludge by ultrasonic treatment. *Chemical Engineering Journal*, **106**, 163–169.

Bundgaard E. and Saabye A. (1992). State of the Art on Sewage Sludge: Environmental Aspects and Regulations of Common Sludge Disposal Methods, in International Exhibition Congress of Solid Waste, ISWA, Madrid, Spain.

Campos J. R., Reali M. A. P., Rossetto R. and Sampaio J. (2009). A wastewater treatment plant composed of UASB reactors, activated sludge with DAF and UV disinfection, in series. *Water Practice & Technology*, **4**(1), doi: 10.2166/WPT.2009.008.

Carrere H., Dumas C., Battimelli A., Batstone D. J., Delgenes J. P., Steyer J. P. and Ferrer I. (2010). Pretreatment methods to improve sludge anaerobic degradability: a review. *Journal of Hazardous Materials*, **183**, 1–15.

Chan Y. J., Chong M. F., Law C. L. and Hassell D. G. (2009). A review on anaerobic–aerobic treatment of industrial and municipal wastewater. *Chemical Engineering Journal*, **155**(1–2), 1–18.

Dai X., Duan N., Dong B. and Dai L. (2013). High-solids anaerobic co-digestion of sewage sludge and food waste in comparison with mono digestions: Stability and performance. *Waste Management*, **33**(2), 308–316.

Demirel B. and Yenigün O. (2002). Two-phase anaerobic digestion processes: a review. *Journal of Chemical Technology and Biotechnology*, **77**(7), 743–755.

Dewettinck T., Van Houtte E., Geenens D., Van Hege K. and Verstraete W. (2001). HACCP (hazard analysis and critical control points) to guarantee safe water reuse and drinking water production – a case study. *Water Science and Technology*, **43**(12), 31–38.

Dhar B. R., Nakhla G. and Ray M. B. (2012). Techno-economic evaluation of ultrasound and thermal pretreatments for enhanced anaerobic digestion of municipal waste activated sludge. *Waste Management*, **32**, 542–549.

Diamantis V., Verstraete W., Eftaxias E., Bundervoet B., Vlaeminck S. E., Melidis P. and Aivasidis A. (2013). Sewage pre-concentration for maximum recovery and reuse at decentralized level. *Water Science and Technology*, **67**(6), 1188–1193.

Draaijer H., Maas J. A. W., Schaapman J. E. and Khan A. (1992). Performance of the 5 MLD UASB reactor for sewage treatment at Kanpur, India. *Water Science and Technology*, **25**(7), 123–133.

Elmitwalli T. A., Zandvoort M. H., Zeeman G., Bruning H. and Lettinga G. (1999). Low temperature treatment of domestic sewage in upflow anaerobic sludge blanket and anaerobic hybrid reactors. *Water Science and Technology*, **39**(5), 177–185.

Frijns J., Hofman J. and Nederlof M. (2013). The potential of (waste) water as energy carrier. *Energy Conversion and Management*, **65**, 357–363.

Galitskaya P. Y., Zvereva P. A. and Selivanovskaya S. Y. (2014). The effectiveness of co-digestion of sewage sludge and phytogenic waste. *World Applied Sciences Journal*, **30**(11), 1689–1693.

Gujer W. and Zehnder A. J. B. (1983). Conversion Processes in Anaerobic Digestion. *Water Science and Technology*, **12**, 127–167.

Guo Js. and Xu Yf. (2011). Review of enzymatic sludge hydrolysis. *Journal of Bioremediation & Biodegradation*, **2**, 130.

Harleman D. R. F. and Murcott S. (1999). The role of physical-chemical wastewater treatment in the mega-cities of the developing world. *Water Science and Technology*, **40**(4–5), 75–80.

Harper S. R. and Pohland F. G. (1987). Enhancement of anaerobic treatment efficiency through process modification. *Journal of Water Pollution Control Federation*, **59**, 152–161.

Hidaka T., Arai S., Okamoto S. and Uchida T. (2013). Anaerobic co-digestion of sewage sludge with shredded grass from public green spaces. *Bioresource Technology*, **130**, 667–672.

Hogan F., Mormede S., Clark P. and Crane M. (2004). Ultrasonic sludge treatment for enhanced anaerobic digestion. *Water Science and Technology*, **50**(9), 25–32.

Jenicek P., Kutil J., Benes O., Todt V., Zabranska J. and Dohanyos M. (2013). Energy self-sufficient sewage wastewater treatment plants: is optimized anaerobic sludge digestion the key? *Water Science and Technology*, **68**(8), 1739–1744.

Jolly M. and Gillard J. (2009). The Economics of Advanced Digestion. 14th European Biosolids and Organic Resources Conference and Exhibition, 9–11 November, The Royal Armouries, Leeds, UK.

Karellas S., Boukis I. and Kontopoulos G. (2010). Development of an investment decision tool for biogas production from agricultural waste. *Renewable and Sustainable Energy Reviews*, **14**(4), 1273–1282.

Khalil N., Sinha R., Raghav A. K. and Mittal A. K. (2008). UASB Technology for Sewage Treatment in India: Experience, Economic Evaluation and its Potential in other Developing Countries. Twelfth International Water Technology Conference, IWTC12 2008, Alexandria, Egypt.

Khan A. A., Gaur R. Z., Tyagi V. K., Khursheed A., Lew B., Mehrotra I. and Kazmi A. A. (2011). Sustainable options of post treatment of UASB effluent treating sewage: a review. *Resources Conservation and Recycling*, **55**, 1232–1251.

Kobus Z. and Kusinska E. (2008). Influence of physical properties of liquid on acoustic power of ultrasonic processor. *TEKA Komisji Motoryzacji i Energetyki Rolnictwa –* OL PA 8a, 71–78.

Lettinga G. (1995). Anaerobic digestion and wastewater treatment systems. *Antonie van Leeuwenhoek*, **67**, 3–28.

Lettinga G., Hobma S. W., Klapwijk A., Van Velsen A. F. M. and De Zeeuw W. J. (1980). Use of the Upflow Sludge Blanket (UAS) reactor concept for biological wastewater treatment. *Biotechnology and Bioengineering*, **22**, 699–734.

Lew B., Tarre S., Belavski M. and Green M. (2004). UASB reactor for domestic wastewater treatment at low temperatures: a comparison between a classical UASB and hybrid UASB-filter reactor. *Water Science and Technology*, **49**(11–12), 295–301.

Libhaber M. and Jaramillo A. O. (2012). Sustainable Treatment and Reuse of Municipal Wastewater For Decision Makers and Practicing Engineers, IWA Publishing, UK.

Lin H., Peng W., Zhang M., Chen J., Hong H. and Zhang Y. (2013). A review on anaerobic membrane bioreactors: applications, membrane fouling and future perspectives. *Desalination*, **314**, 169–188.

Lv W., Schanbacher F. L. and Yu Z. (2010). Putting microbes to work in sequence: recent advances in temperature-phased anaerobic digestion processes (review). *Bioresource Technology*, **101**(24), 9409–9414.

Mahmoud N. (2008). High strength sewage treatment in a UASB reactor and an integrated UASB-digester system. *Bioresource Technology*, **99**(16), 7531–7538.

Mahmoud N., Zeeman G., Gijzen H. and Lettinga G. (2004). Anaerobic sewage treatment in a one-stage UASB reactor and a combined UASB-digester system. *Water Research*, **38**(9), 2348–2358.

McCarty P. L., Bae J. and Kim J. (2011). Domestic wastewater treatment as a net energy producer-can this be achieved? *Environmental Science and Technology*, **45**, 7100–7106.

Mills N., Pearce P., Farrow J., Thorpe R. B. and Kirkby N. F. (2014). Environmental & economic life cycle assessment of current & future sewage sludge to energy technologies, *Waste Management*, **34**, 185–195.

Neyens E. and Baeyens J. (2003). A review of thermal sludge pre-treatment processes to improve dewaterability. *Journal of Hazardous Materials*, **B98**, 51–67.

Novotny V. (2011). Water and energy link in the cities of the future – achieving net zero carbon and pollution emissions footprint. *Water Science and Technology*, **63**(1), 184–190.

Novotny V. (2012). Water and energy link in the cities of the future – achieving net zero carbon and pollution emissions footprint. In: Water/Energy Interactions of Water Reuse, V. Lazarova, K. H. Choo and P. Cornel (eds), IWA Publishing, London.

Park B., Ahn J.-H., Kim J. and Hwang S. (2004). Use of microwave pretreatment for enhanced anaerobiosis of secondary sludge. *Water Science and Technology*, **50**(9), 17–23.

Pastor L., Ruiz L., Pascual A. and Ruiz B. (2013). Co-digestion of used oils and urban landfill leachates with sewage sludge and the effect on the biogas production. *Applied Energy*, **107**, 438–445.

Pavlostathis S. G. and Giraldo-Gomez E. (1991). Kinetics of anaerobic treatment: a critical review. *Critical Reviews in Environmental Control*, **21**(5–6), 411–490.

Pérez-Elvira S. I., Fdz-Polanco M. and Fdz-Polanco F. (2011). Enhancement of the conventional anaerobic digestion of sludge: comparison of four different strategies. *Water Science and Technology*, **64**(2), 375–383.

Pilli S., Bhunia P., Yan S., LeBlanc R. J., Tyagi R. D. and Surampalli R. Y. (2011). Ultrasonic pretreatment of sludge: review. *Ultrasonics Sonochemistry*, **18**, 1–18.

Pind P. F., Angelidaki I., Ahring B. K., Stamatelatou K. and Lyberatos G. (2001). Monitoring and control of anaerobic reactors. *Advances in Biochemical Engineering/ Biotechnology*, Springer, Berlin, **82**, 135–182.

Powell N., Broughton A., Pratt C. and Shilton A. (2013). Effect of whey storage on biogas produced by co-digestion of sewage sludge and whey. *Environmental Technology (United Kingdom)*, **34**(19), 2743–2748.

Schellinkhout A. and Collazos C. J. (1992). Full-Scale Application of the UASB Technology for Sewage Treatment. *Water Science and Technology*, **25**(7), 159–166.

Seghezzo L., Zeeman G., van Lier J. B., Hamelers H. V. M. and Lettinga G. (1998). A review: the anaerobic treatment of sewage in UASB and EGSB reactors. *Bioresource Technology*, **65**, 175–190.

Serrano A., Ángel Siles López J., Chica A. F., Ángeles Martin M., Karouach F., Mesfioui A. and El Bari H. (2014). Mesophilic anaerobic co-digestion of sewage sludge and orange peel waste. *Environmental Technology (United Kingdom)*, **35**(7), 898–906.

Stamatelatou K., Antonopoulou G. and Lyberatos G. (2010). Production of biogas via anaerobic digestion. In: Handbook of Biofuels Production: Processes and Technologies, R. Luque, J. Campelo and J. Clark (eds), Woodhead Publishing Series in Energy No. 15, UK.

Stamatelatou K., Antonopoulou G., Ntaikou I. and Lyberatos G. (2012). The effect of physical, chemical and biological pretreatments of biomass on its anaerobic digestibility and biogas production in Biogas Production: Pretreatment Methods in Anaerobic Digestion, Scrivener Publishing, USA.

Stuckey D. C. (2012). Recent developments in anaerobic membrane reactors. *Bioresource Technology*, **122**, 137–148.

Subtil E. L., Cassini S. T. A. and Goncalves R. F. (2012). Sulfate and dissolved sulfide variation under low COD/sulfate ratio in up-flow anaerobic sludge blanket (UASB) treating domestic wastewater. *Ambi-Agua Taubate*, **7**, 130–139.

Takashima M. and Speece R. E. (1989). Mineral requirements for methane fermentation (review). *Critical Reviews in Environmental Control*, **19**(5), 465–479.

Tchobanoglous G., Burton F. and Stensel H. D. (2003). Wastewater Engineering: Treatment and Reuse. 4th edn, Metcalf and Eddy Inc., McGraw Hill, New York, USA.

Tiwari M., Guha S., Harendranath C. and Tripathi S. (2006). Influence of extrinsic factors on granulation in UASB reactor. *Applied Microbiology and Biotechnology*, **71**(2), 145–154.

Turovskiy I. S. and Mathai P. K. (2006). Wastewater Sludge Processing. Wiley-Interscience. John Wiley & Sons, Inc., Hoboken, New Jersey.

Vanwonterghem I., Jensen P. D., Ho D. P., Batstone D. J. and Tyson G. W. (2014). Linking microbial community structure, interactions and function in anaerobic digesters using new molecular techniques, *Current Opinion in Biotechnology*, **27**, 55–64.

Verstraete W. and Vlaeminck S. E. (2011). Zero WasteWater: short-cycling of wastewater resources for sustainable cities of the future, *International Journal of Sustainable Development & World Ecology*, **18**(3), 253–264.

Verstraete W., van de Caveye P. and Diamantis V. (2009). Maximum use of resources present in domestic 'used water'. *Bioresource Technology*, **100**, 5537–5545.

Vieira S. M. M. and Garcia A. D. Jr. (1992). Sewage treatment by UASB reactor. Operation results and recommendations for design and utilization. *Water Science and Technology*, **25**(7), 143–157.

Vieira S. M. M., Carvalho J. L., Barijan F. P. O. and Rech C. M. (1994). Application of the UASB technology for sewage treatment in a small community at Sumare, Sao Paulo state. *Water Science and Technology*, **30**(12), 203–210.

Wang F., Wang Y. and Ji M. (2005). Mechanisms and kinetic models for ultrasonic waste activated sludge disintegration. *Journal of Hazardous Materials*, **123**, 145–150.

Xu H., He P., Yu G. and Shao L. (2011). Effect of ultrasonic pretreatment on anaerobic digestion and its sludge dewaterability. *Journal of Environmental Sciences*, **23**(9), 1472–1478.

Zandvoort M. H., van Hullebusch E. D., Fermoso F. G. and Lens P. N. L. (2006). Trace metals in anaerobic granular sludge reactors: bioavailability and dosing strategies (review). *Engineering in Life Sciences*, **6**(3), 293–301.

Zhang L., Hendrickx T. L. G., Kampman C., Temmink H. and Zeeman G. (2013). Co-digestion to support low temperature anaerobic pretreatment of municipal sewage in a UASB-digester. *Bioresource Technology*, **148**, 560–566.

Chapter 8

Resource recovery from sewage sludge

M. G. Healy[1], R. Clarke[2], D. Peyton[1,3], E. Cummins[2], E. L. Moynihan[4,5], A. Martins[6], P. Béraud[7] and O. Fenton[3]

[1]*Civil Engineering, National University of Ireland, Galway, Co. Galway, Rep. of Ireland*

[2]*School of Biosystems Engineering, University College Dublin, Co. Dublin, Rep. of Ireland*

[3]*Teagasc Environment Centre, Johnstown Castle, Co. Wexford, Rep. of Ireland*

[4]*T.E. Laboratories Ltd., Loughmartin Industrial Estate, Tullow, Co. Carlow, Rep. of Ireland*

[5]*Danone Nutrition Ireland, Rocklands, Co. Wexford, Rep. of Ireland*

[6]*Águas do Algarve S.A., Rua do Repouso, 10, 8000-302, Faro, Portugal*

[7]*AdP Energias, Rua Visconde de Seabra n°3, 1700-421 Lisboa, Portugal*

8.1 INTRODUCTION

More than 10 million tons of sewage sludge was produced in the European Union (EU) in 2010 (Eurostat, 2014). For the disposal of sewage sludge (solid, semisolid, or liquid residue generated during the treatment of domestic sewage), chemical, thermal or biological treatment, which may include composting, aerobic and anaerobic digestion, solar drying, thermal drying (heating under pressure up to 260°C for 30 min), or lime stabilisation (addition of $Ca(OH)_2$ or CaO such that pH is ≥12 for at least 2 h), produces a stabilised organic material.

The Waste Framework Directive (2008/98/EC; EC, 2008) lays down measures to protect the environment and human health by preventing or reducing adverse impacts resulting from the generation and management of waste. Under the directive, a hierarchy of waste is applied: prevention, preparing for re-use, recycling, other recovery and disposal. The objective of the Directive is to maximise the resource value and minimise the need for disposal (EC, 2008). This has prompted efforts within sewage sludge management to utilise sewage sludge as a commodity. The terminology 'biosolids' reflects the effort to consider these materials as potential resources (Isaac & Boothroyd, 1996). Biosolids may be used in the production of energy, bio-plastics, polymers, construction materials and other potentially useful compounds. However, as the disposal of sewage sludge is commonly achieved by recycling treated sludge to land, nutrient recovery, particularly in the context of pressure on natural resources, and potential barriers to its reuse on land (environmental, legislative), deserves particular attention.

The aim of this chapter is to examine the recovery of nutrients and other compounds, such as volatile fatty acids (VFA), polymers and proteins, from sewage sludge. Due to the increasing awareness regarding risks to the environment and human health, the application of sewage sludge, following treatment, to land as a fertilizer in agricultural systems has come under increased scrutiny. Therefore, any potential benefits accruing from the reuse of sewage sludge are considered against possible adverse impacts associated with its use. Finally, the potential costs and benefits arising from its re-use are examined.

8.2 DEFINING TRENDS FOR MUNICIPAL SLUDGE TREATMENT

The amount of sewage sludge produced in Europe has generally increased (EC, 2011), which is mainly attributable to implementation of the Urban Waste Water Treatment Directive 91/271/EC (EC, 1991) and other legislative measures.

The treatment and disposal of sewage sludge presents a major challenge in wastewater management. As seen over the last decade, the upgrading and development of effective treatment plants has facilitated efforts to improve the quality of the effluent (i.e., removal of microorganisms, viruses, pollutants). Subsequently, legislation regarding sewage sludge in the EU (Sewage Sludge Directive 86/278/EEC; EEC, 1986) and the USA (40 CFR Part 503; USEPA, 1994) has focused on effluent quality and potential contamination. Within the EU, treated sewage sludge is defined as having undergone biological, chemical or heat treatment, long-term storage, or any other appropriate process so as to significantly reduce fermentability and any health hazards resulting from its use (EC, 2012). Physical-chemical treatment of wastewater has been widely practiced, introducing biodegradation and chemical advanced oxidation for biological treatment (Mouri *et al.* 2013). In the treatment of wastewater, biological treatments, such as aerobic and anaerobic digestion, appear to be the more favoured option. Aerobic treatment has a high degree of treatment efficiency, whilst anaerobic biotechnology has significantly progressed, offering resource recovery and utilization while still achieving the objective of waste control (Chan *et al.* 2009). A variety of sewage sludge treatment technologies can be employed and are implemented according to regulations. As can be seen from Table 8.1, significant differences in sewage sludge treatments can be observed between the EU, USA and Canada. With regards to sludge stabilization, aerobic and anaerobic treatments are the most widely used methods of sewage sludge treatment. Within the EU, anaerobic and aerobic wastewater treatments appear to be the most common methods, with 24 countries out of 27 applying this method (Kelessidis & Stasinakis, 2012).

Anaerobic digestion (AD) is most commonly used in Spain, Italy, United Kingdom and Czech Republic (Table 8.1). Within the USA and Canada, biosolids are classed according to their pathogenic levels. Class A biosolids contain minute levels of pathogens and must undergo heating, composting, digestion, or increased pH. Thus, these methods are more commonly employed (Table 8.1). Class B biosolids have less stringent parameters for treatment and contain small, but compliant, amounts of bacteria (USEPA, 2011). In order to achieve Class A biosolids, the sewage sludge must undergo stringent treatment. Stabilization methods such as aerobic, anaerobic, liming and composting, are the recommended options in both the USA and Canada.

8.3 SEWAGE SLUDGE AS A RESOURCE

The two components in sewage sludge that are technically and economically feasible to recycle are nutrients (primarily nitrogen (N) and phosphorus (N)) and energy (carbon) (Tyagi & Lo, 2013). As sewage sludge contains organic matter, energy can be recovered whilst treating it. There are a considerable amount of nutrients within sewage sludge, especially P and N. However, P is fast becoming the most significant nutrient due to depleting sources. Emerging technologies have been developed to extract this valuable resource including KREPO, Aqua-Reci, Kemicond, BioCon, SEPHOS and SUSAN, and are based on physical-chemical and thermal treatment to dissolve the P, with final recovery by precipitation (Cordell *et al.* 2011; Tyagi & Lo, 2013). Other resources include the reuse of sludge for construction materials, heavy metals, polyhydroxyalkanoates (PHA), proteins, enzymes and VFA. Table 8.2 gives an overview of resource recovery products from sewage sludge, their typical values and uses. Apart from the recovery products mentioned in Table 8.2, advances in technology have revealed innovative emerging products from treated sewage sludge and include VFA, polymers, and proteins in the form of worms, larvae and fungi. A short review regarding production, processes and further use is provided on each emerging product.

8.3.1 Nutrient recovery from sewage sludge

Treated sewage sludge may be used as an agricultural fertiliser, as they contain organic matter and inorganic elements (Girovich, 1996). The recycling of treated sewage sludge to agriculture as a source of the fundamental nutrients and metals required for plant growth is going to be essential for future sustainable development, as it is estimated that there are only reserves of 50–100 years of

Table 8.1 Global municipal sewage sludge treatment processes.

	Denmark[a]	France[a]	Germany[a]	Greece[a,b]	Ireland[a]	Italy[a]	Spain[a]	Sweden[a]	UK[a]	Czech Rep.[a]	Poland[a]	Portugal[d]	USA[c]
Stabilisation													
Aerobic	✓	✓		✓✓	✓	✓	✓	✓	✓	✓	✓✓	✓	✓
Anaerobic	✓	✓	✓	✓	✓	✓✓	✓✓	✓	✓✓	✓✓	✓	✓	✓
Lime	✓	✓		✓	✓	✓	✓	✓	✓		✓	✓	✓
Composting	✓	✓		✓	✓	✓	✓	✓	✓	✓✓	✓	✓	✓
Conditioning													
Lime	✓					✓							
Inorganics						✓		✓					
Polymers				✓									
Thermal			✓			✓		✓			✓		
Drying belts				✓				✓					
Dewatering													
Filter press		✓		✓	✓	✓			✓		✓		
Centrifuges		✓						✓			✓	✓	
Belt filter press	✓			✓✓	✓	✓		✓	✓		✓	✓	
Others													
Thermal		✓	✓✓	✓	✓	✓	✓	✓	✓				✓
Solar drying	✓	✓				✓	✓						
Pasteurisation													✓
Long-term storage				✓	✓	✓	✓	✓	✓		✓		
Cold fermentation bag filling											✓		

✓ Common use ✓✓ most common use

[a]Kelessidis and Stasinakis (2012); [b]Tsagarakis *et al.* (1999); [c]Lu *et al.* (2012); [d]Martins and Béraud, pers. comm.

P depending on future demand (Cordell *et al.* 2009). When spread on arable or grassland, and provided that it is treated to the approved standards, treated sewage sludge may offer an excellent source of nutrients and metals required for plant and crop growth (Jeng *et al.* 2006). Treated sewage sludge can also contribute to improving soil physical and chemical characteristics (Mondini *et al.* 2008). It increases water absorbency and tilth, and may reduce the possibility of soil erosion (Meyer *et al.* 2001).

Table 8.2 Resource recovery products from sewage sludge.

Products	Typical values and uses	Reference
Nitrogen	2.4–5% total solids	Tchobanoglous *et al.* (2003)
Phosphorus	0.5–0.7% total solids	Tchobanoglous *et al.* (2003)
Heavy metals	Typical recovery values: Ni 98.8%; Zn 100.2%; Cu 93.3%	Pérez-Cid *et al.* (1999)
Construction materials	Dried sludge or incinerator ash. Biosolid ash is used to make bricks	Tay and Show (1997)
Bio-plastic	Microorganisms in activated sludge can accumulate PHAs ranging from 0.3 to 22.7 mg polymer/g sludge	Yan *et al.* (2008)
Enzymes	Protease, dehydrogenese, catalase, peroxidase, α-amylase, α-glucosidase	Tyagi and Surampalli (2009)

Land application of treated sewage sludge to agricultural land can be relatively inexpensive in countries in which it is considered to be a waste material. An alternative, but costly, option in such countries is to pay tipping fees for its disposal (Sonon & Gaskin, 2009). However, in some countries sewage sludge is seen not as a waste but instead as a product containing valuable nutrients (e.g., the U.K) with an associated fertiliser replacement value (FRV) and cost for its usage.

As the world population increases, pressure on natural resources, especially food, oil and water, will increase. Inorganic fertilizer prices are tied to crude oil prices globally and demand (Bremer, 2009): when prices of oil are high, inorganic fertilizer prices also climb. For instance, in Ireland, the cost of inorganic fertilisers has continually increased, with the cost of a mean kg of N, P and potassium (K) rising from €0.41, 1.06 and 0.23 in 1980 to €103, 203, 105 in 2011 (Figure 8.1). Similar price increases of 13% were seen in the U.K. in 2010 (Tasker, 2010). Recent fertiliser increases since 2008 can be attributed to increases in both energy costs and global demand for fertilisers. Increased prices and volatility are important considerations, as they lead to volatility in farm input costs and profit margins, and make farm planning more difficult and risky (Lalor *et al.* 2012).

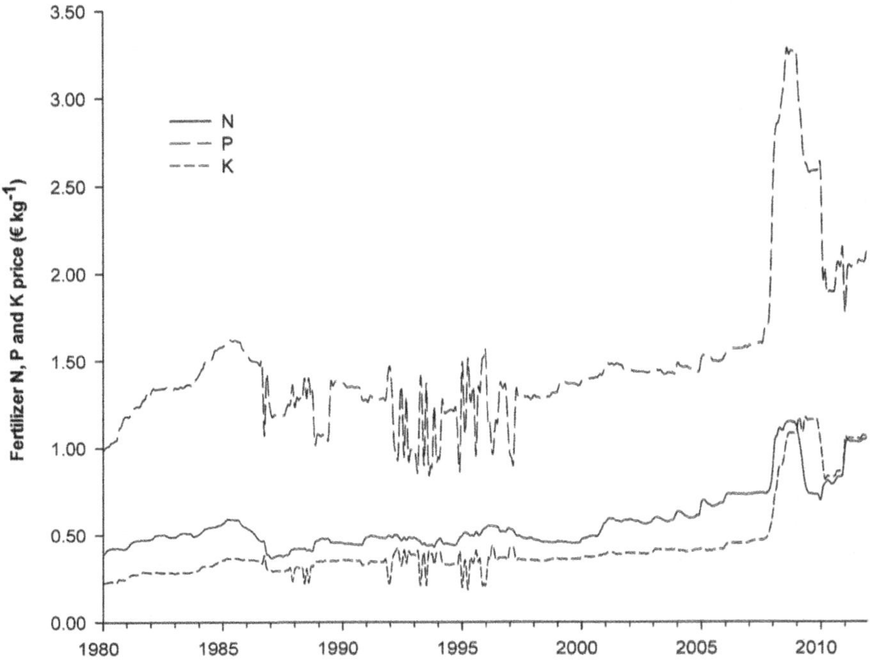

Figure 8.1 Trends in unit cost of nitrogen (N), phosphorus (P) and potassium (K) in chemical fertilisers in Ireland from 1980 to 2011 (Lalor *et al.* 2012).

Nutrient price equivalents of sewage sludge will depend on the nutrient availability and the FRV of the nutrients in the sludge. The FRV of nutrients in cattle slurry over time was calculated in Lalor *et al.* (2012) assuming a total N, P and K content in slurry of 3.6, 0.6 and 4.3 kg m^{-3}, respectively, and an assumption of respective FRV of 25%, 100% and 100% (Coulter, 2004). Of course in treated sewage sludge as in other nutrient streams, micronutrients used by the plant give added value to the product. In addition, factors such as transport and land application costs would also need to be considered in an overall assessment. It is therefore essential that such data are known for treated sewage sludge.

There is a good body of literature that has examined its fertilisation potential (Smith & Durham, 2002; Epstein, 2003; Singh & Agrawal, 2008). Siddique and Robinson (2004) mixed AD-treated sewage sludge, poultry litter, cattle slurry and an inorganic P fertiliser with five soil types at rates equivalent to 100 mg P kg^{-1} soil and, following incubation at 25°C for 100 d, found that AD-treated sewage sludge and poultry litter had a slower rate of P release compared with cattle slurry and inorganic P fertiliser. This may indicate that it may have good long-term fertilisation potential.

One of the main concerns associated with the use of treated sewage sludge as an organic fertiliser on grassland are the loss of nutrients, metals and pathogens along a transfer continuum (Wall *et al.* 2011) to a waterbody *via* direct discharges, surface and near surface pathways and/or groundwater discharge. More recently, so-called 'emerging contaminants', which may include antibiotics, pharmaceuticals and other xenobiotics, have been considered, as they have health risks associated with them. Therefore, nutrient recovery from treated sewage sludge must be considered against possible adverse impacts associated with its use.

8.3.2 Volatile fatty acids

Volatile fatty acids are short-chained fatty acids consisting of six or fewer carbon atoms which can be distilled at atmospheric pressure (Lee *et al.* 2014). Proteins and carbohydrates in sewage sludge can be converted into VFA to enhance methane, hydrogen and poly-hydroxyalkanoate production (Yang *et al.* 2012). The production of VFA from biosolids is an anaerobic process involving hydrolysis and acidogenesis (or dark fermentation) (Su *et al.* 2009). In hydrolysis, complex polymers in waste are broken down into similar organic monomers by the enzymes excreted from the hydrolytic microorganisms. Subsequently, acidogenesis ferment these monomers into mainly VFA such as acetic, propionic and butyric acids. Both processes involve a conglomerate of obligate and facultative anaerobes such as Bacteriocides, Clostridia, Bifidobacteria, Streptococci and Enterobacteriaceace (Lee *et al.* 2014).

8.3.3 Polymers

Extracellular polymeric substances (EPS) are the major constituents of organic matter in sewage sludge floc, which comprises polysaccharides, proteins, nucleic acids, lipids and humic acids (Jiang *et al.* 2011). They occur in the intercellular space of microbial aggregates, more specifically at or outside the cell surface (Neyens *et al.* 2004), and can be extracted by physical (centrifugation, ultrasonication and heating, for example) or chemical methods (using ethylenediamine tetraacetic acid, for example), although formaldehyde plus NaOH has proven to be effective in extracting EPA from most types of sludge (Liu & Fang, 2002). Extracellular polymeric substances perform an important role in defining the physical properties of microbial aggregates (Seviour *et al.* 2009). There are many biotechnical uses of EPS, including the production of food, paints and oil drilling 'muds'; their hydrating properties are also used in cosmetics and pharmaceuticals. Furthermore, EPS may have potential uses as biosurfactants for example, in tertiary oil production, and as biological glue. Extracellular polymeric substances are an interesting component of all biofilm systems and still hold large biotechnological potential (Flemming & Wingender, 2001). A relatively new method for treatment of sewage sludge is aerobic granular sludge technology (Morgenroth *et al.* 1997). A special characteristic of AGS is the high concentration of alginate-like exopolysaccharides (ALE) with different properties compared to converted activated sludge. Aerobic granular sludge technology produces a compound with similar characteristics as alginate, which is a polymer normally harvested from brown seaweed. Alginate-like exopolysaccharides can be harvested and used as a gelling agent in textile printing, food preparation and the paper industry (Hogendoorn, 2013). Lin *et al.* (2010) demonstrated that the potential yield of extractable alginate-like exopolysaccharides reached 160 ± 4 mg/g (VSS ratio). It was also found that they were one of the dominant exopolysaccharides in aerobic granular sludge.

8.3.4 Proteins

Vermicomposting (sludge reduction by earthworms) is a relatively common technology, especially in developing countries with small scale settings. The main product of this process is vermicompost, which consists of earthworm faeces that can be used as a fertilizer due to its high N content, high microbial activity and lower heavy metal content (Ndegwa & Thompson, 2001). Vermicomposting results in bioconversion of the waste streams into two useful products: the earthworm biomass and the vermicompost. In a study by Elissen *et al.* (2010), aquatic worms grown on treated municipal sewage sludge, produced high protein values with a range of amino acids. These proteins can be used as animal feed for non-food animals, such as aquarium fish or other ornamental aquatic fish. Other outlets for the protein could be technical applications such as coatings, glues and emulsifiers. The study also revealed that the dead worm biomass can be utilized as an energy source in anaerobic digestion. Experiments have shown that biogas production of worms is three times that of sewage sludge. Other applications include fats and fatty acid extraction. Treatment of sewage sludge using earthworms has been well documented; however, research studies on protein extraction of earthworms grown on sewage sludge are very limited.

Bioconversion of biosolids using fly larvae has been studied for years. Organic waste has a high nutritional and energy potential and can be used as a feed substrate for larvae. Apart from significantly reducing organic waste, grown larvae make an excellent protein source in animal feed. The insect protein could be used in animal feed to replace fishmeal (Lalander *et al.* 2013). One of the most studied species is the larvae of the Black Soldier fly (*Hermetia illucens* L.). The larvae of this non-pest fly feed on, and thereby degrade, organic material of different origin (Diener *et al.* 2011a). The 6th instar, the prepupa, migrates from the sludge to pupate and can therefore easily be harvested. Since prepupae contain on average 44% crude protein and 33% fat, it is an appropriate alternative to fishmeal in animal feed (St-Hilaire *et al.* 2007). Proposals for other uses for the pupae other than animal feed have been put forward. The other components of the pupae (protein, fat, and chitin) could be fractioned and sold separately. The extracted fat can be converted to biodiesel; chitin is of commercial interest due to its high percentage of N (6.9%) compared to synthetically substituted cellulose (1.25%) (Diener *et al.* 2011b). There has been ample research on the *H. illucens* and its contribution to significantly reducing organic wastes; however, there are several knowledge gaps on the potential utilization of the pupae in terms of protein, fat and chitin.

Filamentous fungi are often cultivated in food industries as a source of valuable products such as protein and a variety of biochemicals, using relatively expensive substrates such as starch or molasses (More *et al.* 2010). The biomass produced during fungal wastewater treatment has potentially a much higher value in the form of valuable fungal by-products such as amylase, chitin, chitosan, glucosamine, antimicrobials and lactic acids, than that from bacterial activated sludge process (van Leeuwen *et al.* 2012). The use of fungi for the production of value added products has been presented by several researchers (Molla *et al.* 2012).

8.4 LEGISLATION COVERING DISPOSAL OF BIODEGRADABLE WASTE ON LAND

Recent estimates of the disposal methods of sewage sludge in EU Member States indicate that although the amount of sewage sludge being applied to land in the EU has dramatically increased, landfill and incineration are still common (EC, 2010), particularly in countries where land application is banned. Less common disposal routes are silviculture, land reclamination, pyrolysis, and reuse as building materials. The drive to reuse sewage sludge has been accelerated by, amongst other legislation, the Landfill Directive, 1999/31/EC (EC, 1999), the Urban Wastewater Treatment Directive 91/271/EEC (EC, 1991), the Waste Framework Directive (2008/98/EC; EC, 2008), and the Renewable Energy Directive (2009/28/EC; EC, 2009), which places an increased emphasis on the production of biomass-derived energy.

The application of treated sewage sludge to agricultural land is governed in Europe by EU Directive 86/278/EEC (EEC 1986), which requires that sewage sludge undergoes biological, chemical or heat treatment, long-term storage, or any other process to reduce the potential for health hazards associated with its use. In the EU, land application of treated sewage sludge is typically based on its nutrient and metal content, although individual member states often have more stringent limits than the Directive (EC, 2010; Milieu *et al.* 2013a, b, c). Generally, when applying treated sewage sludge based on these guidelines and depending on the nutrient and metal content of the treated sewage sludge, P becomes the limiting factor for application. In the USA, the application of treated sewage sludge to land is governed by *The Standards for the Use or Disposal of Sewage Sludge* (USEPA, 1993), and is applied to land based on the N requirement of the crop being grown and is not based on a soil test (McDonald & Wall, 2011). Therefore, less land is required for the disposal of treated sludge than in countries where it is spread based on P content. Evanylo *et al.* (2011) suggests that when soil P poses a threat to water quality in the USA, the application rate could be determined on the P needs of the crop.

8.5 EXISTING AND EMERGING ISSUES CONCERNING THE RE-USE OF BIODEGRADABLE WASTE ON LAND

8.5.1 Societal issues

One of the major stumbling blocks in the use of treated sewage sludge as a low-cost fertiliser is the issue of public perception (Apedaile, 2001). Concerns have been raised over potential health, safety, quality of life and environmental impacts that the land spreading of sludge may have (Robinson *et al.* 2012). This perception could be, in part, due to the fact that treated sewage sludge is heavily regulated or that animal manure is more commonly seen and used. In many countries such as Ireland, for example, companies that produce products for the food and drinks industry will not allow the use of the raw materials produced from agricultural land which has been treated with treated sewage sludge (FSAI, 2008). This limits their use as a fertiliser at the current time.

8.5.2 Nutrient and metal losses

Phosphorus and reactive N losses to a surface waterbody originate from either the soil (chronic) or in runoff where episodic rainfall events follow land application of fertiliser (incidental sources) (Brennan *et al.* 2012). Such losses to a surface waterbody occur *via* primary drainage systems (end of pipe discharges, open drain networks (Ibrahim *et al.* 2013), runoff and/or groundwater discharges. Application of treated sewage sludge to soils may also contribute to soil test phosphorus build-up in soils, thereby contributing to chronic losses of P, metal and pathogen losses in runoff (Gerba & Smith, 2005). Dissolved reactive P losses may also be leached from an agricultural system to shallow groundwater (Galbally *et al.* 2013) and, where a connectivity exists, may affect surface water quality for long periods of time (Domagalski & Johnson, 2011; Fenton *et al.* 2011).

The metal content of treated sludge and of the soil onto which it can be spread is also regulated by legislation in Europe (86/278/EEC; EEC, 1986). However, guidelines governing the application of treated sewage sludge to land (e.g., Fehily Timoney & Company, 1999) mean that is frequently the case that application rates are determined by the nutrient content of the sludge and not its metal content (Lucid *et al.* 2013). Regardless, concerns have been raised about the potential for transfer of metals into water bodies, soil structures and, consequently, the food chain (Navas *et al.* 1999). In countries such as the USA, where in the majority of states, treated sewage sludge is applied to land based on the N requirement of the crop being grown and not on a soil-based test (McDonald & Wall, 2011), excessive metal losses may potentially occur.

8.5.3 Pathogens

During wastewater treatment, the sludge component of the waste becomes separated from the water component. As the survival of many microorganisms and viruses in wastewater is linked to the solid fraction of the waste, the numbers of pathogens present in sludge may be much higher than the water component (Straub *et al.* 1992). Although treatment of municipal sewage sludge using lime, AD, or temperature, may substantially reduce pathogens, complete sterilisation is difficult to achieve (Sidhu & Toze, 2009) and some pathogens, particularly enteric viruses, may persist. Persistence may be related to factors such as temperature, pH, water content (of treated sludge), and sunlight (Sidhu & Toze, 2009). Also, there is often resurgence in pathogen numbers post-treatment, known as the 'regrowth' phenomenon. This may be linked to contamination within the centrifuge, reactivation of viable, but non-culturable, organisms (Higgins *et al.* 2007), storage conditions post-centrifugation (Zaleski *et al.* 2005), and proliferation of a resistant sub-population due to newly available niche space associated with reduction in biomass and activity (McKinley & Vestal, 1985).

The risk associated with sludge-derived pathogens is largely determined by their ability to survive and maintain viability in the soil environment after landspreading. Survival is determined by both soil and sludge characteristics. The major physico-chemical factors that influence the survival of microorganisms in soil are currently considered to be soil texture and structure, pH, moisture, temperature, UV radiation, nutrient and oxygen availability, and land management regimes (reviewed in van Elsas *et al.* (2011)), whereas survival in sludge is primarily related to temperature, pH, water content (of treated sewage sludge), and sunlight (Sidhu & Toze, 2009). Pertinent biotic interactions include antagonism from indigenous microorganisms, competition for resources, predation and occupation of niche space (van Elsas *et al.* 2002). Pathogen-specific biotic factors that influence survival include physiological status and initial inoculum concentration (van Veen *et al.* 1997).

Following landspreading, there are two main scenarios which can lead to human infection. First, pathogens may be transported *via* overland or sub-surface flow to surface and ground waters, and infection may arise via ingestion of contaminated water or accidental ingestion of contaminated recreational water (Jaimeson *et al.* 2002; Tyrrel & Quinton, 2003). Alternatively, it is possible that viable pathogens could be present on the crop surface following biosolid application, or may become internalised within the crop tissue where they are protected from conventional sanitization (Itoh *et al.* 1998; Solomon *et al.* 2002). In this case, a person may become infected if they consume the contaminated produce. Therefore, it is critical to accurately determine the pathogen risk associated with land application of sewage sludge to fully understand the potential for environmental loss and consequently, human transmission.

However, survival patterns of sludge-derived pathogens in the environment are complex, and a lack of a standardised approach to pathogen measurement makes it difficult to quantify their impact. For example, Avery *et al.* (2005) spiked treated and untreated sludge samples with a known concentration of *E. coli* to quantify the time taken to achieve a decimal reduction. The pathogen response was variable and ranged from 3 to 22 days, depending on sludge properties. Lang and Smith (2007) investigated indigenous *E. coli* survival in dewatered, mesosphilic anaerobically digested (MAD) sludge, and in different soil types post MAD sludge application. Again, decimal reduction times proved variable, ranging from 100 days when applied to air-dried sandy loam, to 200 days in air-dried, silty clay. This time decreased to 20 days for both soil types when field moist soil was used, demonstrating the importance of water content in regulating survival behaviour. Therefore, in order to quantify pathogen risk in a relevant, site-specific manner, it is necessary to incorporate both soil and treated sewage sludge characteristics in risk assessment modelling. This has been done previously by conducting soil, sludge and animal slurry incubation studies, where pathogens are often spiked to generate a survival response (Vinten *et al.* 2004; Lang & Smith, 2007; Moynihan *et al.* 2013). Pathogen decay rate is then calculated based on decimal reduction times, or a first-order exponential decay model previously described by Vinten *et al.* (2004), and has been shown to be highly contingent on soil type and sludge or slurry combinations. Currently, the Safe Sludge Matrix provides a legal framework for grazing animals and harvesting crops following landspreading of treated sewage sludge, and stipulates that a time interval of three weeks and 10 months should be enforced to ensure safe practice, respectively (ADAS, 2001). However, further work is required to determine if these regulations are overly stringent, particularly in light of the comparatively higher pathogen concentrations reported for animal manures and slurries. For example, *E. coli* concentrations ranged from 3×10^2 to 6×10^4 CFU g^{-1} in sludge (Payment *et al.* 2001), compared to 2.6×10^8 to 7.5×10^4 CFU g^{-1} in fresh and stored cattle slurry, respectively (Hutchison *et al.* 2004). Therefore, environmental losses associated with treated sewage sludge application may not be as extensive as previously thought, and further comparisons on pathogen risk should form the basis of future research.

8.5.4 Pharmaceuticals

Pharmaceuticals comprise a diverse collection of thousands of chemical substances, including prescription and over-the-counter therapeutic drugs and veterinary drugs (USEPA, 2012). Pharmaceuticals are specifically designed to alter both biochemical and physiological functions of biological systems in humans and animals (Walters *et al.* 2010). Pharmaceuticals are referred to as 'pseudo-persistent' contaminants (i.e., high transformation/removal rates are compensated by their continuous introduction into the environment) (Barceló & Petrovic, 2007). Pharmaceuticals are likely to be found in any body of water influenced by raw or treated waste water, including river, lakes, streams and groundwater, many of which are used as a drinking water source (Yang *et al.* 2011). Between 30 and 90% of an administered dose of many pharmaceuticals ingested by humans is excreted in the urine as the active substance (Cooper *et al.* 2008). In a survey conducted by the US Environmental Agency (see McClellan & Halden, 2010), the mean concentration of 72 pharmaceuticals and personal care products were determined in 110 treated sewage sludge samples. Composite samples of archived treated sewage sludge, collected at 94 U.S. wastewater treatment plants from 32 states and the District of Columbia were analysed by liquid chromatography tandem mass spectrometry using EPA Method 1694. The two most abundant contaminants found in the survey were the disinfectants triclocarban and triclosan. The second most abundant class of pharmaceuticals found were antibiotics, particularly Ciprofloxain, Ofloxacin, 4-epitetra-cycline, tetracycline, minocycline, doxycycline and azithromycin (McClellan & Halden, 2010). It was concluded that the recycling of treated sewage sludge was a mechanism for the release of pharmaceuticals in the environment.

Pharmaceuticals have received increasing attention by the scientific community in recent years, due to the frequent occurrence in the environment and associated health risks (Chen *et al.* 2013). In 2007, the European Medicines Agency (EMEA) issued a guidance document (ERApharm) on environmental risk assessment of human medicinal products. It relies on the risk quotient approach used in the EU and is also used for industrial chemicals and biocides where the predicted environmental concentration is compared to the predicted no-effect concentration. The overall objective of ERApharm is to improve and complement existing knowledge and procedures for environmental risk of human and veterinary pharmaceuticals. The project covers fate and exposure assessment, effects assessment and environmental risk assessment (Lienert *et al.* 2007). A considerable amount of work focused on three case studies. Two of the case studies focused on human pharmaceuticals, β-blocker atenolol and the anti-depressant fluoxetine, and the third on a veterinary parasiticide ivermectin. Atenolol did not reveal any unacceptable risk to the environment but cannot be representative for other β-blockers, some of which show significantly different physiochemical characteristics and varying toxicological profiles in mammalian studies (Knacker & Metcalfe, 2010). Although found in trace levels (several nanograms per litre), some therapeutic compounds such as synthetic sex hormones and antibiotics, have been found to cause adverse effects on aquatic organisms (Chen *et al.* 2013). Therefore, understanding their environmental behaviour and impact has recently become a topic of interest for many researchers.

8.6 QUANTIFICATION OF COSTS AND BENEFITS FROM RE-USE OF SEWAGE SLUDGE

The main pathways for the disposal of sewage sludge in Europe is re-use in agriculture, landfill and incineration. The implementation of the Landfill Directive means that in the coming years, re-use in agriculture or incineration will become common pathways. In countries that preclude the re-use of treated sewage sludge in agriculture, incineration or alternative disposal methods, such as pyrolysis (used in the creation of biochar), the creation of engineering products (e.g., building materials; Hytiris *et al.* 2004), or reuse in power stations, may be alternative options. Landspreading is estimated to be the most cost-effective means of disposal of treated sewage sludge (Table 8.3); however, this does not take into account factors such as legislative requirements, potential savings to the farmer through the use of a low-cost fertiliser, or environmental benefits (or drawbacks) accruing from its use.

Table 8.3 Some treatment and disposal routes for sewage sludge, capital and operating costs, and benefits and drawbacks (adapted from RPA, Milieu Ltd. And WRc., 2008; Fytili & Zabaniotou, 2008; Astals *et al.* 2012; Cao & Pawłowski, 2012).

Treatment/ disposal route for sludge	Costs			Benefits	Drawbacks
	Capital	Operating	Overall cost € per ton DM		
On-site treatment					
Thermal drying			90–160		
Anaerobic digestion			90–160	Biogas produced has a high calorific value (15.9–27.8 MJ m^{-3}).	Elevated heating requirements to heat digester, odour potential.
Lime stabilisation			90–160		
Composting			90–160		
Solar drying	Land acquisition and construction	Labour	30–70	Low investment and operation costs. Final product is useful for industrial valorisation. Sewage volume is reduced.	It depends on sunlight/air temperature. Large areas are required for the greenhouses. Odour emissions.
Landfill	Land acquisition and construction	Labour Vehicle fuel Electricity Landfill tax and gate fees	309	Energy production from gas capture	Leachate production GHG emission (may be reduced in capture) Noise, odour, dust generation

Re-use in agriculture (landspreading)		Labour Regulatory testing of soil	126–280[1]	Potential yield improvement Less reliance on chemical fertiliser	Potential application of emerging contaminants to soil. Potential for leaching, runoff and volatilisation. Potential for introduction of contaminants into food chain.
Thermal (incl. incineration, wet oxidation, gasification and pyrolysis)	Land acquisition and construction	Labour Transport to site Quality control	332–411[2]	Energy production (but less than is used within the process) Large reduction in sludge volume. Thermal destruction of toxic compounds. Pyrolysis can be used to maximize production of chars.	Emissions to air, soil, water. Noise, dust generation. Visual intrusion. Possible impact on human health. Incomplete disposal –30% of solids remains as ash. In pyrolysis, majority of energy consumption is used to reduce sludge moisture content.
Forestry and silviculture		Labour Regulatory testing of soil	210–250[3]	Increased tree growth Nutrient input to soil	Leaching of nutrients to groundwater. Impact on ecosystems.

[1]About €40 ton^{-1} DM in Portugal (Martins & Béraud, pers. comm.) [2]Cost for incineration (RPA, Milieu Ltd. And WRc., 2008) [3]From Anderson and SEDE (2002).

Depending on the type of treatment applied, costs associated with the re-use of sewage sludge may include, amongst other issues, drying, lime amendment, thermal drying costs, along with costs of installation of storage facilities in which to carry out these treatments; labour, energy and transport costs; and where the treated sewage sludge is re-used on land, soil and sewage sludge analysis costs and other professional service costs (Table 8.3). Potential benefits accruing from the land application of treated sewage sludge may be enhanced nutrient availability to crops and enhanced crop yield, and in countries where sewage sludge, treated or untreated, is considered a waste material (e.g., Ireland), there is a substantial saving for the farmer.

8.6.1 Impact of nutrient recovery, energy/product generation on energy and cost savings in a sewage treatment plant

It is well known that the potential energy available in the raw wastewater influent significantly exceeds the electricity requirements of the treatment processes. Energy captured in organics entering the plant can be related to the chemical oxygen demand load of the influent flow. Based on calorific measurements, a capita-specific energy input of 1760 KJ per population equivalent (PE) in terms of 120 g chemical oxygen demand of organic matter can be calculated (Wett *et al.* 2007). This specific organic load is subjected to aerobic and anaerobic degradation processes, partly releasing the captured energy. Traditional wastewater treatment plants (WWTP) have unusually high energy demands and create problems associated with the disposal of sewage sludge and chemical residues. It is estimated that wastewater treatment accounts for about 3–5% of the electrical energy load in many developed and developing countries (Chen & Chen, 2013). Kapshe *et al.* (2013) demonstrated how energy generation in four WWTPs in India can utilize the methane recovery through anaerobic digestion to produce 1.5 to 2.5 million kWh electricity for captive use every year. An additional benefit is the reduction of 80,000 tonnes of CO_2 emission per year.

Dewatered sludge (15–35% D.S.) has a very low Lower Heating Value (LHV), so its use in energy recovery or incineration is not currently feasible. Dried sludge (about 70–75% D.S.), however, may be a valuable energy source, if mixed with fuels (e.g., natural gas) and/or other waste with a high calorific value (e.g., Residue Derived Fuels, RDF), as its LHV may reach up to 16 MJ/kg, allowing its use as a secondary fuel in, for example, the cement industry. The reader is referred to Tsagarakis and Papadogiannis (2006) for further information on energy recovery from sewage sludge in a treatment plant in Greece.

Within Germany, 344 WWTPs in North Rhine Westphalia (NRW) have undergone energy analysis (Wett *et al.* 2007) comprising two stages: a first stage, where operational data are collected and energy consumption rates and biogas yields are targeted; and a second stage, where optimization measures are adopted.

By application of this protocol, energy costs can be reduced. Through the re-use of energy produced during wastewater treatment, the long-term sustainability of the WWTPs is enhanced, while also contributing to offset installation and on-going operational costs.

In Southern European countries, including the Mediterranean area, cultural, social and economic reasons means that the management of the sewage sludge is not necessarily the same as in other EU countries. Here, recycling to agriculture is the main route for final disposal. For example, in Portugal and Spain about 50% of the sewage sludge is recycled in agriculture (Milieu *et al.* 2013a, 2013b, 2013c). Therefore, sewage sludge management in these countries should be governed by the following objectives (Martins & Béraud, pers. comm.): (1) provision of solutions that are technically and economically adapted to the economic realities of these countries (lower investment and operating costs); (2) full legal compliance, including the ability to adapt to future restrictions, which may be placed on the disposal of treated sludge in agriculture; (3) diversification of the final disposal of sludge with new sludge treatment systems; (4) reduction in the quantity of sewage sludge to be disposed of; (5) optimization of the utilisation of weather conditions for sludge treatment, which makes solar drying an appealing solution.

8.7 ACKNOWLEDGEMENTS

The authors wish to acknowledge funding from the Irish EPA (Project reference number 2012-EH-MS-13). M.G. Healy and A. Martins are members of EU COST Action Water_2020.

8.8 REFERENCES

ADAS (2001). Safe Sludge Matrix – Guidelines for the Application of Sewage Sludge to Agricultural Land, UK.

Alam M. Z., Molla A. H. and Fakhru'l-Razi A. (2005). Biotransformation of domestic wastewater treatment plant sludge by two-stage integrated processes – LSB & SSB. *IIUM Engineering Journal*, **6**, 23–40.

Andersen A. and SEDE (2002). Disposal and Recycling Routes for Sewage Sludge. Synthesis Report (22 February 2002). http://ec.europa.eu/environment/waste/sludge/pdf/synthesisreport020222.pdf (accessed 17 September 2013).

Apedaile E. (2001). A perspective on biosolids management. *Canadian Journal of Infectious Diseases*, **12**(4), 202–204.

Astals S., Venegas C., Peces M., Jofre J., Lucena F. and Mata-Alvarez J. (2012). Balancing hygienization and anaerobic digestion of raw sewage sludge. *Water Research*, **46**(19), 6218–6227.

Avery L. M., Killham K. and Jones D. L. (2005). Survival of E. coli O157: H7 in organic wastes destined for land application. *Journal of Applied Microbiology*, **98**(4), 814–822.

Barceló D. and Petrovic M. (2007). Pharmaceuticals and personal care products (PPCPs) in the environment. *Analytical and Bioanalytical Chemistry*, **387**(4), 1141–1142.

Bremer J. (2009). Fertilizer Prices Continue to Rise. http://www.hpj.com/archives/2009/jan09/jan19/Fertilizerpricescontinuetor.cfm (accessed 14 September 2013).

Brennan R. B., Healy M. G., Grant J., Ibrahim T. G., Fenton O. (2012). Incidental phosphorus and nitrogen loss from grassland plots receiving chemically amended dairy cattle slurry. *Science of the Total Environment*, **441**, 132–140.

Cao Y. and Pawłowski A. (2012). Sewage sludge-to-energy approaches based on anaerobic digestion and pyrolysis: brief overview and energy efficiency assessment. *Renewable and Sustainable Energy Reviews*, **16**(3), 1657–1665.

Chan Y. J., Chong M. F., Law C. L. and Hassell D. G. (2009). A review on anaerobic–aerobic treatment of industrial and municipal wastewater. *Chemical Engineering Journal*, **155**(1–2), 1–18.658.

Chen S. and Chen B. (2013). Net energy production and emissions mitigation of domestic wastewater treatment system: a comparison of different biogas–sludge use alternatives. *Bioresource Technology*, **144**, 296–303.

Chen Y., Yu G., Cao Q., Zhang H., Lin Q. and Hong Y. (2013). Occurrence and environmental implications of pharmaceuticals in Chinese municipal sewage sludge. *Chemosphere*, **93**(9), 1765–1772.

Cooper E. R., Siewicki T. C. and Phillips K. (2008). Preliminary risk assessment database and risk ranking of pharmaceuticals in the environment. *Science of The Total Environment*, **398**(1–3), 26–33.

Cordell D., Schmid-Neset T., White S. and Drangert J. O. (2009). Preferred future phosphorus scenarios: a framework for meeting long-term phosphorus needs for global food demand, In: International Conference on Nutrient Recovery From Wastewater Streams, K. Ashley, D. Mavinic and F. Koch (eds), Vancouver, British Columbia, Canada, 2009.

Cordell D., Rosemarin A., Schröder J. J. and Smit A. L. (2011). Towards global phosphorus security: a systems framework for phosphorus recovery and reuse options. *Chemosphere*, **84**(6), 747–758.

Coulter B. S. (2004). Nutrient and Trace Element Advice for Grassland, Tillage, Vegetable and Fruit Crops. Teagasc, Johnstown Castle, Wexford, p. 95.

Diener S., Studt Solano N., Roa Gutiérrez F., Zurbrügg C. and Tockner K. (2011a). Biological treatment of municipal organic waste using black soldier fly larvae. *Waste Biomass Valorization*, **2**(4), 357–363.

Diener S., Zurbrügg C., Gutiérrez F. R., Nguyen D. H., Morel A. and Koottatep T. (2011). Black soldier fly larvae for organic waste treatment – prospects and constraints. In: Proceedings of the WasteSafe – 2nd International Conference on Solid Waste Management in the Developing Countries, Khulna, Bangladesh, M. Alamgir, Q. H. Bari, I. M. Rafizul, S. M. T. Islam, G. Sarkar and M. K. Howlader (eds), pp. 1–8.

Domagalski J. L. and Johnson H. M. (2011). Subsurface transport of orthophosphate in five agricultural watersheds, USA. *Journal of Hydrology*, **409**, 157–171.

EC (1991). Council Directive Concerning Urban Waste-Water Treatment. http://eur-lex.europa.eu/LexUriServ/LexUriServ.do?uri = OJ:L:1991:135:0040:0052:EN:PDF (accessed 11 June 2014).

EC (1999). Council Directive 1999/31/EC of 26 April 1999 on the Landfill of Waste. http://www.central2013.eu/fileadmin/user_upload/Downloads/Document_Centre/OP_Resources/Landfill_Directive_1999_31_EC.pdf (accessed 11 June 2014).

EC (2008). Waste Framework Directive. http://ec.europa.eu/environment/waste/legislation/a.htm (accessed 18 September 2013).

EC (2010). Environmental, Economic and Social Impacts of the Use of Sewage Sludge on Land. Summary Report 1. http://ec.europa.eu/environment/waste/sludge/pdf/part_iii_report.pdf (accessed 11 June 2014).

EC (2012). Sewage Sludge. http://ec.europa.eu/environment/waste/sludge/index.htm (accessed 11 June 2014).

EEC (1986). Council Directive of 12 June 1986 on the Protection of the Environment, and in Particular of the Soil, when Sewage Sludge is used in Agriculture (86/278/EEC). http://www.efma.org/PRODUCT-STEWARDSHIP-PROGRAM-10/images/86278EEC.pdf (accessed 11 June 2014).

Elissen H. J. H., Mulder W. J., Hendrickx T. L. G., Elbersen H. W., Beelen B., Temmink H. and Buisman C. J. N. (2010). Aquatic worms grown on biosolids: biomass composition and potential applications. *Bioresource Technology*, **101**(2), 804–811.

Epstein E. (2003). Land Application of Sewage Sludge and Biosolids. Lewis, Boca Raton, Fl.

European Commission (2009). Directive 2009/28/EC of the European Parliament and of the Council of 23 April 2009 on the promotion of the use of energy from renewable sources and amending and subsequently repealing Directives 2001/77/EC and 2003/30/EC.

European Commission (2011). EUROSTAT. http://epp.eurostat.ec.europa.eu/tgm/table.do?tab=table&init=1&language=en&pcode=ten00030&plugin=0 (accessed 19 November 2013).

Eurostat (2014). Sewage Sludge Production and Disposal. http://appsso.eurostat.ec.europa.eu/nui/show.do?dataset=env_ww_spd&lang=en (accessed 11 June 2014).

Evanylo G. K., Haering K. C., Sherony C. and Christian A. H. (2011). Mid-Atlantic composting directory. Virginia Cooperative Extension Publication, Virginia Tech, Blacksburg, VA, 452–230. http://pubs.ext.vt.edu/452/452-230/452-230.pdf (accessed 21 November 2014).

Fehily Timoney and Company (1999). Codes of good practice for the use of biosolids in agriculture – guidelines for farmers. http://www.environ.ie/en/Publications/Environment/Water/FileDownLoad,17228,en.pdf (accessed 11 June 2014).

Fenton O., Schulte R. P. O., Jordan P., Lalor S. T. J. and Richards K. G. (2011). Time lag: a methodology for the estimation of vertical and horizontal travel and flushing timescales to nitrate threshold concentrations in Irish aquifers. *Environmental Science and Policy*, **14**, 419–431.

Flemming H. C. and Wingender J. (2001). Relevance of microbial extracellular polymeric substances (EPSs) – Part II: Technical aspcets, **43**, 9–16.

Food Safety Authority of Ireland (2008). Food safety implications of land-spreading agricultural, municipal and industrial organic materials on agricultural land used for food production in Ireland. www.fsai.ie/WorkArea/DownloadAsset.aspx?id=8226 (accessed 11 June 2014).

Fytili D. and Zabaniotou A. (2008). Utilization of sewage sludge in EU application of old and new methods – a review. *Renewable and Sustainable Energy Reviews*, **12**(1), 116–140.

Galbally P., Ryan D., Fagan C. C., Finnan J., Grant J. and McDonnell K. (2013). Biosolid and distillery effluent amendments to Irish short rotation coppiced willow plantations: impacts on groundwater quality and soil. *Agricultural Water Management*, **116**, 193–203.

Gerba C. P. and Smith J. E. (2005). Sources of pathogenic microorganisms and their fate during land application of wastes. *Journal of Environmental Quality*, **34**, 42–48.

Girovich M. J. (1996). Biosolids Treatment and Management: Processes for Beneficial Use. CRC Press, New York.

Higgins M. J., Chen Y.-C., Murthy S. N., Hendrickson D., Farrel J. and Schafer P. (2007). Reactivation and growth of non-culturable indicator bacteria in anaerobically digested biosolids after centrifuge dewatering. *Water Research*, **41**, 665–673.

Hogendoorn A. (2013). 'Enhanced digestion and alginate-like-exopolysaccharides extraction from Nereda sludge'. Masters in Science Civil Engineering, University of Technology, Delft. http://www.citg.tudelft.nl/fileadmin/Faculteit/CiTG/Gezondheidstechniek/doc/Afstudeerrapporten/final_thesis_anthonie_hogendoorn.pdf (accessed 21 November 2014).

Hutchison M. L., Walters L. D., Avery S. M., Synge B. A. and Moore A. (2004). Levels of zoonotic agents in british livestock manures. *Letters in Applied Microbiology*, **39**, 207–214.

Hytiris N., Kapellakis I. E., La Roij de R. and Tsagarakis K. P. (2004). The potential use of olive mill sludge in solidification process. *Resources, Conservation and Recycling*, **40**(2), 129–139.

Ibrahim T. G., Fenton O., Richards K. G., Fealy R. M. and Healy M. G. (2013). Loads and forms of nitrogen and phosphorus in overland flow and subsurface drainage on a marginal land site in south east Ireland, *Biology and Environment: Proceedings of the Royal Irish Academy*, **113B**(2), 169–186.

Isaac R. A. and Boothroyd Y. (1996). Beneficial use of biosolids: Progress in controlling metals. *Water Science and Technology*, **34**(3–4), 493–497.

Itoh Y., Sugita-Konishi Y., Kasuga F., Iwaki M., Hara-Kudo Y., Saito N., Noguchi Y., Konuma H. and Kumagai S. (1998). Enterohemorrhagic Escherichia coli O157: H7 present in radish sprouts. *Applied and Environmental Microbiology*, **64**, 1532–1535.

Jamieson R. C., Gordon R. J., Sharples K. E., Stratton G. W. and Madani A. (2002). Movement and persistence of fecal bacteria in agricultural soils and sub-surface drainage water: a review. *Canadian Biosystems Engineering*, **44**, 1–1.9.

Jeng A. S., Haraldsen T. K., Grønlund A. and Pedersen P. A. (2006). Meat and bone meal as nitrogen and phosphorus fertilizer to cereals and rye grass. *Nutrient Cycle in Agroecosystems*, **76**, 183–191.

Jiang J., Zhao Q., Wei L., Wang K. and Lee D.-J. (2011). Degradation and characteristic changes of organic matter in sewage sludge using microbial fuel cell with ultrasound pretreatment. *Bioresource Technology*, **102**(1), 272–277.

Kapshe M., Kuriakose P. N., Srivastava G. and Surjan A. (2013). Analysing the co-benefits: case of municipal sewage management at Surat, India. *Journal of Cleaner Production*, **58**, 51–60.

Kelessidis A. and Stasinakis A. S. (2012). Comparative study of the methods used for treatment and final disposal of sewage sludge in European countries. *Waste Management*, **32**(6), 1186–1195.

Knacker T. and Metcalfe C. (2010). Introduction to the special issue on environmental risk assessment of pharmaceuticals. *Integrated Environmental Assessment and Management*, **6**(S1), 511–513.

Lalander C., Diener S., Magri M. E., Zurbrügg C., Lindström A. and Vinnerås B. (2013). Faecal sludge management with the larvae of the black soldier fly (hermetia illucens) — from a hygiene aspect. *Science of the Total Environment*, **458–460**, 312–318.

Lalor S. T. J., Hoekstra N. J., Murphy P. N. C., Richards K. G. and Lanigan G. J. (2012). Practical Advice for Slurry Application Strategies for Grassland Systems. Proceedings 712, International Fertiliser Society, Cambridge, 6 December 2012, UK.

Lang N. L., Smith S. R. (2007). Influence of soil type, moisture content and biosolids application on the fate of escherichia coli in agricultural soil under controlled laboratory conditions. *Journal of Applied Microbiology*, **103**, 2122–2131.

Lee W. S., Chua A. S. M., Yeoh H. K. and Ngoh G. C. (2014). A review of the production and applications of waste-derived volatile fatty acids. *Chemical Engineering Journal*, **235**, 83–99.

Lienert J., Güdel K. and Escher B. I. (2007). Screening Method for Ecotoxicological Hazard Assessment of 42 Pharmaceuticals Considering Human Metabolism and Excretory Routes. *Environmental Science and Technology*, **41**(12), 4471–4478.

Lin Y., de Kreuk M., van Loosdrecht M. C. M. and Adin A. (2010). Characterization of alginate-like exopolysaccharides isolated from aerobic granular sludge in pilot-plant. *Water Research*, **44**(11), 3355–3364.

Liu, H. and Fang, H. H. P. (2002). Extraction of extracellular polymeric substances (EPS) of sludges. *Journal of Biotechnology*, **95**(3), 249–256.

Lu Q., He Z. L. and Stoffella P. J. (2012). Land application of biosolids in the USA: a review. *Applied and Environmental Soil Science*, **2012**, 1–11.

Lucid J. D., Fenton O., Healy M. G. (2013). Estimation of maximum biosolids and meat and bone meal application to a low P index soil and a method to test for nutrient and metal losses. *Water, Air and Soil Pollution*, **224**, 1464–1476.

McClellan K. and Halden R. U. (2010). Pharmaceuticals and personal care products in archived U.S. biosolids from the 2001 EPA national sewage sludge survey. *Water Research*, **44**(2), 658–668.

McDonald N., Wall D. (2011). Soil Specific N Advice – Utilising our Soil Nitrogen Resources. National Agri-environment Conference 2011, 10 November 2011, Athlone. http://www.teagasc.ie/publications/2011/1050/Agrienvironment_Proceedings.pdf (accessed 11 June 2014).

McKinley V. L. and Vestal J. R. (1985). Physical and chemical correlates of microbial activity and biomass in composting municipal sewage sludge. *Applied and Environmental Microbiology*, **50**, 1395–1403.

Meyer V. F., Redente E. F., Barbarick K. A. and Brobst R. (2001). Biosolids applications affect runoff water quality following forest fire. *Journal of Environmental Quality*, **30**, 1528–1532.

Milieu, WRC, RPA (2013a). Environmental, Economic and Social Impacts of the Use of Sewage Sludge on Land. Final Report – Part I: Overview Report. Service contract No 070307/2008/517358/ETU/G4.

Milieu, WRC, RPA (2013b). Environmental, Economic and Social Impacts of the Use of Sewage Sludge on Land. Final Report – Part II: Report on Options and Impacts. Service contract No 070307/2008/517358/ETU/G4.

Milieu, WRC, RPA (2013c). Environmental, Economic and Social Impacts of the Use of Sewage Sludge on Land. Final Report – Part III: Project Interim Reports. Service contract No 070307/2008/517358/ETU/G4.

Mondini C., Cayuela M. L., Sinicco T., Sánchez-Monedero M. A., Bertolone E. and Bardi L. (2008). Soil application of meat and bone meal. Short-term effects on mineralization dynamics and soil biochemical and microbiological properties. *Soil Biology and Biochemistry*, **40**, 462–474.

Morgenroth E., Sherden T., van Loosdrecht M. C. M., Heiinen J. J. and Wilderer P. A. (1997). Aerobic granular sludge in sequencing batch reactors. *Water Research*, **31**(12), 3191–3194.

More T. T., Yan S., Tyagi R. D. and Surampalli R. Y. (2010). Potential use of filamentous fungi for wastewater sludge treatment. *Bioresource Technology*, **101**(20), 7691–7700.

Mouri G., Takizawa S., Fukushi K. and Oki T. (2013). Estimation of the effects of chemically-enhanced treatment of urban sewage system based on life-cycle management. *Sustainable Cities and Society*, **9**(0), 23–31.

Moynihan E. L., Richards K. G., Ritz K., Tyrrel S. and Brennan F. P. (2013). Impact of soil type, biology and temperature on the survival of non-toxigenic Escherichia coli O157. *Biology and Environment*, **113**(B), 1–6.

Navas A., Machn J., and Navas B. (1999). Use of biosolids to restore the natural vegetation cover on degraded soils in the badlands of Zaragoza (NE Spain). *Bioresource Technology*, **69**, 199–205.

Ndegwa P. M. and Thompson S. A. (2001). Integrating composting and vermicomposting in the treatment and bioconversion of biosolids. *Bioresource Technology*, **76**(2), 107–112.

Neyens E., Baeyens J., Dewil R. and De heyder B. (2004). Advanced sludge treatment affects extracellular polymeric substances to improve activated sludge dewatering. *Journal of Hazardous Materials*, **106**(2–3), 83–92.

Payment P., Plante R. and Cejka P. (2001). Removal of indicator bacteria, human enteric viruses, giardia cysts, and cryptosporidium oocysts at a large wastewater primary treatment facility. *Canadian Journal of Microbiology*, **47**,188–93.

Pérez-Cid B., Lavilla I. and Bendicho C. (1999). Application of microwave extraction for partitioning of heavy metals in sewage sludge. *Analytica Chimica Acta*, **378**(1–3), 201–210.

Robinson K. G., Robinson C. H., Raup L. A. and Markum T. R. (2012). Public attidudes and risk perception toward land application of biosolids within the south-eastern United States. *Journal of Environmental Management*, **98**, 29–36.

RPA, Milieu Ltd. And WRc. (2008). Environmental, Economic and Social Impacts of the use of Sewage Sludge on Land. Final report. Part II: report on options and impacts. http://ec.europa.eu/environment/waste/sludge/pdf/part_ii_report.pdf (accessed 6 November 2013).

Seviour T., Pijuan M., Nicholson T., Keller J. and Yuan Z. (2009). Gel-forming exopolysaccharides explain basic differences between structures of aerobic sludge granules and floccular sludges. *Water Research*, **43**(18), 4469–4478.

Siddique M. T. and Robinson J. S. (2004). Differences in phosphorus retention and release in soils amended with animal manures and sewage sludge. *Soil Science Society of Am Journal*, **68**, 1421–1428.

Sidhu J. P. S., and Toze S. G. 2009. Human pathogens and their indicators in biosolids: a literature review. *Environment International*, **35**, 187–201.

Singh R. P. and Agrawal M. (2008). Potential benefits and risks of land application of sewage sludge. *Wate Management*, **28**, 347–358.

Smith S. R. and Durham E. (2002). Nitrogen release and fertiliser value of thermally-dried biosolids. *Water and Environment Journal*, **16**(2), 121–126.

Solomon E. B., Yaron S. and Matthews K. R. (2002). Transmission of Escherichia coli O157: H7 from contaminated manure and irrigation water to lettuce plant tissue and its subsequent internalization. *Applied and Environmental Microbiology*, **68**, 397–400.

Sonon L. S. and Gaskin J. (2009). Metal concentration standards for land application of biosolids and other by-products in Georgia. Learning for Life Bulletin 1353.

Straub T. M., Pepper I. L. and Gerba C. P. (1992). Persistence of viruses in desert soils amended with anaerobically digested sewage sludge. *Applied Environmental Microbiology*, **58**, 636–641.

St-Hilaire S., Sheppard C., Tomberlin J. K., Irving S., Newton L., McGuire M. A., Mosley E. E., Hardy R. W. and Sealey W. (2007). Fly prepupae as a feedstuff for rainbow trout, oncorhynchus mykiss'. *Journal of the World Aquaculture Society*, **38**(1), 59–67.

Su H., Cheng J., Zhou J., Song W. and Cen K. (2009). Improving hydrogen production from cassava starch by combination of dark and photo fermentation. *International Journal of Hydrogen Energy*, **34**(4), 1780–1786.

Tasker J. (2010). Frustration at 13% fertiliser price hike. Farmers Weekly. http://www.fwi.co.uk/Articles/03/09/2010/123163/Frustration-at-13-fertiliser-price-hike.htm (accessed 14 September 2013).

Tchobanoglous G., Barton F. and Stensel H. (2003). Wastewater Engineering Treatment and Reuse, 4th edn, McGraw Hill, New York.

Tsagarakis K. P. and Papadogiannis Ch. (2006). Technical and economic evaluation of the biogas utilization for energy production at Iraklio municipality, Greece. *Energy Conservation and Management*, **47**(7–8), 844–857.

Tsagarakis K. P., Horan N. J., Mara D. D. and Angelakis A. N. (1999). Management of Biosolids from Municipal Wastewater Treatment Plants in Greece. Fourth European Biosolids and Organic Residuals Conference, Wakefield, 15–17 November 1999. Paper no. 35.

Tyagi V. K. and Lo S.-L. (2013). Sludge: a waste or renewable source for energy and resources recovery? *Renewable and Sustainable Energy Reviews*, **25**, 708–728.

Tyagi R. D. and Surampalli S. Y. E. (2009). Sustainable Sludge Management: Production of Value Added Products. American Society of Civil Engineers, Reston, VA, USA.

Tyrrel S. F. and Quinton J. N. (2003). Overland flow transport of pathogens from agricultural land receiving faecal wastes. *Journal of Applied Microbiology*, **94**, 87–93.

U.S. EPA (1993). Standards for the use or Disposal of Sewage Sludge; Final Rules. http://water.epa.gov/scitech/wastetech/biosolids/upload/fr2-19-93.pdf (accessed 11 June 2014).

USEPA (1994). A Plain Guide to the EPA Part 503 Biosolids Rule. http://water.epa.gov/scitech/wastetech/biosolids/upload/2002_06_28_mtb_biosolids_503pe_503pe_toc.pdf (accessed 6 November 2013).

USEPA (2011). Problem Formulation for Human Health Risk Assessments of Pathogens in Land-Applied Biosolids, Cincinnatti OH. EPA/600/R-08/035F.

USEPA (2012). Pharmaceuticals and Personal Care Products. http://www.epa.gov/ppcp/ (accessed 20 September 2013).

van Elsas J. D., Garbeva P. and Salles J. (2002). Effects of agronomical measures on the microbial diversity of soils as related to the suppression of soil-borne plant pathogens. *Biodegradation*, **13**(1), 29–40.

van Elsas J. D., Semenov A. V., Costa R. and Trevors J. T. (2011). Survival of escherichia coli in the environment: fundamental and public health aspects. *The ISME Journal*, **5**(2), 173–183.

van Leeuwen J. H., Rasmussen M. L., Sankaran S., Koza C. R., Erickson D. T., Mitra D. and Jin B. (2012). Fungal treatment of crop processing wastewaters with value-added co-products. In: Sustainable Bioenergy and Bioproducts, K. Gopalakrishnan, J. (Hans) van Leeuwen and Robert C. Brown (eds), Springer, Germany, pp. 13–44.

van Veen J. A., van Overbeek L. S., and van Elsas J. D. (1997). Fate and activity of microorganisms introduced into soil. *Microbiology and Molecular Biology Reviews*, **61**(2),121–135.

Vinten A. J. A., Douglas J. T., Lewis D. R., Aitken M. N. and Fenlon D. R. (2004). Relative risk of surface water pollution by *E. coli* derived from faeces of grazing animals compared to slurry application. *Soil Use and Management*, **20**(1), 13–22.

Wall D., Jordan P., Melland A. R., Mellander P.-E., Buckley C., Reaney S. M. and Shortle G. (2011). Using the nutrient transfer continuum concept to evaluate the European Union Nitrates Directive National Action Programme. *Environmental Science and Policy*, **14**(6), 664–674.

Walters E., McClellan K. and Halden R. U. (2010). Occurrence and loss over three years of 72 pharmaceuticals and personal care products from biosolids–soil mixtures in outdoor mesocosms. *Water Research*, **44**(20), 6011–6020.

Wett B., Buchauer K. and Fimml C. (2007). Energy self-sufficiency as a feasible concept for wastewater treatment systems. In: IWA Leading Edge Technology Conference, Singapore.

Yan S., Subramanian S., Tyagi R. and Surampalli R. (2008). Bioplastics from waste activated sludge-batch process. *Practice Periodical of Hazardous, Toxic, and Radioactive Waste Management*, **12**(4), 239–248.

Yang X., Flowers R. C., Weinberg H. S. and Singer P. C. (2011). Occurrence and removal of pharmaceuticals and personal care products (PPCPs) in an advanced wastewater reclamation plant. *Water Research*, **45**(16), 5218–5228.

Yang X., Du M., Lee D.-J., Wan C., Zheng L. and Wan F. (2012). Improved volatile fatty acids production from proteins of sewage sludge with anthraquinone-2,6-disulfonate (AQDS) under anaerobic condition. *Bioresource Technology*, **103**(1), 494–497.

Zaleski K. J., Josephson K. L, Gerba C. P. and Pepper I. L. (2005). Survival, growth and regrowth of enteric indicator and pathogenic bacteria in biosolids, compost, soil and land applied biosolids. *Journal of Residuals Science and Technology*, **2**(1), 49–63.

Chapter 9

Odour abatement technologies in WWTPs: energy and economic efficiency

José M. Estrada[1], Raquel Lebrero[1], Guillermo Quijano[1], N. J. R. Bart Kraakman[2] and Raúl Muñoz[1]

[1]Department of Chemical Engineering and Environmental Technology, Dr Mergelina Street s/n, 47011, University of Valladolid, Spain

[2]CH2M Hill, Level 7, 9 Hope Street, Chatswood NSW 2067, Australia and Laboratory of Biotechnology, Faculty of Applied Sciences, Julianalaan 67, 2628 BC Delft, Delft University of Technology, Netherlands

9.1 INTRODUCTION

Over the past few decades, atmospheric pollution has become more important since recent investigations have consistently demonstrated it poses a threat for human health and natural ecosystems. In October 2013, the specialized cancer agency of the World Health Organization (WHO) classified outdoor air pollution as a human carcinogenic and related it to lung and bladder cancer (WHO, 2013). The harmful consequences derived from polluted gas emissions have resulted in an increasing public concern and the enforcement of the stricter environmental legislations (Stuetz *et al.* 2001; Inrapour *et al.* 2005).

Odorous emissions constitute an important contributor to atmospheric pollution and represent a significant contribution to photochemical smog formation and particulate secondary contaminant emission (Sucker *et al.* 2008). Moreover, at wastewater treatment plants (WWTPs), the accumulation of specific odorants such as H_2S in confined spaces may reach lethal concentrations, entailing a severe occupational risk to the operators (Vincent, 2001). H_2S also leads to the corrosion of valuable assets, reducing the life of the WWTP's infrastructure.

Malodors are the main cause of the public complaints received by environmental regulatory agencies worldwide (Kaye & Jiang, 2000). Malodorous emissions from WWTPs rank among the most unpleasant ones and their nuisance on the nearby population is increasing as a result of the encroachment of residential areas on WWTPs. These odours are caused by volatile organic compounds (aromatics, aldehydes, alcohols, ketones), sulphur compounds (H_2S, mercaptans) and ammonia, which are produced throughout the WWTP, with the headworks and the sludge treatment lines being the most important sources of malodours. Whereas they are not a direct cause of disease, they negatively affect human health causing nausea, headache, insomnia, loss of appetite, respiratory problems, irrational behavior, and so on (Sucker *et al.* 2008; Zarra *et al.* 2008: Jehlickova *et al.* 2008).

Since WWTP operating companies are increasingly concerned about their public image, malodorous emission control has become a mandatory and challenging task for WWTP operators due to the particular characteristics of odourous emissions: high air flow rates containing a wide range of chemicals from many different sources at trace level concentrations (in the order of $\mu g\ m^{-3}$–$mg\ m^{-3}$), highly variable in time and with climatological conditions (Iranpour *et al.* 2005; Zarra *et al.* 2008). Depending on the type of source, the nature of the odorous compounds emitted and the level of odour reduction required, different control approaches might be adopted. Therefore, a detailed characterization of the key sources, the odours emitted and their impact is essential prior to the implementation of any odour management strategy (Capodaglio *et al.* 2002).

In this context, odour management at WWTPs should always consider hierarchically emission prevention at the source, impact mitigation on the nearby community and finally, odorant removal. Hence, a correct design, operation and maintenance of the WWTP facilities must be ensured in the first place. Good operation practices in WWTPs and modifications in the process or plant configuration might prevent odour formation at source. This can be done, for instance, by frequently cleaning the grit chambers and screening units, by minimizing the sludge retention time in the sludge handling line or by operating under adequate aeration and mixing in the wastewater treatment process units. The installation of covers on specific process units or enclosures not only prevents odorant emission, but also assists odour abatement by capturing the odorous emission (Tchobanoglous *et al.* 2003). However, unwanted side effects from covering process units, such as corrosion and reduced operator access, usually require proper ventilation strategies.

The nuisance caused on the nearby population can be mitigated by enhancing the dispersion and dilution of the emission. Thus, the implementation of buffer zones (separation between the odour source and the potentially affected population), turbulence-inducing structures such as trees or high barrier fences, and chimneys can dilute the emission and reduce the odour impact (Capodaglio *et al.* 2002; Tchobanoglous *et al.* 2003; Estrada *et al.* 2013a). Minimization of the odour annoyance can also be accomplished by odour masking or neutralization, a strategy commonly applied in extended or intermittent sources. However, the use of masking and inhibitory agents is relatively controversial. Some studies have demonstrated that, in spite of reducing the unpleasantness (the hedonic tone) of the emission, they might increase the odour concentration of the emission even above regulatory limits (Decottignies *et al.* 2007; Bilsen & De Fre, 2008).

Finally, odour control technologies are implemented when neither prevention nor mitigation are viable or sufficiently efficient. It is important to highlight that the capital and operating costs associated with odour treatment using traditional technologies (biofilters, biotrickling filters, adsorption and chemical scrubbing) might represent from 5% up to 15% of the total costs of WWTPs (Kiesewetter *et al.* 2012). Odour abatement often involves covering and extracting foul air from the odour emission source and subsequently treating this air using a specific process unit prior to atmospheric discharge. These treatment technologies are based either on physical-chemical or biological principles (Lebrero *et al.* 2011).

In this chapter, an energy/economic efficiency analysis of the most typically employed odour treatment technologies (chemical scrubbing, activated carbon filtration, biofiltration and biotrickling filtration) together with three hybrid systems (chemical scrubbing with activated carbon filtration, biotrickling filtration with activated carbon filtration and biotrickling filtration with chemical scrubbing) and three emerging technologies recently applied for odour abatement (activated sludge diffusion, activated sludge recycling and step-feed biofiltration) will be presented. The sensitivity of this energy/economic efficiency towards design parameters such as the length-to-diameter ratio (L/D) in packed based systems will be also evaluated.

9.2 ODOUR ABATEMENT TECHNOLOGIES

Technologies based on physical and chemical odorant removal mechanisms, such as chemical scrubbers or adsorption filters, are reliable and well-established techniques. Chemical scrubbing constitutes nowadays one of the most commonly implemented technologies in the odour control market due to its reasonably high performance, lower operating costs compared to other physical-chemical technologies and extensive experience in design and operation (Card, 2001; Sanchez *et al.* 2006). Typically implemented in packed towers, odorants are transferred from the gas phase to an aqueous solution containing a chemical oxidant (sodium hypochlorite, sodium hydroxide or hydrogen peroxide) where they are destroyed (Figure 9.1a). On the other hand, odour adsorption is based on odorant trapping onto a fixed bed of adsorbent (commonly activated carbon, zeolites or silica-gel) by intermolecular forces. Adsorption systems usually require a minimum of two beds, alternating in operation for adsorbent replacement or regeneration (Figure 9.1b) (Turk & Bandosz, 2001).

The odour removal efficiencies (REs) of these technologies depend on the hydrophobicity of the odorants. Chemical scrubbers offer REs > 99% for water soluble odorants, but as low as 50% for the highly hydrophobic ones. On the other hand, activated carbon adsorption (AC) presents the highest REs for hydrophobic odorants (>99%), while supporting REs from 80 to 90% for those smaller pollutants with higher water affinity (Lebrero *et al.* 2011; Estrada *et al.* 2013a).

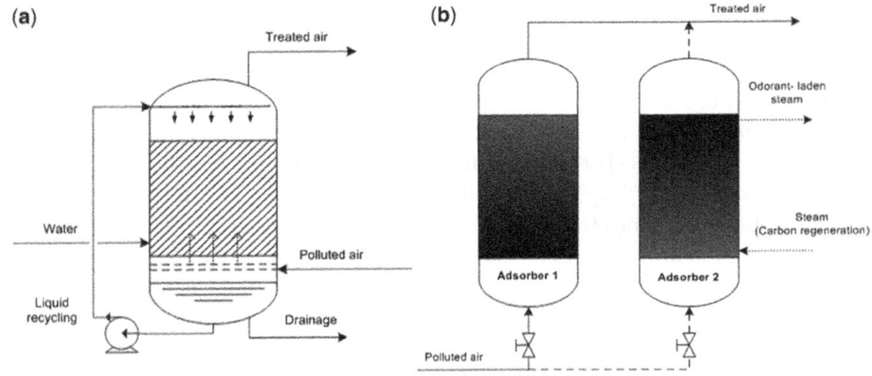

Figure 9.1 Schematic illustration of a chemical scrubber (a) and an activated carbon adsorption unit with steam regeneration (b).

Biological technologies such as biofilters (BF), biotrickling filters (BTF) or activated sludge diffusion systems (ASD), are gaining ground as a result of their high cost-effectiveness and low environmental impact derived from their reduced energy or chemical requirements (Shareefdeen & Singh, 2005; Estrada *et al.* 2011, 2012). Biofilters are the most commonly implemented biotechnology for odour abatement. The wide experience acquired over the past 30 years on their design and operation, their relatively easy operation and maintenance and their moderate investment and operating costs make biofilters often an adequate technology for a cost-effective odour management (Estrada *et al.* 2013a; Prado *et al.* 2009). They consist of a fixed bed packed with an organic or organic/inorganic material, which hosts the microbial population responsible for the odorant biodegradation (Figure 9.2a). The malodorous air is forced through the packed bed where odorants are transferred to the biofilm and subsequently biologically oxidized under aerobic conditions (Lebrero *et al.* 2011; Estrada *et al.* 2013a). In biotrickling filters, microorganisms grow attached to an inert packing material, and a nutrient aqueous solution is continuously irrigated through the bed (Figure 9.2b) (Lebrero *et al.* 2011; Estrada *et al.* 2013a). The main difference between biofilters and biotrickling filters is the continuous recycling of the aqueous solution in the latter, while biofilters are only irrigated sporadically to maintain moisture, pH and nutrient levels. Odorant mass transfer from the gas phase to the biofilm usually determines the maximum REs achieved in these biotechnologies: RE > 99% can be reached for water soluble biodegradable compounds, decreasing to ~75% and 50–80% in BFs and BTFs, respectively, for the most hydrophobic odorants (Delhomenie & Heitz, 2005; Kraakman *et al.* 2011). On the other hand, in ASD systems the malodourous emission is employed for wastewater oxygenation by directly sparging the emission into the aeration tank of the plant. Odorants diffuse into the activated sludge broth together with oxygen and are biodegraded by the activated sludge (Figure 9.2c). Odour REs > 99% have been recorded in large-scale WWTPs treating the odorous emissions in their aeration tanks (Kiessewetter *et al.* 2012; Lebrero *et al.* 2011; Barbosa *et al.* 2006). ASD has been applied for more than 30 years mainly in North America, and only in the past decade started to be perceived as a real engineered alternative for odour abatement all over the world. Recent works have ruled out the traditional concerns of detrimental effects on wastewater treatment caused by the sparging of malodours in the aeration basin due to pH modification by the H_2S present in the malodorous stream or to possible alterations in the structure of the biological communities responsible for wastewater treatment (Barbosa & Stuetz, 2013). A more detailed description of these odour control technologies can be found elsewhere (Delhomenie & Heitz, 2005; Lebrero *et al.* 2011; Estrada *et al.* 2013a).

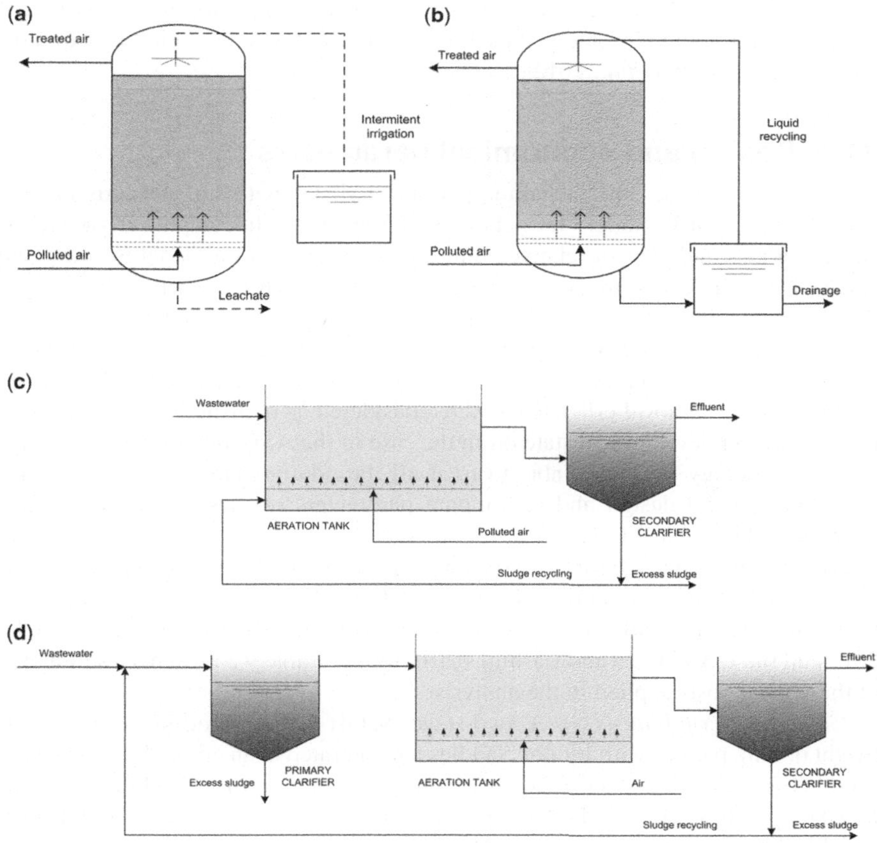

Figure 9.2 Schematic illustration of a biofilter (a), a biotrickling filter (b) an activated sludge diffusion system (c) and activated sludge recycling (d).

Finally, activated sludge recycling (ASR) is an innovative odour treatment approach consisting of the recycling of return activated sludge from the secondary clarifier or aerobic activated sludge from the aeration basin to the head works of the WWTP in order to aerobically or anoxically degrade potentially odorous compounds or prevent their release (Figure 9.2d). The main advantages of ASR are its low investment costs and the absence of covers and ductwork. Not much information is yet available regarding this odour abatement method, but preliminary studies have shown the potential for its application in WWTPs. A similar and innovative strategy to reduce odour impact is based on the recycling of effluents with high nitrate content, such as centrates from sludge dewatering, to the headworks of the WWTP in order to promote anoxic conditions, thus allowing the oxidation of potential odorants. The only study regarding this strategy estimated savings of 310 USD per day in odour treatment by the implementation of nitrified centrate recycling (Husband *et al.* 2010). ASR constitutes the most recent approach to odour control and has emerged as a promising candidate for future research in the field (Kiesewetter *et al.* 2012).

9.2.1 Design and economical parameters

A WWTP of about 300 megaliter per day (MLD) with an air emission of 50000 $m^3\ h^{-1}$ with 39 model VOC odorants, H_2S and methylmercaptan at trace level concentrations and 40% of relative humidity was selected as a model malodorous emission for the economical and energy efficiency analysis (Zarra *et al.* 2008; Barbosa *et al.* 2002; Estrada *et al.* 2011). All technologies evaluated were designed to support removal efficiencies >99% for H_2S and >95% for odour concentration, except for activated sludge diffusion and activated sludge recycling, where 75% and 50% odour removal efficiencies were considered as realistic values according to their current development state or, in the case of the ASD, due to the possibility of the technology not being able to treat all the odorous emission produced in the plant. Typical design and operational parameters of these technologies are summarized in Table 9.1.

All investment and operating costs of the most commonly used odour abatement technologies were based on Estrada *et al.* (2011) and (2012) and updated according to the most recent available Chemical Engineering Process Cost Index (CEPCI) 2012 and the UBS Prices and Earnings 2012 report. Table 9.2 presents a summary of the typical costs applied in the analysis.

Chemical scrubbing (CS). A two-stage NaOH-NaClO process with a total height of 2 m, packed with Intalox Saddles and operated at an empty bed residence time (EBRT) of 4 s (2 s per stage), was considered in this study. Packing material lifespan, purchase costs and disposal costs were set up at 10 years, 1370 EUR m^{-3} and 137 EUR m^{-3}, respectively. The total pressure drop in the system, including ductwork, was estimated at 1000 Pa, and water was recycled at a rate of 180 L $m^{-3}\ min^{-1}$. Labour costs of 20700 EUR per media substitution were considered.

Table 9.1 Data compilation for the design parameters and operating costs for the different technologies evaluated.

Technology	Height	Packing material	EBRT	Packing material lifespan	Packing material purchase costs	Packing material disposal costs	Labour, transport and handling costs	Pressure drop	Others
Chemical scrubbing (CS) two-stage NaOH-NaClO	2 m	Intalox Saddles	4 s (2 s per stage)	10 years	1370 EUR m^{-3}	137 EUR m^{-3}	20700 EUR per media substitution	1000 Pa	Liquid recirculation: 180 L m^{-3} min^{-1}
Activated carbon filtration (AC)	0.6 m	Granular impregnated AC (450 kg m^{-3})	2.5 s	6 months	5.5 EUR kg^{-1}	137 EUR m^{-3}	20700 EUR $year^{-1}$	2250 Pa	No regeneration of the activated carbon was considered in the estimation of lifespan and costs
Biofiltration (BF)	1 m	Compost	60 s	2 years	82 EUR m^{-3}	48 EUR m^{-3}	33130 EUR $year^{-1}$	1500 Pa	Humidification: 0.02 kg water (kg air)$^{-1}$ Irrigation by means of 2 water nets with 49 drips m^{-2} and each drip irrigating 1.9 L h^{-1} for 3 min day^{-1}
Biotrickling Filtration (BTF)	4 m (2 m per stage)	Inert polyurethane foam (PUF)	15 s	10 years	1370 EUR m^{-3}	137 EUR m^{-3}	20700 EUR per media substitution	1000 Pa	Liquid recirculation: 7.2 L m^{-3} min^{-1}. Liquid renewal rate: 2.5 L per g of H_2S removed

(Continued)

Table 9.1 Data compilation for the design parameters and operating costs for the different technologies evaluated (*Continued*).

Technology	Height	Packing material	EBRT	Packing material lifespan	Packing Material Purchase Costs	Packing Material Disposal Costs	Labour, transport and handling costs	Pressure Drop	Others
1 stage BTF + AC	2 m + 0.6 m	PUF + Standard AC	9 s + 2.5 s	PUF: 10 years. AC: 2 years	1370 EUR m^{-3} for PUF. 4.1 EUR kg^{-1} for AC	137 EUR m^{-3}	22770 EUR $year^{-1}$	2500 Pa	Extended lifespan of 2 years for AC due to the lower concentration of odorants to be treated in the adsorption unit.
1 stage CS + AC	2 m + 0.6 m	Intalox Saddles + Standard AC	2 s + 2.5 s	Intalox Saddles: 10 years. AC: 2 years	1370 EUR m^{-3} for Intalox Saddles. 4.1 EUR kg^{-1} for AC	137 EUR m^{-3}	22770 EUR $year^{-1}$	2500 Pa	Extended lifespan of 2 years for AC due to the lower concentration of odorants to be treated in the adsorption unit.
1 stage BTF + 1 stage CS	2 m + 2 m	PUF + Intalox Saddles	9 s + 2 s	10 years	1370 EUR m^{-3}	137 EUR m^{-3}	20700 EUR per media substitution	1500 Pa	
Step Feed BF	1 m	Compost	60 s	2.5 years	82 EUR m^{-3}	48 EUR m^{-3}	24850 EUR $year^{-1}$	1250 Pa	BF modified by supplying the odorous emission in three different locations along the BF height (Estrada *et al.* 2013b).

Table 9.2 Data compilation for the operating costs based on previous studies by Estrada *et al.* 2012 (updated to 2012).

Source		Value	Remarks
Energy		EUR 0.105 kW^{-1}	
Chemicals	Caustic soda	EUR 0.175 kg^{-1}	50% w/w, density 1.53 kg l^{-1}
	Hypochlorite	EUR 0.210 kg^{-1}	12.5% w/w, density 1.22 kg l^{-1}
Activated Carbon	Virgin	EUR 4.11 kg^{-1}	density 0.45 kg l^{-1}
	Impregnated	EUR 5.50 kg^{-1}	density 0.45 kg l^{-1}
Water	Potable	EUR 1.12 m^{-3}	
	secondary effluent	EUR 0.56 m^{-3}	

Activated carbon filtration (AC). The adsorbent selected for the filtration process was granular impregnated activated carbon, characterized by a density of 450 kg m^{-3}, purchase cost of 5.5 EUR kg^{-1} and a lifespan of 6 months (no regeneration of the activated carbon was considered). The adsorption filter consisted of a 0.6 m height column operated at an EBRT of 2.5 s, resulting in a total pressure drop of 2250 Pa including grease filters and ductwork. The disposal costs of activated carbon were 137 EUR m^{-3}, while AC transport and renewal costs added up to 20700 EUR $year^{-1}$.

Biofiltration (BF). A 1 m biofilter packed with compost with a lifespan of 2 years was selected in this study. The system operated at an EBRT of 60 s and a total pressure drop of 1500 Pa (including the pressure drop of the humidifier and ductwork). Irrigation of the biofilter was performed by means of 2 water nets located at the top of the unit, each of them provided with 49 drips m^{-2} and each drip irrigating 1.9 L h^{-1} for 3 min day^{-1}. Total humidification requirements were estimated to be 0.02 kg water (kg air)$^{-1}$. The packing purchase costs were 82 EUR m^{-3}, while the costs associated to its disposal and transport-handling were 48 EUR m^{-3} and 33130 EUR $year^{-1}$, respectively.

Biotrickling filtration (BTF). A two-stage BTF operated at acid (~2) and neutral pH, respectively, with a total height of about 4 m (2 m per stage) was considered as model BTF. Inert PUF, with a lifespan of 10 years and a cost of 1370 EUR m^{-3}, was selected as the packing material. Disposal and labour costs were estimated at 137 EUR m^{-3} and 20700 EUR per media substitution, respectively. The total pressure drop in the system and ductwork was 1000 Pa. The EBRT, liquid recycling and liquid renewal rate were set at 15 s, 7.2 L m^{-3} min^{-1} and 2.5 L $(g_{H2S\ removed})^{-1}$, respectively.

BTF + AC. This hybrid technology consisted of a single stage BTF and an AC filter acting as a polishing step. Similar operating parameters as in stand-alone technologies were used in the hybrid technology except for a shorter EBRT of 9 s in the BTF and the use of standard activated carbon with a price of 4.1 EUR kg^{-1} and an extended lifespan of the packing material of up to 2 years due to the lower concentration of odorants to be treated in the unit after the treatment in the BTF. The total pressure drop of this combined system was 2500 Pa.

CS + AC. A single-stage CS operated at an EBRT of 2 s was coupled with an AC as a polishing step. The rest of the operating parameters were maintained as in the stand-alone technologies except for the use of standard activated carbon packing with a price of 4.1 EUR kg^{-1} and extended lifespan of 2 years due to the lower concentration of odorants to be treated in the unit. A total pressure drop of 2500 Pa was considered.

BTF + CS. A single-stage CS acts as the polishing stage after a single-stage BTF in this hybrid technology. EBRTs in the BTF and the CS stages can be set at 9 s and 2 s, respectively, to fulfil the target odour and H_2S REs. The rest of the operating parameters were maintained as in the stand-alone technologies. The total pressure drop of this combined system was 1500 Pa.

Step-feed biofilter (Step BF). The operation of the standard BF was modified by supplying the odourous emission in three different locations along the BF height (Estrada *et al.* 2013b). This configuration allows for an increased packing lifespan of 25% compared to the conventional BF, while reducing the overall pressure drop of the bed by 25% (total pressure drop in the step-feed BF of 1250 Pa).

Activated sludge diffusion (ASD). The odorous emission is sparged into an activated sludge tank (devoted to wastewater treatment) with a depth of 4 m. An additional pressure drop of 500 Pa was considered to take into account the ducting required to conduct the emission to the aeration basin. Grease filters, corrosion resistant blower, upgrade to fine bubble diffusers and instrumentation were included as capital costs.

Activated sludge recycling (ASR). A sludge flowrate of 625 m^3 h^{-1} is pumped from the secondary settler of the activated sludge tank to the head of the WWTP, representing 5% of the total wastewater flowrate treated in the plant (a model WWTP treating 300 megaliter per day was considered, typical size of a WWTP with approximately 50000 m^3 h^{-1} malodorous air emission). Piping, sludge pumps, dispensers, valves, instrumentation and automation needed were included in the costs of this technology.

The CO_2 footprint for each technology was calculated according to the data shown in Table 9.3. The following transportation distances were assigned to the different materials required in the technologies evaluated: 50 km of road transportation to the compost needed in BF based on the possibility of locally purchasing this packing material; 500 km of road transportation to the polyurethane foam (PUF) and Intalox Saddles required in the BTF and CS, respectively, based on the possibility of purchasing these materials inside the country; 5000 km of sea transportation + 200 km of road transportation to the activated carbon, according the present trend of activated carbon purchase from Asian manufacturers. Finally, 15 km of road transportation were considered for the disposal of all spent packing materials in local landfills.

Table 9.3 Data compilation for the calculation of CO_2 footprint.

Source		CO_2 equivalence	Unit	Reference
Energy consumption		250	kg CO_2-eq GJ^{-1}	IChemE (2002)
Chemical usage	Caustic soda	1376	kg CO_2-eq $t_{chemical}^{-1}$	Owen (1982)
	Hypochlorite	1065		Owen (1982)
Packing material usage	PUF	4300		Mattinen and Nissinen (2011)
	Intalox saddles	2950	kg CO_2-eq $t_{packing}^{-1}$	Mattinen and Nissinen (2011)
	Compost	600		Boldrin *et al.* (2009)
	Activated carbon	1000		Agentschap (2012)
Transport	Road (Truck)	0.127	kg CO_2-eq $t_{transported}^{-1}$ km^{-1}	IMO (2009)
	Sea (Ship)	0.015		IMO (2009)

9.3 COMPARATIVE PARAMETRIC EFFICIENCY ANALYSIS

9.3.1 Energy consumption

A simple analysis of the technologies can be initially made in terms of their energy consumption during operation to achieve the target H_2S and odour REs (99 and 95%, respectively). The energy requirements entail a significant share of the annual operating costs and are responsible for a part of the environmental impacts in odour treatment technologies (Alfonsín *et al.* 2013).

According to Figure 9.3, the hybrid technologies (BTF + AC, CS + AC, BTF + CS) and the physical/chemical techniques presented the highest energy consumptions (50–80 kW) among the odour abatement techniques evaluated. This can be explained by the high pressure drop across the activated carbon beds in AC, BTF + AC and CS + AC. On the other hand, chemical scrubbers presented relatively high power requirements associated to their high liquid recycling rates. This accounts for the high power consumptions in CS and BTF + CS, despite the low pressure drops across their tailored packing material.

Conventional biological technologies such as BF and BTF exhibited moderate energy consumptions. Innovative technologies such as Step BF and ASR also presented moderate power consumptions of ≈30 kW. Biofiltration (BF) was the technology with the highest energy requirements among the biological techniques, mainly due to the significant pressure drop in the system. Step feed biofiltration, where the total gas flow is split and fed at different heights along the biofilter bed, was shown to reduce the overall pressure drop across the bed without significantly impacting BF performance, which resulted in similar pressure drops to those recorded in BTFs (Estrada *et al.* 2013b). In ASR, the energy requirements were exclusively devoted to the pumping of the return activated sludge from the secondary settler at a flow rate of ≈5% of the total wastewater volume treated in the plant. This rate was estimated based on recent experimental results and should be able to reduce the hydrogen sulphide concentration in the inlet wastewater by 80–95% (Zhang *et al.* 2011).

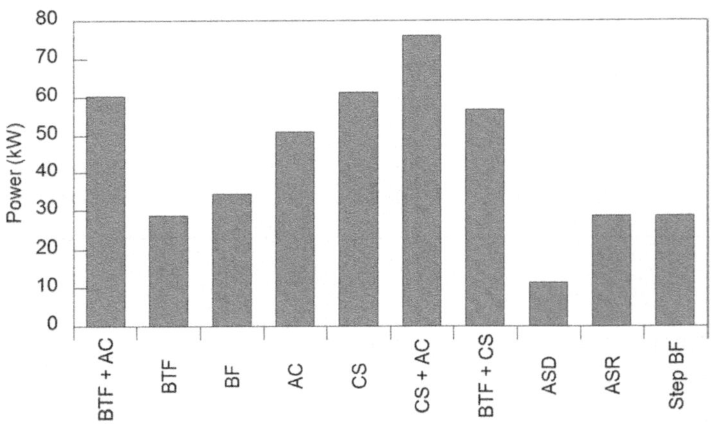

Figure 9.3 Power consumption of the evaluated odour treatment technologies.

Finally, ASD presented the lowest energy requirements among the technologies evaluated, since it only accounts for the energy needed to overcome the pressure drop of the piping to conduct the air to the aeration basin (500 Pa). The energy consumption associated to the bubbling of the air emission into the reactor was not considered as an extra energy need for the technology, since these energy requirements are usually already taken into account in the overall consumption in the water treatment line of the WWTP (Estrada *et al.* 2011).

9.3.2 Energy efficiency parameter

Despite the overall energy consumption of the technologies for odour abatement in WWTPs constitutes a relevant economic and environmental information, many other parameters must be taken into account for technology selection. Economic criteria such as investment and operating costs, together with the degree of abatement that a technology can offer, are usually most relevant when selecting technologies for full-scale applications. Therefore, the application of 'cost/benefit parameters' accounting for different aspects of the odour abatement process and lumping information into a single parameter would allow for a fairer comparison among odour abatement technologies.

In this context, the cost/benefit Energy Efficiency Parameter (EEP) was devised to account for three key features of the evaluated techniques: one cost and two benefits. The Net Present Value in 20 years (NPV_{20}, EUR) was considered as the 'technology cost'. The NPV_{20} accounts for the total amount of money spent in the installation and operation of a technology for a period of 20 years and includes both investment and operating costs as defined elsewhere (Estrada *et al.* 2011). Any change reducing the operating or investment costs, such as more economic construction or packing materials, lower salaries in work costs or an increased packing material lifespan would reduce the NPV_{20} values as described in depth in Estrada *et al.* (2012). The first 'benefit' considered was the odour abatement performance quantified as odour RE for the corresponding technology. The second relevant 'benefit' selected was the inverse of the power required (P, kW), considering a low energy consumption as a potential benefit in a technology. Thus, the EEP was defined as follows:

$$\mathrm{EEP} = \frac{\mathrm{NPV}_{20}}{RE \cdot (P)^{-1}} \tag{9.1}$$

A technology with a low EEP value will be therefore preferred since ideally this would entail a low NPV_{20}, high odour REs and low energy consumption.

The main conclusions derived from the EEP analysis were in agreement with those drawn from the total energy consumption data (Figure 9.3), with physical/chemical and the hybrid technologies also exhibiting the highest EEP values (Figure 9.4). Compared to the overall energy consumption, the differences among technologies got sharpened using this parameter, CS + AC being the less efficient technology in terms of costs and energy use. This finding confirmed previous research on the field that concluded that physical/chemical technologies are both more expensive and energy demanding than their biological counterparts (Estrada *et al.* 2011). In the particular case of CS + AC, the high operating costs of both individual technologies combined with their intrinsic high energy requirements, ranked this technology as the less preferred with an EEP value of 3.37×10^8 EUR kW. AC and CS ranked as the second and third less efficient technologies (EEP values of 2.35×10^8 and 2.08×10^8 EUR kW, respectively) based on the previously mentioned rationales. In this context, the hybrid technologies involving BTF performed better than the above discussed technologies despite exhibiting similar power requirements. Both BTF + AC and BTF + CS exhibited EEP values below 1.5×10^8 (Figure 9.4) even though they involved similar energy requirements to CS, AS and CS + AC (Figure 9.3).

This enhanced EEP derived from the lower NPV_{20} of these technologies, since the upfront biological technology significantly counter-balanced the overall costs of the AC or the CS employed as polishing steps (mainly due to the lower frequency of activated carbon purchase and replacement, and lower chemical usage, respectively).

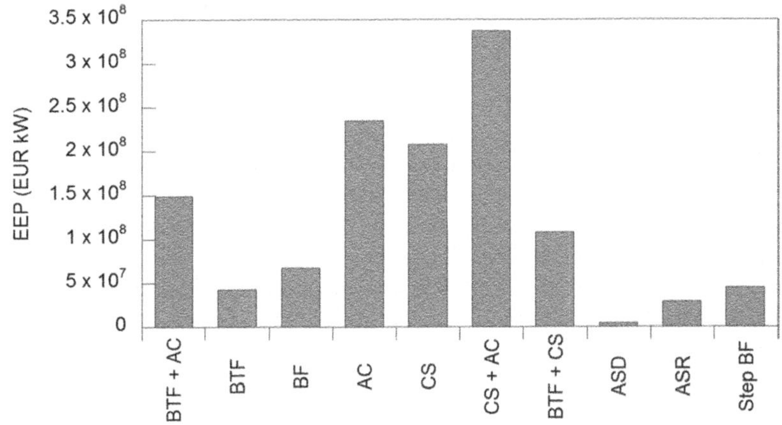

Figure 9.4 Energy Efficiency Parameter for the evaluated odour treatment technologies.

Interestingly, BF was the less efficient technology among the purely biological techniques due to its high operating costs caused by the low lifespan of the organic packing material (EEP value of 6.75×10^7 EUR kW). BTF and step feed BF exhibited similar EEPs of 4.26×10^7 and 4.48×10^7 EUR kW, respectively, as a result of their low operating costs and low energy requirements. ASR and ASD emerged as the most efficient technologies in terms of economic and energy efficiency. Despite their lower odour REs (75% for ASD and 50% for ASR vs 95% considered for the rest of technologies), the low NPV_{20} inherent to ASD and ASR ranked these techniques as the top performing technologies for odour abatement in terms of EEP. However, it is important to remark that they might not be able to achieve the odour removal required in some particular full scale applications, and in these scenarios complementary solutions to odour control in WWTP should be considered. In this context, ASR has been scarcely tested to date and ASD might not be able to treat the whole malodorous emission when its flow rate exceeds the aeration requirements of the wastewater aerobic treatment (Kiesewetter *et al.* 2012), which would entail the implementation of an additional treatment. In brief, these wastewater treatment-associated technologies should be considered in future WWTP designs and emerge as promising techniques, with limitations that must be gradually overcome in on-site research programs.

9.3.3 Sustainability efficiency parameter

Economic criteria have traditionally driven process selection at full-scale. However, nowadays standards are gradually shifting towards more holistic approaches including environmental aspects and process sustainability. Tools such as the IChemE Sustainability Metrics and Life Cycle Assessment have been recently applied to the most commonly implemented odour abatement technologies in order to quantify their environmental and social impacts (Estrada *et al.* 2011; Alfonsín *et al.* 2013). These comparative analyses provided a wider picture of the economic/environmental/social performance of each specific technology.

In this context, a cost/benefit parameter can be defined in order to include environmental aspects in combination with process economics. In our particular case, a low CO_2 footprint was considered to be a benefit anticipating any imminent change in atmospheric legislation to prevent the accumulation of greenhouse gases. Thus, the inverse of the CO_2-footprint produced (CO_2ftp, t CO_2-eq $m^{-3}_{treated}$) along with the odour abatement efficiency were employed to define the Sustainability Efficiency Parameter (SEP):

$$SEP = \frac{NPV_{20}}{RE \cdot (CO_2 ftp)^{-1}} \tag{9.2}$$

A technology with a low SEP value will be therefore preferred since ideally this would entail a low NPV_{20}, high odour REs and low CO_2 footprint. Any parameter reducing the NPV_{20} would therefore be beneficial for the SEP. In this context, the use of renewable energy sources would decrease the CO_2 footprint also improving the SEP results, specially benefiting those technologies with higher energy demands.

The SEP (Figure 9.5) was strongly correlated to the EEP previously reported (Figure 9.4) since the main contributor to the CO_2 footprint was the CO_2 associated to energy consumption, and most technologies were designed to achieve similar odour and H_2S REs. The CO_2 footprint associated to energy usage was at least one order of magnitude higher than the CO_2 footprint from other sources, except for BF and Step BF, where it was only 3 and 5 times higher, respectively (Table 9.4). Packing material manufacturing, transport and the untreated odour emission contributed marginally to the overall CO_2 footprint among the technologies evaluated. On the other hand, chemicals manufacture became the second most important contributor in CS and CS + AC. In technologies with low packing material lifespan such as AC, BF and Step BF, packing material manufacture constituted the second contributor to the total CO_2 footprint. Finally, the untreated odorous emission became the second most important contributor to the total CO_2 footprint in technologies with low packing material and transport requirements and/or low RE, such as BTF, ASD and ASR (Table 9.4).

The CS + AC hybrid technology was the least efficient technology due to its high energy requirements and chemical usage, contributing to a higher CO_2

footprint (SEP value of 7.34 EUR t CO_2-eq $m^{-3}_{treated}$). CS and AC ranked second and third in terms of sustainability efficiency with SEP values of 4.76 and 4.63 EUR t CO_2-eq $m^{-3}_{treated}$, respectively, confirming the relatively poor performance of physical/chemical techniques in terms of environmental impact. Despite the absence of CO_2 footprint associated to chemical usage, AC presented the highest CO_2 footprint associated to packing material manufacture due to its low lifespan.

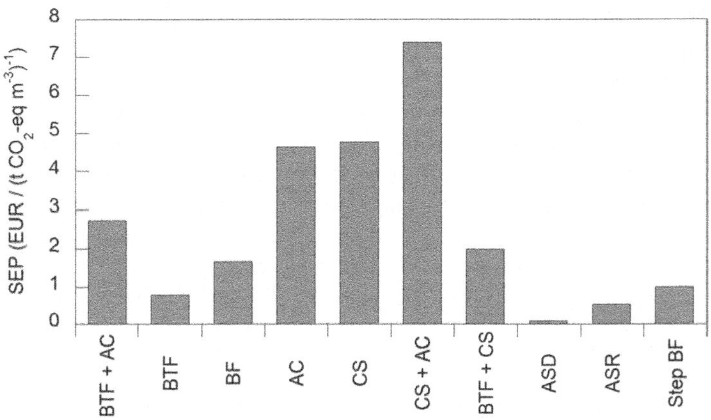

Figure 9.5 Sustainability Efficiency Parameter for the odour treatment technologies evaluated.

Table 9.4 CO_2 footprint of the different technologies evaluated.

	Footprint and Source (t CO_2-eq h^{-1})					
Technology	**Energy consumption**	**Untreated odour emission**	**Chemical manufacture**	**Packing material manufacture**	**Transport**	**Total**
BTF + AC	0.0543	0.0000	0.0000	0.0010	0.0001	**0.0554**
BTF	0.0257	0.0007	0.0000	0.0002	0.0000	**0.0266**
BF	0.0308	0.0006	0.0000	0.0106	0.0001	**0.0422**
AC	0.0459	0.0005	0.0000	0.0036	0.0004	**0.0503**
CS	0.0552	0.0012	0.0133	0.0001	0.0005	**0.0702**
CS + AC	0.0685	0.0000	0.0133	0.0010	0.0005	**0.0833**
BTF + CS	0.0512	0.0007	0.0001	0.0007	0.0000	**0.0527**
ASD	0.0102	0.0007	0.0000	0.0000	0.0000	**0.0109**
ASR	0.0257	0.0007	0.0000	0.0000	0.0000	**0.0264**
Step BF	0.0257	0.0006	0.0000	0.0053	0.0001	**0.0317**

Among biological technologies, standard BF and step feed BF exhibited higher SEP than BTF (1.66 and 0.95, respectively vs 0.79 EUR t CO_2-eq $m^{-3}_{treated}$) due to the short lifespan of their organic packing material, which increased packing material transportation frequency and compost utilization, with the subsequent increase in the CO_2 emissions associated to those aspects. In our particular analysis, BF exhibited a SEP value of 1.66 EUR t CO_2-eq $m^{-3}_{treated}$, very close to that estimated for the hybrid BTF + CS system (SEP = 1.99 EUR t CO_2-eq $m^{-3}_{treated}$). BTF constituted the preferred option in terms of sustainability among the conventional end-of-the-pipe odour treatment technologies, with a SEP value of 0.79 EUR t CO_2-eq $m^{-3}_{treated}$. However, the wastewater treatment-associated technologies ASD and ASR were indeed the most cost-sustainable with SEP values of 0.53 and 0.09 EUR t CO_2-eq $m^{-3}_{treated}$, respectively. This confirmed ASD and ASR are extremely cost-efficient, while entailing a low environmental impact in terms of energy use and CO_2 emissions.

Despite providing a limited knowledge of the overall environmental impact of odour abatement technologies, the use of CO_2 footprint as environmental benefit parameter confirmed the widely accepted best performance of biological techniques for odour abatement (Estrada *et al.* 2011). In addition, the SEP could be formulated to include other environmental or social impacts relevant in the future. Similarly, weighted factors could be added to tune the relevance of a certain cost or benefit in the SEP definition. For instance, in a sensitive scenario where the odour RE constitutes the most important parameter, the RE could be multiplied by a sensitivity factor.

9.3.4 Robustness efficiency parameter

Process robustness constitutes a key issue in odour abatement technology selection once the economic and environmental aspects have been evaluated. Under real scenarios, any technology has to cope with odorant concentration fluctuations, technical problems or shutdowns. Process robustness (R) can be quantified according to the methodology proposed by Kraakman (2003), where the probability of an operational upset or event to happen is multiplied by the negative effect caused in the overall odour abatement performance of the technology evaluated. In this simple way, the lower the values of R, the higher the robustness of a technology (Table 9.5).

Table 9.5 Robustness evaluation according to the methodology proposed by Kraakman (2003) and the semi-quantitative evaluation in Estrada *et al.* (2012).

Technology	BTF + AC	BTF	BF	AC	CS	CS + AC	BTF + CS	ASD	ASR	Step BF
Robustness value (R)	15	31	35	13	39	16	29	23	14	35

This estimated robustness (R) can be then included as a benefit in the previous cost/ benefit parameter SEP in order to have a more complete comparative parameter.

This new parameter, defined as Sustainability and Robustness Efficiency Parameter (SREP), includes process economic, odour abatement performance, environmental impact (and energy, indirectly) and process robustness (Eq. 9.3):

$$\text{SREP} = \frac{\text{NPV}_{20}}{RE \cdot (\text{CO}_2\text{ftp})^{-1} \cdot R^{-1}} \tag{9.3}$$

Chemical scrubbing was the least efficient technology according to the SREP as a result of its relatively low robustness due to the key importance of water and chemicals for its correct operation (Figure 9.6). The high robustness of adsorption systems derived from its relative simplicity and the fact that it does not rely on water or process control to operate may explain the worldwide acceptance of AC and all the AC-involving technologies among the WWTP operators and resulted in SREPs comparable to biofiltration. The consideration of process robustness among technology selection criteria entailed that the hybrid technology CS + AC would be preferred over CS (SREP values of 118 vs 185 EUR t CO_2-eq $m^{-3}_{treated}$, respectively). The hybrid technology BTF + AC, combining the low energy consumption and environmental impact of BTF and the robustness of AC, remained at the level of the step BF in terms of SREP (41 vs 35 respectively EUR t CO_2-eq $m^{-3}_{treated}$). BTF would be the preferred conventional technology according to the SREP analysis due to its balanced costs and sustainability despite not showing the highest robustness.

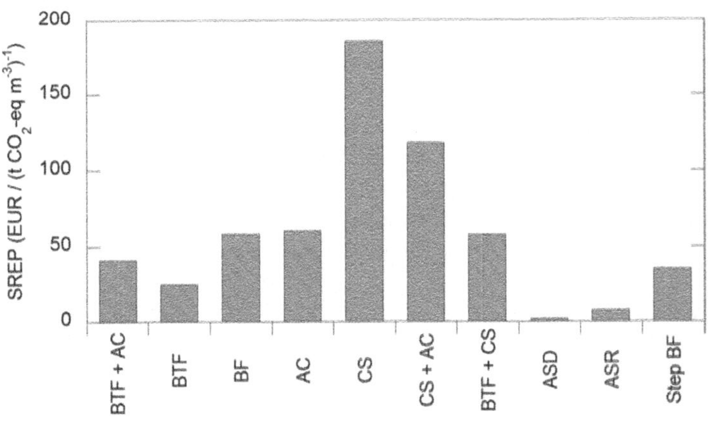

Figure 9.6 Sustainability and Robustness Efficiency Parameter for the evaluated odour treatment technologies.

ASD and ASR also showed a proficient performance in terms of SREP. In addition to their previously discussed economic and environmental advantages, these activated sludge-based technologies benefit from a high robustness derived from their relative simplicity and ease of operation. Despite practical experience available is scarce, especially for ASR, the high robustness of ASD for odour abatement was recently confirmed (Lebrero *et al.* 2010). In addition, both pilot and field studies suggest that the diffusion of malodorous emissions into the activated sludge process does not negatively affect the efficiency of wastewater treatment (Barbosa & Stuetz, 2013), but ASD might promote the development of filamentous bacteria and induce a poor biomass sedimentation under high H_2S concentration scenarios (Kiesewetter *et al.* 2012).

9.3.5 Influence of the H_2S concentration

The above defined parametric indicators represented a useful tool for comparative technology evaluation since they allowed considering simultaneously process economics, environmental impacts and robustness. However, the variable nature of the odorous emissions must be taken into account in order to adapt the parameters to different scenarios. Previous works have shown the significant impact of the different operational and design parameters on both the investment and operating costs of most odour abatement technologies (Estrada *et al.* 2011, 2012). H_2S concentration is a key parameter due to the wide range of concentrations and variability found in malodorous emissions depending on the origin and treatment of the wastewater. Moreover, the concentration of this compound is commonly employed as a reference for the calculation of main design parameters of odour abatement technologies, such as the estimation of chemical consumption in scrubbers, water needs in biotrickling filters or packing material lifespan in adsorption filters and biofilters (Estrada *et al.* 2011). Therefore, a comprehensive understanding of the influence of H_2S concentration on the different efficiency parameters above discussed is also crucial for a successful technology selection.

The SREP was calculated for all the evaluated technologies at three typical H_2S concentrations in WWTPs: 5, 15 and 45 ppm (Figure 9.7). ASD and ASR technologies were not included in the present analysis since not enough data were available to evaluate the influence of H_2S concentration on their performance. A recent sensitivity analysis conducted by the authors highlighted the high sensitivity of physical/chemical technologies towards H_2S (Estrada *et al.* 2012). In the present study, the SREP corresponding to the CS was the most impacted by variations in H_2S concentration. Indeed, despite the requirements of packing material in the CS were not affected by H_2S concentration, more water and chemicals (=operating costs) were needed for the correct operation of the system. In addition, the increase in chemical requirements implies higher CO_2 emissions in their manufacture and transportation. The second most sensitive technology to H_2S was AC, where higher concentrations of H_2S reduced the lifespan of the AC (thus increasing its NPV_{20}), entailing a more frequent packing replacement (thus increasing the CO_2 associated to its manufacture and transportation). However, it must be mentioned that AC presented a similar SREP value to that of BF at low H_2S concentrations.

On the other hand, the combination of increased packing material and chemical consumption at 45 ppm of H_2S boosted the SREP value of the CS + AC. Finally, the good performance of BTF + AC and BTF + CS even at high H_2S concentration must be highlighted, which confirmed that physical/chemical technologies highly reduced their overall costs and environmental impacts when implemented in combination with a biological technology.

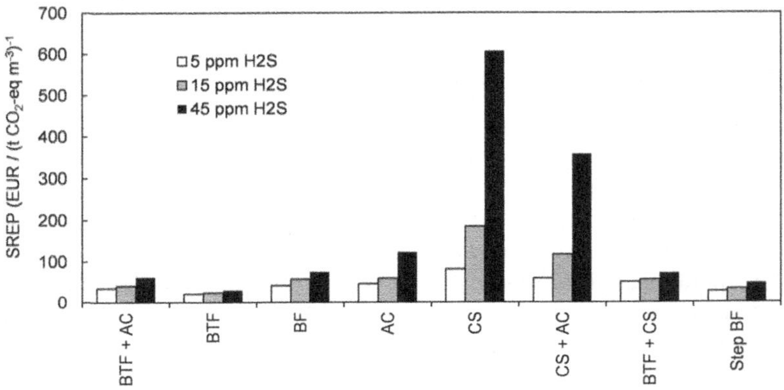

Figure 9.7 Influence of H_2S concentration on the Sustainability and Robustness Efficiency Parameter of the odour abatement technologies evaluated.

Overall, biotechnologies presented the lowest sensitivity towards fluctuations in H_2S concentration, with BTF exhibiting the lowest SREP variations when H_2S increased from 5 to 45 ppm. This sensitivity analysis revealed that despite the differences in SREP might not be significant at low H_2S concentrations, the selection of the optimum technology might result in important economic and environmental savings in WWTPs with higher H_2S concentrations.

9.3.6 Exploring alternatives to increase technology efficiency: L/D ratio

The length-to-diameter ratio (L/D) ratio is an important reactor design parameter to be analysed in order to improve the energy, economic and environmental efficiency of odour treatment technologies. The height and diameter of the reactors commonly employed are often fixed by conventional design criteria. L/D ratios are usually high due to various reasons: (i) a lower diameter means a lower land footprint, which might be a key issue when odour abatement techniques are installed in pre-existing WWTPs with limited space availability or in scenarios with high land costs (Estrada *et al.* 2011); (ii) the higher investment cost of units with high area, whose components would have to be built on-site or transported in special trucks; (iii) the high gas velocities achieved in units with high L/D ratios mediate an enhanced odorant mass transfer from the gas to the aqueous phase or to the adsorbent surface, which is a key operational issue since mass transfer limitations are commonly encountered in odour treatment applications (Kim & Deshusses, 2008; Iranpour *et al.* 2005). Energy efficiency considerations apply also in the selection of the optimum L/D ratio in biofilters, where a packed bed height of approximately 1 m is commonly accepted as a maximum in order to limit the pressure drop across the bed, and the diameter is then determined by the EBRT (Iranpour *et al.* 2005).

However, in the current scenario of increasing energy prices around the world, a modification in conventional design parameters such as the L/D ratio could be considered. In this regard, technologies exhibiting the highest pressure drops (i.e., AC, BF, CS + AC and BTF + AC) presented the highest benefits in terms of reductions in power consumption derived from a decrease in the L/D ratio (Figure 9.8). For instance, a reduction of 30% in the L/D ratio resulted in energy savings of up to 22% in AC, 20% in both BF and Step-feed BF, 18% in CS + AC and 21% in BTF + AC, mainly due to a decreased pressure drop in each reactor configuration. In the particular case of CS, the benefits derived mainly from the savings in energy consumption for liquid recycling. However, the energy savings in a CS accounted only for 11% when the L/D ratio is reduced by a 30% due to the low pressure drop across the packing material and the lower size of the reactors employed. BTFs exhibited moderate savings of 15% when the L/D ratio was reduced by 30%.

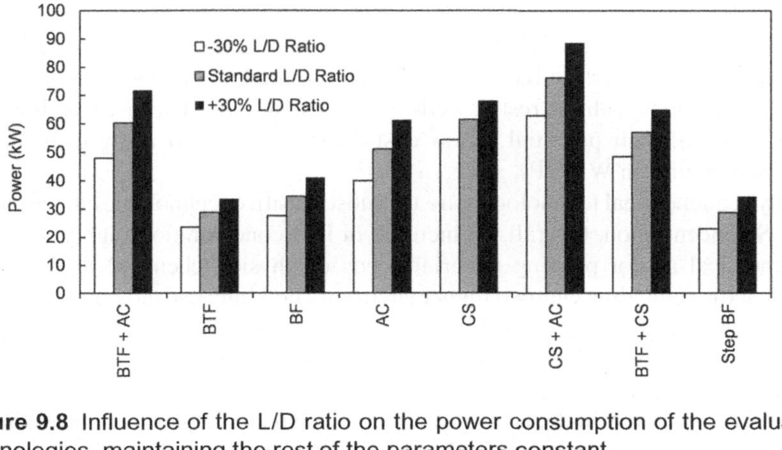

Figure 9.8 Influence of the L/D ratio on the power consumption of the evaluated technologies, maintaining the rest of the parameters constant.

These results suggest that relatively small variations in the L/D ratio can strongly impact on the energy consumption of the treatment technologies. Moreover, unnecessary increases in L/D ratios, even by a moderate 30%, can derive in a superfluous energy consumption of approximately 20% for most odour abatement technologies. This analysis was carried out assuming that odour REs remained constant by changing the L/D ratio by +/– 30%. Based on the fact that not much data are available in literature about this issue, more research will be necessary in order to better understand the influence of such design parameter on odour abatement.

9.4 CONCLUSIONS

In brief, physical/chemical technologies exhibited an inferior performance than their biological counterparts in terms of energy consumption, economic and environmental efficiency.

Among biological technologies, biotrickling filtration (BTF) exhibited one of the lowest energy requirements and better overall efficiency when process economic, environmental impacts and robustness are taken into account.

Biotrickling filtration technology can be backed up by an activated carbon filtration unit (AC) in order to increase its robustness, providing better sustainability than AC in standalone applications. When backed-up by chemical scrubbing (CS), the hybrid BTF + CS technology showed similar sustainability results to those of BTF + AC, but a lower process robustness.

Standard biofiltration (BF) showed low energy requirements and environmental impacts, however, its sustainability and robustness efficiency decreased to values similar to those of AC when robustness was considered. However, biofiltration performance can be increased in both economic and environmental terms when a step-feed configuration is employed.

The lack of sufficient reliable data on activated sludge diffusion (ASD) and activated sludge recycling (ASR) performance limited a complete comparison of both technologies with the rest of well-established systems. However, preliminary results showed their potential as low cost environmentally-friendly technologies for odour control in WWTPs.

Physical/chemical technologies are the most sensitive technologies to variations in H_2S concentration. Overall, an increase in H_2S concentration highly impacted the chemical and/or packing material needs of physical/chemical technologies, which at the same time entailed higher environmental impacts and operating costs. On the other hand, biological techniques were less influenced by H_2S concentration fluctuations, which rendered biotechniques more predictable over time in economic and environmental terms.

Finally, variations in the design of conventional technologies must be evaluated to reduce power consumption and increase efficiency based on the increasing energy prices. Limited reductions of 30% in the L/D ratio can yield reductions in the energy requirements of ≈20% for most of the odour treatment technologies here evaluated. However, there is still a lack of experimental data on the influence of L/D ratio on RE of odour treatment technologies.

9.5 REFERENCES

Agentschap (2012). Translation of I&E Monitoring Protocol 12-014.

Alfonsín C., Hernández J., Omil F., Prado Ó. J., Gabriel D., Feijoo G. and Moreira M. T. (2013). Environmental assessment of different biofilters for the treatment of gaseous streams. *Journal of Environmental Management*, **129**(0), 463–470.

Barbosa V. L. and Stuetz R. M. (2013). Performance of activated sludge diffusion for biological treatment of hydrogen sulphide gas emissions. *Water Science and Technology*, **68**(9), 1932–1939.

Barbosa V. L., Burgess J. E., Darke K. and Stuetz R. M. (2002). Activated sludge biotreatment of sulphurous waste emissions. *Reviews in Environmental Science and Biotechnology*, **1**, 345–362

Barbosa V. L., Hobbs P. J., Sneath R. W., Callen J. and Stuetz R. M. (2006). Investigating the capacity of an activated sludge process to reduce odour emissions. *Water Environment Research*, **78**, 842–851.

Bilsen I. and De Fré R. (2008). Evaluation of a Neutralizing Agent Applied on a Cooling Tower at a Citric Acid Production Plant. 3rd IWA International Conference on Odour and VOCs, Barcelona, Spain.

Boldrin A., Andersen J. K., Møller J., Christensen T. H. and Favoino E. (2009). Composting and compost utilization: accounting of greenhouse gases and global warming contributions. *Waste Management & Research*, **27**(8), 800–812.

CEPCI Chemical Engineering Plant Cost Index (2012). Chemical Engineering, New York.

Capodaglio A. G., Conti F., Fortina L., Pelosi G. and Urbini G. (2002). Assessing the environmental impact of WWTP expansion: Odour nuisance and its minimization. *Water Science and Technology*, **46**(9), 339–346.

Card T. (2001). Chemical odour scrubbing systems. In: odours in Wastewater Treatment: Measurement, Modelling and Control, R. Stuetz and F.-B. Frenchen (eds), IWA Publishing, Padstow, Cornwall, UK.

Decottignies V., Filippi G. and Bruchet A. (2007). Characterisation of odour masking agents often used in the solid waste industry for odour abatement. *Water Science and Technology*, **55**(5), 359–364.

Delhoménie M. C. and Heitz M. (2005). Biofiltration of air: a review. *Critical Reviews in Biotechnology*, **25**(1), 53–72.

Estrada J. M., Kraakman N. J. R., Muñoz R. and Lebrero R. (2011). A comparative analysis of odour treatment technologies in wastewater treatment plants. *Environmental Science and Technology*, **45**(3), 1100–1106.

Estrada J. M., Kraakman N. J. R., Lebrero R. and Muñoz R. (2012). A sensitivity analysis of process design parameters, commodity prices and robustness on the economics of odour abatement technologies. *Biotechnology Advances*, **30**(6), 1354–1363.

Estrada J. M., Lebrero R., Quijano G., Kraakman N. J. R. and Muñoz R. (2013a). Strategies for odour control. In: Odour Impact Assessment Handbook, V. Belgiorno, V. Naddeo and T. Zarra (eds), John Wiley & Sons, Ltd., Hoboken, NJ, USA.

Estrada J. M., Quijano G., Lebrero R. and Muñoz R. (2013b). Step-feed biofiltration: a low cost alternative configuration for off-gas treatment. *Water Research*, **47**(13), 4312–4321.

Husband J. A., Phillips J., Coughenour J. R., Walz T. and Blatchford G. (2010). Innovative approach to centrate nitrification accomplishes multiple goals: nitrogen removal and odour control. *Water Science and Technology*, **61**(5), 1097–1103.

IChem (2002). The Sustainability Matrics. The Institution of Chemical Engineers, Rugby, UK.

IMO (2009). Second IMO GHG Study. International Maritime Organization, London, UK.

Iranpour R., Cox H. H. J., Deshusses M. A. and Schroeder E. D. (2005). Literature Review of air pollution control biofilters and biotrickling filters for odor and volatile organic compound removal. *Environmental Progress*, **24**(3), 254–267.

Jehlickova B., Longhurst P. J. and Drew G. H. (2008). Assessing effects of odour: a critical review of assessing annoyance and impact on amenity. 3rd IWA International Conference on Odour and VOCs, Barcelona, Spain.

Kaye R. and Jiang K. (2000). Development of odour impact criteria for sewage treatment plants using odour complaint history. *Water Science and Technology*, **41**(6), 57–74.

Kiesewetter J., Kraakman N. R. J., Cesca J., Trainor S. and Witherspoon J. (2012). Expanding the use of Activated Sludge at Biological Wastewater Treatment Plants for Odor Control. Proceedings of the WEF Odors and Air Pollutants Conference. Louiseville, Kentucky, USA.

Kim S. and Deshusses M. A. (2008). Determination of mass transfer coefficients for packing materials used in biofilters and biotrickling filters for air pollution control. 1. Experimental results. *Chemical Engineering Science*, **63**(4), 841–855.

Kraakman N. J. R. (2003). Robustness of a full-scale biological system treating industrial CS2 emissions. *Environmental Progress*, **22**(2), 79–85.

Kraakman N. J. R., Roche Rios J. and van Loosdrecht M. C. M. (2011). Review of mass transfer aspects for biological gas treatment. *Applied Microbiology and Biotechnology*, **91**, 873–886.

Lebrero R., Rodríguez E., Martin M., García-Encina P. A. and Muñoz R. (2010). H2S and VOCs abatement robustness in biofilters and air diffusion bioreactors: a comparative study. *Water Research*, **44**(13), 3905–3914.

Lebrero R., Bouchy L., Stuetz R. and Muñoz R. (2011). Odor assessment and management in wastewater treatment plants: a review. *Critical Reviews in Environmental Science and Technology*, **41**(10), 915–950.

Mattinen M. and Nissinen A. (2011). Carbon Footprint Calculators for Public Procurement. The Finnish Environment 36/2011, Helsinki, Finland.

Owen W. F. (1982). Energy in Wastewater Treatment. Prentice Hall, USA.

Prado Ó. J., Gabriel D. and Lafuente J. (2009). Economical assessment of the design, construction and operation of open-bed biofilters for waste gas treatment. *Journal of Environmental Management*, **90**(8), 2515–2523.

Sanchez C., Couvert A., Laplanche A. and Renner C. (2006). New compact scrubber for odour removal in wastewater treatment plants. *Water Science and Technology*, **54**(9), 45–52.

Shareefdeen Z. and Singh A. (2005). Biotechnology for Odor and Air Pollution Control. Springer-Verlag, Heidelberg, Germany.

Stuetz R. M., Gostelow P. and Burgess J. E. (2001). Odour perception. In: Odours in Wastewater Treatment: Measurement, Modeling and Control, R. Stuetz and F. B. Frechen (eds), IWA Publishing, London, UK.

Sucker K., Both R. and Winneke G. (2008). Review of adverse health effects of odours in field studies. 3rd IWA International Conference on Odour and VOCs, Barcelona, Spain.

Tchobanoglous G., Burton F. L. and Stensel H. D. (2003). Wastewater Engineering: Treatment and Reuse. McGraw Hill, New York, USA.

Turk A. and Bandosz T. J. (2001). Adsorption systems for odour treatment. In: Odours in Wastewater Treatment: Measurement, Modeling and Control, R. Stuetzand and F. B. Frechen (eds), IWA Publishing, London, UK.

UBS – United Bank of Switzerland (2012). Prices and Earnings: A Comparison of Purchasing Power Around the Globe. UBS CIO WM Research, Switzerland.

Vincent A. J. (2001). Sources of odours. In: Odours in Wastewater Treatment: Measurement, Modeling and Control, R. Stuetzand and F. B. Frechen (eds), IWA Publishing, London, UK.

WHO, World Health Organization International Angency for Research on Cancer (2013). Press Release Number 221, 17 October.

Zarra T., Naddeo V., Belgiorno V., Reiser M. and Kranert M. (2008). Odour monitoring of small wastewater treatment plant located in sensitive environment. *Water Science and Technology*, **58**(1), 89–94.

Zhang L., De Gusseme B., Cai L., De Schryver P., Marzorati M., Boon N., Lens P. and Verstraete W. (2011). Addition of an aerated iron-rich waste-activated sludge to control the soluble sulphide concentration in sewage. *Water and Environment Journal*, **25**(1), 106–115.

Chapter 10

Instrumentation, monitoring and real-time control strategies for efficient sewage treatment plant operation

Sergio Beltrán, Ion Irizar and Eduardo Ayesa

CEIT and Tecnun (University of Navarra), Manuel de Lardizábal 15, 20018 San Sebastián, Spain

10.1 INTRODUCTION

Over the last decades, the operation of urban wastewater treatment plants (WWTPs) has been mainly focused on maintaining the amount of active biomass required for removing the influent organic matter within the system. For this purpose, the most common operation and control strategies were oriented toward regulating the suspended solids and dissolved oxygen concentrations needed to guarantee biodegradation (Olsson, 1987). However, stricter requirements in effluent quality (especially for nutrient removal in sensitive areas) have been promoting new, more flexible and efficient WWTP plant layouts. New treatment plants incorporate complex combinations of aerobic, anoxic, anaerobic and facultative tank reactors, and their optimum operation cannot be based on the traditional strategies (Andrews, 1993), which are frequently based on rigid and conservative rules that do not consider the interactions between processes and their dynamic characteristics. For these reasons, new Instrumentation, Control and Automation (ICA) tools are being progressively incorporated into plants in order to optimally govern (both from an environmental and economic point of view) all the dynamic and interrelated mechanisms that appear in an urban WWTP.

The need to meet stricter effluent requirements at a minimal cost is not the only reason for the significant increase in ICA implementation in recent years. New advances in monitoring and actuation equipment are also crucial. On the one hand, new on-line monitoring sensors and analysers are more reliable and economical, and on the other hand the new actuators incorporate enough flexibility for implementing more efficient control strategies. In this sense, it is very important to note that the dynamic behaviour of the WWTP processes should be taken into account in the design and dimensioning stages in order to optimize future operational costs.

Computation and data acquisition tools have also progressed significantly in recent years. However, the lack of appropriate data management tools is presently a limiting factor for a broader implementation and a more efficient use of sensors and analysers, monitoring systems and process controllers in WWTPs. Therefore, there is the challenge to develop new tools that are able to synthesize useful information from the processing and analysis of combined data gathered from several interrelated processes and deal with their heterogeneous characteristics and storage decentralization (Beltrán *et al.* 2012).

It is clear that optimizing WWTP operation using automatic controllers makes it possible to reduce operational costs. However, this reduction cannot be generically evaluated because it depends on many factors like the size and flexibility of the plant, the amount of available information, the wastewater characteristics and the effluent requirements. This chapter focuses firstly on the state-of-the-art of instrumentation for monitoring and control purposes. Then it describes the main real-time control strategies for efficient sewage treatment plant operation: control of the aeration system, control of chemical addition, control of the internal, external and wastage flow-rates, and control of the anaerobic processes. Finally, plant-wide control is addressed and several conclusions are highlighted.

10.2 INSTRUMENTATION FOR MONITORING AND CONTROL PURPOSES

During the last 20–25 years most wastewater treatment plants have been replacing the old control and monitoring systems with new SCADA systems (Supervisory Control and Data Acquisition), which allow both the central monitoring of a plant and its controllers and the storage of plant data (Garrett, 1998; Olsson *et al.* 1998; Lynggaard-Jensen, 1999). However, independently of the introduction of SCADA systems, the control of plants remained mainly focused on strategies based on actuation times (with a few exceptions such as control based on on-line measurements of hydraulic flow-rates and dissolved oxygen concentrations). It was not until the last half of the 1990s when the SCADA systems were completely utilised, since the instrumentation needed for extracting plant information was unavailable or considered too unreliable to be used in practical applications (Bourgeois *et al.* 2001).

Nowadays, it is possible to measure in a continuous or semi-continuous way nearly all the physical and chemical parameters relevant for the monitoring and control of WWTPs (Olsson, 2012). For example, Table 10.1 shows the most commonly used measurements taken by instrumentation.

Table 10.1 Commonly used measurements taken by instrumentation in WWTPs (from Olsson *et al.* 2004)

Properties	Constituents
Conductivity	Ammonium
pH	Biogas production
Redox	Dissolved oxygen
Sludge blanket	Nitrate
Temperature	Organic matter
Turbidity	Phosphate
	Solids concentration

Aeration is one of the key systems for WWTP stability, performance and operational costs, and consequently dissolved molecular oxygen (DO) sensors are considered particularly important. The first DO determination was performed by L. W. Winkler in 1888 by using a colorimetric method based on titration of oxygen with thiosulfate ($S_2O_3^{-2}$) and iodine (I^2) (Winkler, 1988). The amount of the DO is proportional to the amount of tetrathionate ($S_4O_6^{-2}$) generated, which is determined by reducing I_2 to iodide (I^-). In spite of being difficult to use for on-line sensing purposes, automatic measurement of DO based on potentiometric determination of the I^- produced has also been developed (Orellana *et al.* 2011) and it is still employed as a reference method for calibrating DO sensors since it is the most precise and reliable titrimetric procedure for DO analysis (Standard Methods, 2012). Commercial electrochemical (e.g., polarographic, galvanic) DO sensors were already being used for control on a routine basis from the early 1980s. These sensors are protected by an oxygen-permeable plastic membrane that serves as a diffusion barrier against impurities, which makes electrochemical DO sensors suitable for analysis in situ and particularly convenient for field applications such as the continuous monitoring of DO in activated sludge. However, DO control was still far from fully utilised in the early 2000s (Ingildsen *et al.* 2002b; Jeppsson *et al.* 2002). Nowadays, even though galvanic cells are the dominant electrochemical technology, the control of aeration systems based on luminescence-based oxygen sensors is growing rapidly. The advantages over electrochemical devices include the ease of miniaturization, the lack of chemical reactive agent consumption, faster response (<60 s), robustness and insensitivity to interfering agents (e.g., H_2S, CO_2 or NH_X-N). The low maintenance, extended operational lifetime and reliability of optic oxygen sensors are so notable that every major manufacturer of environmental monitors is currently offering at least one model for in situ DO measurements in water, rapidly phasing out the membrane electrode sensors.

Nutrient analysers (e.g., NH_X-N, NO_X-N, PO_4-P) were emerging at WWTPs in the early 2000s (Olsson, 2005) and they now have developed into in situ sensors (e.g., ion-selective electrodes probes for ammonia and ultraviolet probes for nitrate and nitrite) for reasons of maintenance, costs, measurement delays and sensor dynamics, resulting in easier control and better performance (Kaelin *et al.* 2008).

In the case of biological parameters, the on-line information is more reduced and the most used analysers are the respirometers (Spanjers *et al.* 1998). Despite the great expectations already present in the 1970s, there are not many controls based on respirometry implemented in full scale (Trillo, 2004). Still, respirometry is a viable method for in-stream early warning systems.

In anaerobic sludge digestion operation, gas production, gas quality, volatile acid content, temperature, sludge feed rate, alkalinity and total organic carbon (TOC) are usually measured (Spanjers & van Lier, 2006). The sludge blanket height in the clarifiers is monitored today by reliable sensors that are used on a routine basis, sometimes for the dynamic control of the return sludge (Vanrolleghem *et al.* 2006).

Further information of on-line measuring equipment in WWTPs is listed in Rieger *et al.* (2003) and Vanrolleghem and Lee (2003). In addition, the Water Environment Federation publishes an annual literature review about instrumentation for monitoring and control purposes in WWTPs (e.g., Sweeney & Kabouris, 2013).

In summary, it can be stated that instrumentation, including sensors, analysers and other measuring instruments, is no longer the bottleneck for the control of wastewater systems (Jeppsson *et al.* 2002). In fact, instrumentation for control is no longer the main focus for international research. Currently, according to a recent industrial marketing analysis, there are almost 100 sensor companies in the world working with water (Olsson *et al.* 2013). Indeed, increased confidence in instrumentation is now driven by the fact that clear definitions of performance characteristics and standardised tests for instrumentation have become available (ISO 15839:2003).

In a short time the research on sensors will become more focused by providing relevant and reliable data on the problem at hand and they will deal with the painstaking fouling problems whilst at the same time minimising maintenance requirements. In addition, there is an exponential growth in the use of soft-sensors for estimating process variables that considerably influence process behaviour but cannot be measured on-line in a successful manner (e.g., active biomass, soluble substrate). Such tools are being used as an effective utilisation for advanced control and for the development of new optimization strategies, helping the operator or a supervision system to take the appropriate actions to maintain the process in good operating conditions, diagnose possible process failures or prevent accidents (Haimi *et al.* 2013).

Finally, it can be said that although the instrumentation in wastewater treatment systems has increased almost exponentially over the past decades, there is still no standardized way to check data quality (fault detection) and know the source of the error (diagnosis), even if a lot of progress has been made (Olsson *et al.* 2013). While some years ago the scarcity of data due to the lack of reliable measuring devices was a major limitation, at present WWTPs deal with extremely large volumes of data. Data logging and SCADA tools manage thousands of data from all points of the plant on a daily basis, meaning that the likelihood that errors or faults will occur has greatly increased. Processing and managing these heterogeneous, incomplete and frequently inconsistent data appropriately often exceeds the capacity of the WWTP staff. Consequently, valuable plant information for diagnosis and optimisation continues to be limited, in this case due to the excess of data rather than the previous shortage. Although several methodologies and tools have been developed in recent years to support decision making at WWTPs (Beltrán *et al.* 2012), at present it is clear that the optimum operation of WWTPs urgently requires advanced data management algorithms and tools to optimize the global operation of wastewater systems by adequately managing and using all the information available in the plant at every moment.

10.3 CONTROL OF AERATION SYSTEMS

The level of aeration of biological reactors in WWTPs may be one of the first manipulated variables that was automatically regulated (more than 40 years ago). This early development was motivated by the advances made in dissolved oxygen sensors (see Section 10.2) and because it plays a crucial role in the stability, performance and operational costs of the process, accounting up to 50%–60% of a plant's total energy demand (Brandt *et al.* 2011).

The basic level of aeration control consists of keeping a stable dissolved oxygen concentration in the presence of the disturbances associated with the load fluctuations, manipulating the air flow-rate supplied to the bioreactors (regulating the compressors' speed or the opening degree of the airflow valves) or, in the case where mechanical aerators are used, their speed or submergence. This is usually done by a simple proportional-integral-derivative (PID) feedback controller operated at constant set-point (reference value) or by an on-off control (see Figure 10.1). Generally speaking, it can be said that if the controllers have been properly designed and tuned, these control loops are a mature and proven technology that offers satisfactory results (Olsson, 2012). In fact, PI control or variations thereof are today the most common strategies in full-scale.

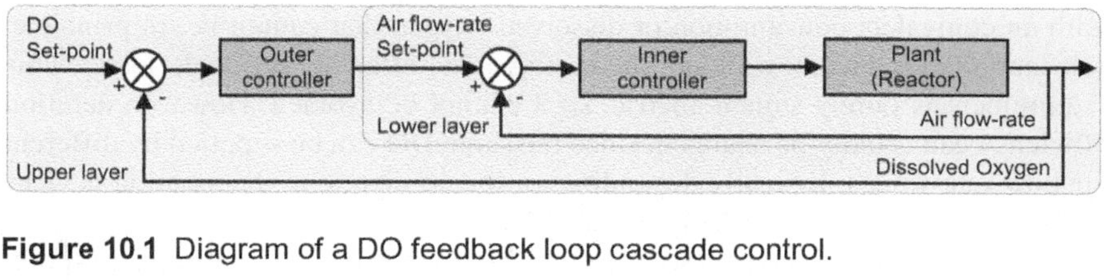

Figure 10.1 Diagram of a DO feedback loop cascade control.

The most frequent problems of DO control at a constant set-point are related to aeration systems of reduced flexibility, which are not able to adapt to the high fluctuations in oxygen demand that appear through time and at the different treatment zones. Another common limitation in keeping DO constant is motivated by the need to avoid the sedimentation of the suspended solids in the bioreactors, which requires a minimum air flow-rate value that, in some cases, could be excessive during the low load periods, and above all at the final zones of the aerated bioreactors. In addition to the extra energy cost, the peaks in DO associated with these situations can be detrimental for the anoxic or anaerobic bioreactors because of the highly oxygenated water recirculations.

The reduction of the air flow-rate that is achieved with the incorporation of a simple feedback control loop at a constant DO set-point is difficult to evaluate in a general way due to the high dependence of the influent load fluctuations. In any case, a reduction of between 20% and 40% in the air flow-rate is very common (Åmand *et al.* 2013), in addition to there being other possible advantages from the point of view of process stability and possible nutrient removal.

A second level of aeration control, which is a bit more sophisticated, consists of varying the DO concentration with the aim of optimising costs and nutrient removal. The DO concentration in the aerated zones of the biological reactors is the manipulated variable with quicker time response when it comes to regulating biological activity. On the one hand, the DO concentration in the aerated zones must be kept high enough to favour the growth of non-filamentous microorganisms, to guarantee the needed nitrifying speed and to maintain the adequate mixing characteristics in order to maintain solids in suspension and to assure that the oxygen is well distributed throughout the reactors. On the other hand, the DO concentration must be low enough to avoid wasting energy and an excessive agitation that could fragment the biological flocs, and to minimise as much as possible the oxygenation induced by the internal recirculations in the anoxic and anaerobic zones. It is clear that the nitrification is enhanced for high DO concentrations, but operating a plant with relatively low DO concentrations not only promotes denitrification in the anoxic zones, but it also encourages a certain simultaneous denitrification-nitrification in the aerated zones (Olsson & Newell, 1999).

When it comes to assessing the oxygen requirements and thus the reduction in energy costs that can be achieved by manipulating the aeration, it is important to consider that a specific load to be oxidised in aerobic conditions is associated with an equivalent consumption of dissolved oxygen that cannot be, in principle, substantially reduced without increasing the effluent residual load. This requirement is purely stoichiometric, so it cannot be avoided. However, aeration efficiency can usually be changed, since the same DO can be supplied by different air flow-rate values, basically depending on the efficiency of the aeration devices and the difference from the DO saturation value.

In this way, a simple control strategy for adapting the DO concentration set-point to the nitrification needs is the incorporation of a control feedback loop (Upper layer, Figure 10.2) that selects the DO reference needed by the lower control loop (Lower layer) on a continuous basis. This is done by using the discrepancy between the NH_X-N set-point selected by the plant operator and the NH_X-N that is measured experimentally in the effluent. In this cascade, the three control loops work at different time scales with the airflow control (see Figure 10.1) being the fastest. This combined control scheme guarantees good performance.

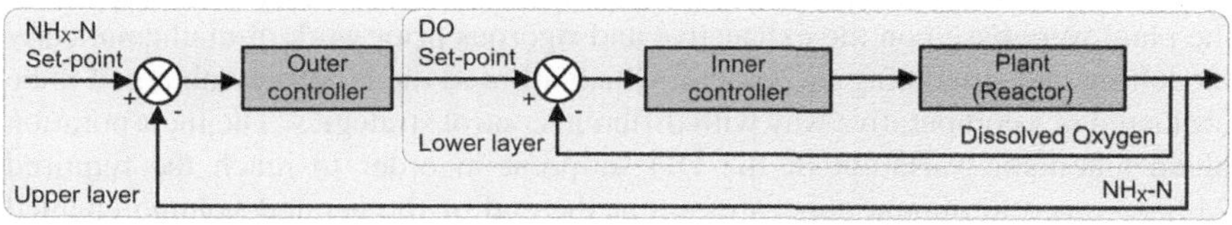

Figure 10.2 Diagram of an NH_X-N feedback loop cascade control.

The discrepancy is usually based on an average value of the experimentally measured NH_X-N (e.g., using a 24-h mobile averaged window, Suescun *et al.* 2001). Consequently, the typical variation in the effluent concentration generated by the daily load profile is filtered, and so the DO set-point selected by the controller moves smoothly according to medium- and long-term disturbances (e.g., dry or wet weather, unexpected load variations, changes in water temperature, weekly and seasonal load variations, etc.). The width of the mobile average window can also be used to 'distribute' these cyclical perturbations between the effluent ammonia and the DO set-point. For example, a narrower mobile average filter (8 h or 12 h) would reduce both the controller's time-response and the height of the peaks in the instantaneous effluent ammonia, but it does so at the expense of increasing the short-term fluctuations in the DO set-point selected by the controller. Therefore, the plant operator can select the most appropriate size of the mobile average window for each particular plant. However, narrowing the mobile-averaged windows for effluent ammonia will cause higher short-term fluctuations in the DO set-point, and therefore there will be a net increase in the total air consumption for a similar mass of ammonia that is removed. This negative effect is due to the nonlinear relationship between the DO concentration and the nitrification rate combined with the decrease in oxygen transfer efficiency at higher DO levels.

The advantages of applying this concept at full-scale has been widely studied with numerical simulation, leading to an average energy savings of around 10%–20% (Åmand *et al.* 2013). However, in spite of the excellent results predicted by simulation, the implantation of this control strategy at full-scale is still very limited.

One of the first examples of full-scale application of this control concept was developed in the Källby WWTP (100 kPE, Sweden, Ingildsen *et al.* 2002a). The plant incorporated a PI control loop with a limited range of DO operation, which allowed the plant to achieve an effluent NH_X-N concentration around the selected set-point. The installation of the controllers at full-scale decreased the supplied air flow-rate in the range of 14% to 28%, depending on the NH_X-N concentration that was used as reference. The most likely improvement of the effluent nitrates concentration was not studied.

A successful example of full-scale application was developed at the Galindo-Bilbao WWTP, which has an average design flow-rate of 4 m^3/s (1.5 MPE). The specific design and preliminary verification of the most appropriate controllers for the plant were based on the exhaustive and rigorous prior work of mathematically modelling and simulating the system, which allowed the efficiency obtained to be evaluated in a comparative way with different control strategies. The incorporation of an automatic variation of the DO set-point in order to reach the required 24 h-averaged ammonia concentration at the end of the aerated volume enabled the reduction of energy consumption in the range of 15% to 20% and effluent nitrates (and consequently in total nitrogen) in 2.0 g N/m^3 (Ayesa *et al.* 2006). The different phases of the project lasted 8 years, combining model simulations, pilot-plant experimentation and full-scale validation.

When there is a benefit to reacting quickly to a disturbance (e.g., to the NH_X-N influent load), such as in the case of 'never-to-exceed' effluent limits, feedforward control can be used. In order to do this, more sensors and a mathematical model of the controlled system are needed. Using feedforward without a feedback loop is not recommended since feedback contributes to a more robust performance in light of feedforward model uncertainty and it can compensate for non-modelled disturbances. Essentially, there are two ways to use a feedforward control in an effort to reduce effluent ammonia peaks: changing the aeration intensity (e.g., by varying the DO set-point, see Figure 10.3), or adding more aerated volume (e.g., by switching on aeration in a swing zone). However, the latter is preferred since the nitrification capacity of the system may be exceeded (Rieger *et al.* 2012).

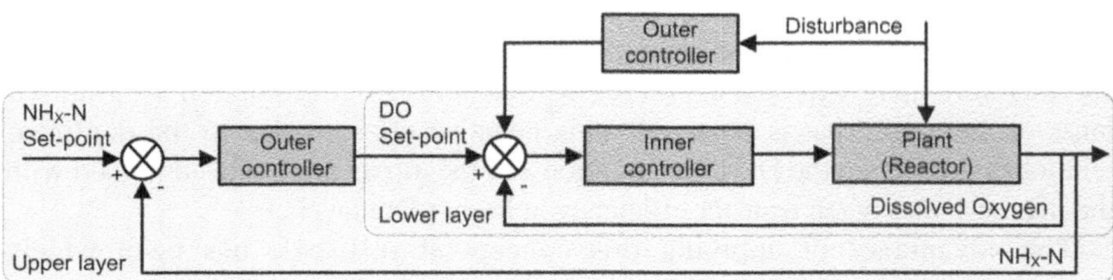

Figure 10.3 Diagram of an NH_X-N feedforward-feedback loop cascade control.

Finally, it is also worth pointing out the possibilities of controlling the aeration in sequential processes such as in sequential batch reactors (SBR) or in intermittent aerated reactors. In these cases, the controllers of the phases' duration are usually based on the interpretation of the trajectories of ORP (Oxidation Reduction Potential) and/or pH. The ORP profiles show inflexion points generated by the transition among the oxic, anoxic and anaerobic phases (e.g., the ORP 'elbows' that appear when the DO grows suddenly at the end of the nitrification process and when the NO_X-N run out at anoxic conditions and anaerobic conditions start).

On the other hand, pH profiles also show variations motivated by the biochemical transformations of nitrification and denitrification (e.g., 'the NH_X-N valley' that appears at the end of the nitrification and 'the NO_X-N peak' that occurs at the end of the denitrification). In addition to the trajectory variations, several controllers use the absolute value of the measurements of ORP or pH, but it is a risky practice because of the difficulty in choosing suitable values in each case. On the other hand, it is clear that optimising the duration of each cycle phase induces an improvement in the effluent quality, an increase in treatment capacity and therefore a reduction in the total operating costs.

10.4 CONTROL OF CHEMICAL ADDITION

The chemical precipitation of phosphate by using aluminium of ferric salts is very fast in comparison with the biological transformations in the tank reactors. Therefore, from the point of view of automatic control, perturbations can be easily overruled with feedback loop controllers. However, depending on the place selected for dosage and measuring, special attention should be paid to the hydraulic response time of the reactors.

Chemical precipitation can be carried out prior to the biological treatment (pre-precipitation), in the biological treatment (simultaneous precipitation) or after the biological treatment (post-precipitation) (Sedlak, 1991). The selection of the most appropriate dosage in time should be adapted to the influent P load, because an excessive dosage increases operational costs (due to the price of chemicals and the increase in sludge production) and it does not improve the removal of phosphate. Therefore, the convenience of introducing automatic control strategies is clear, although the most appropriate strategy should be selected depending on each specific case.

In the case of pre-precipitation, chemicals are added prior to the primary settler and the large time delay from the dosage point to the effluent discharge (several hours) makes it difficult to incorporate feedback control loops. Therefore, most controllers have been based on open loops that regulate the chemical dosage proportionally to influent flow or turbidity.

In the case of simultaneous precipitation, the dosage is directly introduced into the biological reactors (normally at the end of the anoxic volume) and the smaller delay times make it possible to implement feedback loops with reasonable results. Finally, post-precipitation incorporates flocculation and settling chambers after the biological treatment and facilitates very efficient regulation of the feedback control, measuring the phosphate concentration just at the end of the flocculation chamber.

Nowadays, the reliability of the on-line phosphate analysers makes the successful implementation of feedback loops with a return period of the investment smaller than one year possible. A comparative efficiency analysis for different post-precipitation strategies carried out in the WWTP at Källby (100,000 PE, Sweden) (Ingildsen & Wendelboe, 2003) revealed that the relative consumption of chemicals for constant dosage, flow-proportional dosage, load-proportional dosage and PI feedback control were 100/95/80/60, respectively. Finally, it is important to remark that, depending on the characteristics of the sludge treatment, the cost associated with the reduction of sludge production can be comparable to the chemical reduction cost (up to 75%).

10.5 CONTROL OF THE INTERNAL, EXTERNAL AND SLUDGE WASTAGE FLOW-RATES

10.5.1 Control of the nitrates internal flow-rate and the carbon external addition

The most extended configuration for the biological removal of nitrogen is the denitrification-nitrification process (DN), mainly used when a low effluent ammonia concentration (e.g., 2 g N/m^3) is required. The conventional operating strategies for the DN process are focused on the control of the effluent ammonia and/or nitrate concentration, with denitrification being a key stage. For the plants with nitrification and denitrification processes, the internal recirculation between the biological reactors provides the nitrates required as electron acceptors (oxidising agents) instead of oxygen to the anoxic zones for the removal of organic matter. Therefore, the degree of the internal recirculation has to be large enough so that the nitrates concentration in the anoxic zones does not limit the denitrification (e.g., 1.0 g N/m^3 is enough) while at the same time it should not be very high that would cause an excessive pumping cost or an inhibition of the denitrification process because of the dissolved oxygen contribution from the oxic zones.

Some control strategies have been studied for the appropriate regulation of the internal recirculation flow-rate (Yuan *et al.* 2002). In practice, the most implemented solution consists of the regulation of the internal flow-rate between the oxic and the anoxic reactors so that the nitrate concentration at the end of the anoxic zones is kept at a low value (e.g., 0.4 g N/m^3). A simple Proportional-Integral controller would be enough for implementing this strategy using only a nitrates analyser (Figure 10.4). Such a control strategy has been successfully implemented in the Galindo-Bilbao WWTP (Ayesa *et al.* 2006) and in the Mekolalde WWTP (Irizar *et al.* 2014). This strategy maximises the use of the influent COD for the denitrification process, and thus it also maximises the nitrate removal and the denitrification rate. However, it is of limited efficacy in maintaining the effluent nitrate levels because the maximum quantity of nitrates that can be removed is determined by the ratio of COD/N in the influent wastewater.

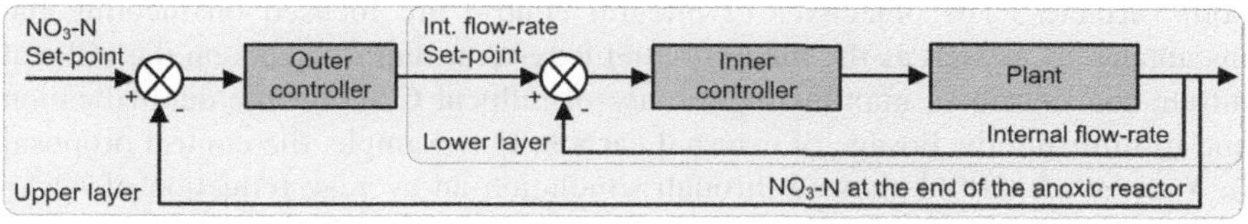

Figure 10.4 Diagram of an NO_3-N feedback loop cascade control.

The economic impact of the pumping costs associated with the internal recirculation is really low because the height elevation of the water is low even though the flow-rate could be high. Moreover, the implementation of this controller does not have a direct economic implication since its effect is more focused on the optimisation of the process and, in particular, on the optimisation of the use of the plant denitrification potential. However, this optimisation always has indirect economic consequences for the aeration costs since all the COD that is used under anoxic conditions will not consume oxygen and will allow savings in the carbon external addition when this is necessary.

When the limited factor of the denitrification process is the organic matter and when internal addition is not possible (sludge supernatants, acid fermentation, etc.), the most widespread practice today consists of applying an external carbon source (e.g., methanol, ethanol, acetic, etc.), which enhances the denitrification rate and the nitrate removal, especially in the case of WWTPs that treat wastewater with an unfavourable COD/N ratio.

The control strategies with constant dosage do not allow the process operation to be optimised since at high nitrogen loads or low temperature periods, an immediate increase of the denitrification rate would be desirable, while at low load periods the external carbon requirements would be lower. Therefore, there is a need for regulating the quantity of the external carbon to be dosed into the system in an automatic way.

When automatic control of nitrate recirculation is not available, the most common strategy consists of regulating the carbon dosage so that the nitrate concentration at the end of the anoxic zone is kept low. In this case, the internal recirculation flow-rate is set to a level in which the average effluent nitrate concentration is kept within the appropriate limits. Since nitrates are in the anoxic zone when the carbon source is added, this strategy guarantees that the carbon source is instantaneously used for the denitrification. However, although an average effluent nitrate concentration can be fixed beforehand, this strategy does not have direct control on the effluent nitrate concentration as it depends on the influent variations of COD and N. Moreover, this strategy does not guarantee the maximum use of the influent COD for denitrification. For example, when the nitrogen load is low, part of the anoxic zone is changed to aerobic, even when external carbon is not added, due to an inappropriate nitrate recirculation.

The disadvantages associated with individual control of the nitrate recirculation and the dosage of external carbon can be overcome by an integral control of both variables. The objectives of integral control are focused on meeting the instantaneous as well as the medium- and long-term requirements on the effluent nitrate concentration, maximising the use of influent COD for the denitrification and minimising the dosage of external carbon. For example, the control proposal in Yong *et al.* (2006) showed through simulation an average reduction of 42%–47% in the effluent nitrate concentration.

It is not possible to make a direct quantitative evaluation of the benefits that are obtained when an external carbon dosage control loop is implemented since they clearly depend on the plant characteristics. It is clear that the automatic control loops allow the dosage to be adjusted to the one that is strictly necessary for meeting the effluent requirements, avoiding excessive dosages that could involve a significant direct cost (because of the additive price) and an additional indirect cost associated with the subsequent aerobic oxidation of the carbon excess.

10.5.2 Control of the external flow-rate or sludge recirculation

The sludge recirculation from the secondary settlers is commonly kept under a constant flow-rate or proportional to the influent flow-rate. The latter, in principle, allows a more stable sludge blanket height to be maintained in the presence of the influent flow-rate variations, although the hydraulic disturbances in the clarifiers can increase in excess. The main restrictions on its operation are the hydraulic loads induced in the clarifiers, the sludge blanket height, the retention time in the settlers and the dilution effects of the recirculated sludge.

From a general point of view, the economic implications of sludge recirculation control are marginal, comparable to the rest of the plant controllers. The only worth mentioning effect on the operation costs comes from the different level of thickening that is achieved for large values of the sludge blanket height. In this case, the sludge is more thickened, so the pumping costs are reduced, and in the case of making the wastage from the thickened sludge, the wastage volume is also reduced. In this way, the costs of the sludge treatment are reduced. However, the economic implications of this effect are not usually significant.

10.5.3 Control of the sludge wastage flow-rate

The sludge wastage flow-rate is one of the most important manipulated variables for the solids content in the process. Its known actuation limits are the impossibility of selecting a specific type of biomass (since it works on the total solids of the system) and its high response time (sludge retention time (SRT) is a stationary parameter that cannot be changed instantaneously). For this reason, its use is generally limited to control in the medium and long term (in response to seasonal variations) with the aim of obtaining the desirable characteristics in the biomass. The sludge age represents the average retention time of the solids in the biological reactors and as a result it has a direct effect on the solids that are attained in the process and in their distribution. Each plant has its specific range of SRT since a longer SRT results in a higher concentration of suspended solids and thus increases the solids loading into the settler. A concentration that is too high could cause an elevated concentration of suspended solids in the effluent, especially when coupled with high hydraulic loads. On the other hand, a minimum SRT is needed to guarantee the necessary active biomass. For example, in the case of nitrogen removal plants, a minimum SRT value is essential in order to achieve nitrification, especially at low temperatures.

In a general way, higher SRTs mean lower sludge production, and therefore the costs associated with sludge treatment are lower. However, the potential benefits that can be achieved by methane production in the anaerobic digestion are also lower. On the other hand, the oxygen required by the microorganisms increases with sludge age. This increase is logically associated with a higher reduction in the contaminant load of COD and nitrogen and with an increase in the biomass endogenous respiration. Selecting the optimum SRT set-point for the different operational conditions of a WWTP should take into account the intrinsic inevitable limitations of the process (nitrification safety and maximum solids load to the settler) and operating costs criteria (sludge treatment, aeration costs, etc.).

There are different strategies for controlling the sludge wastage flow-rate. One option is to maintain a constant SRT via a feedforward loop, complemented with a loop that stops the wastage flow-rate when the average effluent ammonia exceeds a maximum fixed value (Olsson *et al.* 2005). Another option is to keep the solids mass under a fixed range, such as the one that was successfully implemented in the Galindo-Bilbao WWTP (Ayesa *et al.* 2006) and in the Mekolalde WWTP (Irizar *et al.* 2014).

10.6 CONTROL OF ANAEROBIC PROCESSES

At the time of writing this chapter, the status of automatic control in real-world anaerobic processes is in general limited only to those basic operations that are strictly required to guarantee autonomous operation under normal conditions. The characterisation 'basic' involves operations such as: (1) keeping the reactor temperature and the recirculation flow-rate close to pre-set values; (2) feeding the anaerobic reactor with a pre-set volumetric flow-rate; and (3), repeating a fixed sequence of phases when the process operates in a semi-batch mode. The latter combined with a user-friendly software for real-time process supervision (the so-called SCADA software) provides plant operators with an adequate interface for monitoring the process and adjusting the operating point in each situation. This mode of operation in which plants are automated with low-level controllers and operators who decide their reference values, has been usually called 'conventional operation' or 'semi-automatic operation'. Conventional operation has the advantage of minimising the use of on-line instrumentation and thus the sensitivity of the process to sensor failures. In contrast, the efficiency of the plant lies almost completely upon the operators who cannot be in contact with the process continuously.

Because of the over-dependence on plant operators, who should remain in charge of the decision-making, the capacity of the process to accommodate hourly disturbances (e.g., imposed by the continuous variation of the influent waste) is limited. The result is a reduction in the efficiency of the process, which under extreme conditions can even lead to a complete failure of the anaerobic reactor. In order to prevent the above, two different approaches can be followed. The most common solution is to absorb disturbances hydraulically by over dimensioning the volume of the plant. The second approach, as one can guess, calls for the installation of high-level controllers (also known as 'advanced control') that perform the actions that operators are not efficient to carry out. While the first option entails higher construction costs, the second one may require investment in on-line instrumentation. Nonetheless, use of advanced control is preferable since it offers several benefits, namely: (1) the plant volume can be decreased when advanced controllers are present, and (2) by adapting the actuators to the disturbance characteristics, the performance of the plant is optimised in the medium-/long-term.

10.6.1 Technological barriers

Although the benefits of an extensive application of advanced monitoring and control solutions on the anaerobic processes are clear, faster development of control products in this technology has been hampered for a number of reasons:

- Lack of on-line sensors. In the last decade, considerable research has focused on the consolidation of a sensor technology specific to anaerobic digesters. However, the so-called advanced instrumentation is still not sufficiently reliable for use in full-scale real applications (Spanjers & van Lier, 2006).
- The design and development of industrial control products involves real aspects such as the physical and operational constraints of actuators, disturbances in the form of sensor faults, noise, and so on, that need to be considered at the outset (Liu *et al.* 2004a). This means extrapolating control solutions that have been satisfactorily validated at laboratory or pilot scale is not straightforward.
- The intricate non-linear nature of anaerobic digestion phenomena and the intrinsic uncertainty of their mathematical formulations make the design of control solutions a non-trivial issue.

10.6.2 Applications of control in anaerobic digestion

In the last few years automatic control of anaerobic digestion (AD) has become a central research topic in the water field, as corroborated by the multiple works on this subject that have been recently reported (Batstone *et al.* 2004; Liu *et al.* 2004b; Olsson *et al.* 2005). Unfortunately, most of these works on the control of AD have not gone beyond experimental validations at either the lab or pilot scale.

In addition to this, given that the feed flow-rate is the only manipulated variable used for control, a surprising conclusion drawn from all these works is the lack of consensus about which output signals (sensors) must be linked to this actuator. As shown below, the list of suggested signals is relatively long: from pH, intermediate alkalinity, hydrogen gas, dissolved hydrogen, volatile fatty acids (VFAs), effluent COD, to methane gas, and so on.

Cord-Ruwisch *et al.* (1997), for instance, propose a very simple proportional controller that automatically adjusts the feeding flow-rate by trying to keep the dissolved hydrogen close to a pre-set reference. Antonelli *et al.* (2003) suggest a similar control scheme where the dissolved hydrogen signal is replaced by the methane gas flow-rate. In the same way, Rodríguez *et al.* (2006) argue that hydrogen gas, being an intermediate process variable, becomes especially appropriate for the early detection of abnormal situations (overloads, drops in reactor temperature, etc.). On the basis of this argument, they design a very simple control law to keep the hydrogen gas concentration around a pre-set reference. Moreover, the non-linearity of the process is taken into account through a real-time variable gain algorithm dependent on the instantaneous measurements of hydrogen gas concentration and methane gas flow-rate. A more sophisticated cascade control scheme is described in Álvarez-Ramírez *et al.* (2002), where the control signals for the inner and outer loops are the concentration of VFA in the reactor and effluent COD, respectively.

Fuzzy logic is another technique that has received considerable acceptance in control solutions for anaerobic reactors. Just to cite a few examples, Estaben *et al.* (1997) implemented this technique to control the pH and the biogas flow-rate. Similarly, García *et al.* (2007) applied a fuzzy-based real-time system in a pilot plant, which controls the anaerobic digester by monitoring the following process variables: the ratio between the intermediate and total alkalinities (IA/TA), methane gas production, the concentration of hydrogen in the gas headspace and its rate of change.

Adaptive control techniques have also been used in control solutions for anaerobic processes. An example is the adaptive controllers with linearisation designed by Bernard *et al.* (2001) to control the IA/TA ratio and TA in an anaerobic filter. Similarly, by using a simplified model to reproduce plant behaviour, a non-linear adaptive controller to control the effluent COD was developed by Mailleret *et al.* (2004). Méndez-Acosta *et al.* (2007) suggested a model-based non-linear controller to regulate the effluent TOC and the VFA in the reactor. Although the above approaches are of great interest in the academic research, their application in real-world plants currently faces a major barrier: it is extremely difficult to develop reduced models that guarantee good predictions for all operations taking place in a WWTP.

A common feature of all the above control solutions is that the ultimate goal is to keep one or more process variables close to a fixed reference value. Moreover, a different control approach for anaerobic systems also found in the literature focuses its objective on maximising a process variable, usually the methane gas production rate. These strategies, known as extremum-seeking, are especially appropriate when it comes to controlling non-linear systems (Krstic, 2000). Moreover, extremum-seeking controllers are classified in two main groups: (1) those that need a predictive model to optimise the objective variable (Marcos *et al.* 2004; Guay *et al.* 2005); and (2) those that apply external excitation in order to diagnose whether or not the 'objective' has reached its maximum value. It is actually its model-free condition that makes this second group attractive from a practical point of view.

Reported works on model-free extremum-seeking controllers differ from each other in terms of the pattern followed for external excitation. Thus, in Steyer *et al.* (1999) a periodic pulse is superimposed on the feed rate and its effect on the methane production observed in order to determine whether or not the organic loading rate can be increased. In contrast, in Simeonov *et al.* (2004, 2007) a sinusoidal excitation is applied in combination with an adaptation law for the feeding flow-rate that is proportional to the gradient of the biogas production. Liu *et al.* (2004c) proposed a cascade control scheme consisting of an inner loop controlling the pH of the reactor at a pre-set value which is determined by an outer loop. The outer loop compares the biogas production rate of the reactor with a reference value, set by a rule-based supervision module on a regular basis in order to automatically steer the process towards the maximum biogas production. The validation of this control solution in a lab-scale anaerobic reactor fed with synthetic wastewater reveals very good disturbance rejection properties as well as potential to keep the process stable even at organic loading rates above 25 kg COD/m^3 · d. Some years later, the same control architecture was adopted and extended (Figure 10.5) by Alferes and Irizar (2010) to prove that the combination of extremum-seeking controllers and optimum management of the equalisation tanks preceding the anaerobic reactor leads in the medium/long-term to a significant increase in the production of biogas. Comparative simulation results with respect to conventional operation reveal improvement of about 20% in both the effluent quality and the biogas yield. Nonetheless, the extent to which these estimates are valid during the operation of a full scale plant still needs to be verified.

The economic evaluation of the automatic control of anaerobic digesters has to be studied in function of the characteristics of each WWTP. For example, the possibility of including pre-treatment steps that hydrolyse the sludge or digest the sludge with other external substrates confirms the interest of having automatic tools that allow the digester to work at its maximum efficiency. It is clear that maximising methane production yields an important energy recovery (1.0 m_N^3 of methane is equivalent to 8570 kcal, i.e., it is equivalent to 1.15 l of gasoline, 1.3 kg of coil, 0.94 m_N^3 of natural gas or 9.7 kWh of electricity). However, it is also clear that the potential of generating methane in WWTPs is determined by the quantity and type of the available sludge. Therefore, a plant-wide control strategy operating the system as a whole with a global objective should be considered instead of trying to optimise the operation of the individual units separately.

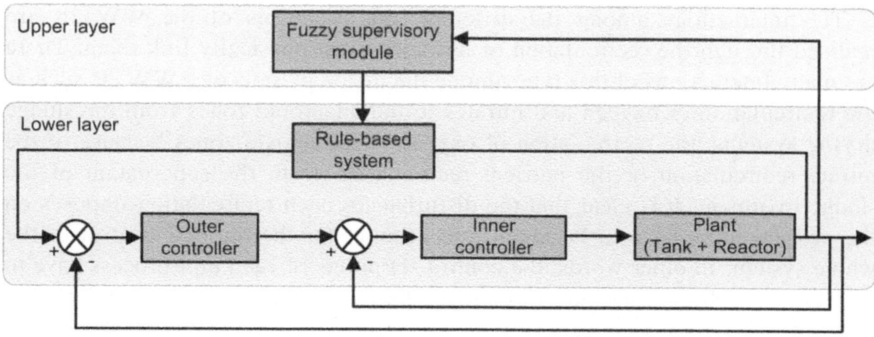

Figure 10.5. Diagram of a control strategy for the maximisation of methane production.

10.7 PLANT-WIDE CONTROL

In order to address the control of WWTPs in a reasonable way, it is necessary to break the problem down into smaller parts, taking advantage of the separation among lines and process units, as well as the different time responses of the physical, chemical and biological mechanisms of the process. This allows the control strategies to be organised in a hierarchy and enables the optimisation of each subprocess to be resolved separately.

However, it is clear that the optimisation of the operation of a WWTP when understood as a global system does not, in principle, have to be equivalent to the result of optimising each one of the elements and process units (e.g., primary settler, secondary biological reactors, anaerobic digesters, etc.). Therefore, the global optimisation of the system has to take the interactions between the different parts of the plant into account and use them to find the operational and control strategies that optimise a global criterion. In some cases, this could mean that some parts of the WWTP could be operated in a suboptimal way, while still contributing into the plant operation an acceptable way or near the global optimum.

Speaking of the integrated control of sanitary systems, the first point of improvement is taking advantage of the relation between the networks of drainage and sanitation (including the sewers, retention tanks and storm tanks) and the WWTPs. Taking the undesirable effects of the load variations in the WWTP into account, an appropriate management of the hydraulic retention capacity of a sewer network may induce a significant increase in the treatment capacity. The experience gained with the combined operation of sewers, storm tanks and WWTPs shows that integrating all these parts in a control strategy has great potential.

However, even in the WWTP itself, there is wide margin for designing operation strategies and loops of automatic control that would comprise an integrated and flexible control system. This is the so-called 'plant-wide control'.

The interactions among the different unit processes of the WWTPs are realised through the recirculation of flow-rates that physically link them. There are many interactions of this type among the different units of a WWTP, such as the recirculation of oxygen and nitrates to the anaerobic zones from the sludge drying systems, the recirculation of oxygen to the anoxic zones because of the nitrate recirculation or the nutrient recirculation from the supernatant of the sludge treatment. It is clear that the disturbances each recirculation imposes on the process units have to be taken into account for the optimal control of the whole system. In other words, the control set-points of each unit process have to be supervised by taking into account the most relevant interactions among the different control loops.

This integral operation also involves rethinking of the objectives, taking into account the whole system. A clear example could be the discussion of the global objective of sludge production (or sludge retention time), considering the requirements needed for appropriate nutrient elimination, the costs of the sludge treatment or energy recovery through methane or incineration. Another point of great interest when global plant operation is considered is the decision regarding the optimum use of organic matter in each unit process, considering, for example, the sedimentation in the primary settlers, the operating costs associated with aeration in the secondary treatment, the need for carbon for nutrient elimination and the possibilities of energy recovery in the anaerobic digestion. These examples clearly show the need for supervision of the local control loops by a global supervision strategy that optimally manages the flexibility of the whole system.

Finally, it is important to note that the integral control of a WWTP is the one that provides real possibilities for increasing the capacity of the global treatment and as a result removing or postponing new investments when the size of the WWTP is not sufficiently large. This can lead to important economic savings.

10.8 CONCLUSIONS

The full-scale implementation of automatic control strategies in WWTPs has been severely limited for years because of a limitation in the sturdiness and reliability of on-line monitoring devices and the operating flexibility of plants. However, in recent years new analysers that are more reliable and require less maintenance have been developed. In this way, the problem of the instrumentation is no longer the bottleneck to progressive implantation of advanced automation in WWTPs. On the other hand, the new treatment technologies, in addition to providing higher efficiency, also provide higher operating flexibility, which can and must be taken advantage by the automatic control strategies.

The concept of automatic control in WWTPs is closely linked to plant-wide optimisation of the process, with all the implications that this has for the improvement of treatment efficiency and the reduction of the operating costs. The results of this optimisation are starting to be evident now. On the one hand, the long list of simulation studies have demonstrated the great advantages that are derived from the implantation of automatic control strategies in WWTPs, which also enable the detection of the most important points of improvement and a comparative analysis of the efficiency of the different control alternatives. On the other hand, the results of applying controllers are no longer limited to simulation studies. Instead the reports that illustrate their successful implementation at full scale are increasing every day, corroborating the improvement in the stability and efficiency of the process. It can be expected that in a few years' time the majority of the controllers whose efficiency has been successfully evaluated by simulation will be experimentally validated.

The economic implications resulting from the implantation of automatic controllers in new WWTPs are increasingly more evident. Some control loops, such as the ones that optimise aeration (at constant or variable set points), the dosing of additives for physical and chemical precipitation and the dosing of methanol for denitrification are demonstrating their capability to not only improve process performance but also substantially reduce the operating costs and recover initial investments in a short time. The cost reduction associated with other automatic control loops is more difficult to quantify, since it significantly depends on the specific characteristics of each plant, such as flexibility, treatment type, size, effluent quality requirements, and so on. Therefore, it is highly recommended that a simulation study be undertaken beforehand, making it possible to quantify the expected benefits of each control strategy and prioritize the different available alternatives, depending on the needs and limitations of each WWTP. It is also important to note the interest in the automatic control loops for improving the stability and general governability of the process, facilitating the final objective of 'global optimisation' of the WWTP.

The lines for the future developments of ICA tools are many and varied. Some examples of the major challenges that have to be faced in the near future and are being actively worked on now include the optimum operation of aeration systems that minimises the energetic costs, the design of the most appropriate control strategies that optimises the biological elimination of phosphorus, the adaptation of the controllers to small-sized plants in which simplicity and toughness must take precedence, and the plant-wide control that integrates the operation of the water and sludge lines. In order to efficiently respond to these challenges it is necessary that the research groups who are experts in the control and operation of plants closely collaborate with the water engineering and control companies and the entities that manage WWTPs.

10.9 REFERENCES

Alferes J. and Irizar I. (2010). Combination of extremum-seeking algorithms with effective hydraulic handling of equalization tanks to control anaerobic digesters. *Water Science and Technology*, **61**(11), 2825–2834.

Álvarez-Ramírez J., Meraz M., Monroy O. and Velasco A. (2002). Feedback control design for an anaerobic digestion process. *Journal of Chemical Technology and Biotechnology*, **77**, 725–734.

Andrews J. F. (1993). Modeling and simulation of wastewater treatment process. *Water Science and Technology*, **28**(11–12), 141–150.

Antonelli R., Harmand J., Steyer J. P. and Astolfi A. (2003). Set-point regulation of an anaerobic digestion process with bounded output feedback. *IEEE Transactions on Control Systems Technology*, **11**(4), 495–504.

Åmand L., Olsson G. and Carlsson B. (2013). Aeration control – a review. *Water Science and Technology*, **67**(11), 2374–2398.

Ayesa E., de la Sota A., Grau P., Sagarna J. M., Salterain A. and Suescun J. (2006). Supervisory control strategies for the new WWTP of Galindo-Bilbao: the long run from the conceptual design to the full-scale experimental validation. *Water Science and Technology*, **53**(4–5), 193–201.

Batstone D. J., Gernaey K. V. and Steyer J. P. (2004). Instrumentation and control in anaerobic digestion In: *Proeedings of the 2nd IWA Leading-Edge Water and Wastewater Treatment Technologies* (*LET 2004*), Prague, Czech Republic, 1–4 June.

Beltrán S., Maiza M., de la Sota A., Villanueva J. M. and Ayesa E. (2012). Advanced data management for optimising the operation of a full-scale WWTP. *Water Science and Technology*, **46**(2), 314–320.

Bernard O., Polit M., Hadj-Sadok Z., Pengov M., Dochain D., Estaben M. and Labat P. (2001). Advanced monitoring and control of anaerobic wastewater treatment plants: software sensors and controllers for an anaerobic digester. *Water Science and Technology*, **43**(7), 175–182.

Bourgeois W., Burgess J. E. and Stuetz R. M. (2001). On-line monitoring of wastewater quality: a review. *Journal of Chemical Technology and Biotechnology*, **76**(4), 337–348.

Brandt M., Middleton R., Wheale G. and Schulting F. (2011). Energy efficiency in the water industry, a global research project. *Water Practice & Technology*, **6**(2).

Cord-Ruwisch R., Mercz T. I., Hoh C. Y. and Strong G. E. (1997). Dissolved hydrogen concentration as an on-line control parameter for the automated operation and optimization of anaerobic digesters. *Biotechnology and Bioengineering*, **56**(6), 626–634.

Estaben M., Polit M. and Steyer J. P. (1997). Fuzzy control for an anaerobic digester. *Control Engineering Practice*, **5**(98), 1303–1310.

García C., Molina F., Roca E. and Lema J. M. (2007). Fuzzy-based control of an anaerobic reactor treating wastewaters containing ethanol and carbohydrates. *Industrial and Engineering Chemistry Research*, **46**, 6707–6715.

Garrett M. T. (1998). Instrumentation, control and automation progress in the United States in the last 24 years. *Water Science and Technology*, **37**(12), 21–25.

Guay M., Dochain D. and Perrier M. (2005). Adaptive extremum-seeking control of nonisothermal continuous stirred tank reactors. *Chemical Engineering Science*, **60**(13), 3671–3680.

Haimi H., Mulas M., Corona F. and Vahala R. (2013). Data-derived soft-sensors for biological wastewater treatment plants: an overview. *Environmental Modelling & Software*, **47**, 88–107.

Ingildsen P., Jeppsson U. and Olsson G. (2002a). Dissolved oxygen controller based on on-line measurements of ammonium combining feed-forward and feedback. *Water Science Technology*, **45**(4–5), 453–460.

Ingildsen P., Lant P. and Olsson G. (2002b). Benchmarking plant operation and instrumentation, control and automation in the wastewater industry. *Water Supply*, **2**(4), 163–171.

Ingildsen P. and Wendelboe H. (2003). Improved nutrient removal using in situ continuous on-line sensors with short response time. *Water Science and Technology*, **48**(1), 95–102.

Irizar I., Beltrán S., Urchegui G., Izko G., Fernández O. and Maiza M. (2014). Lessons learnt from the application of advanced controllers in the Mekolalde WWTP: good simulation practices in control. *Water Science and Technology*, **69**(6), 1289–1297.

ISO 15839 (2003). Water quality – on-line sensors/analysing equipment for water – Specifications and performance tests. International Organization for Standardization.

Jeppsson U., Alex J., Pons M. N., Spanjers H. and Vanrolleghem P. A. (2002). Status and future trends of ICA in wastewater treatment – a European perspective. *Water Science and Technology*, **45**(4–5), 485–494.

Kaelin D., Rieger L., Eugster J., Rottermann K., Banninger C., Siegrist H. (2008). Potential of in-situ sensors with ion-selective electrodes for aeration control at wastewater treatment plants. *Water Science and Technology*, **58**(3), 629–637.

Krstic M. (2000). Performance improvement and limitations in extremum seeking control. *Systems and Control Letters*, **39**, 313–326.

Liu J., Olsson G. and Mattiasson B. (2004a). Monitoring and control of an anaerobic up-flow fixed bed reactor for high loading rate operation and rejection of disturbances. *Biotechnology and Bioengineering*, **87**(1), 43–53.

Liu J., Olsson G. and Mattiasson B. (2004b). Towards an economically competitive anaerobic degradation process – an ICA approach. In: *Proceedings of the 10th IWA Anaerobic Digestion Congress* (*AD10*), Montreal, Canada, 29 Aug.–2 Sept.

Liu J., Olsson G. and Mattiasson B. (2004c). Control of an anaerobic reactor towards maximum biogas production. *Water Science and Technology*, **50**(11), 189–198.

Lynggaard-Jensen A. (1999). Trends in monitoring of waste water systems. *Talanta*, **50**(4), 707–716.

Mailleret L., Bernard O. and Steyer J. P. (2004). Non-linear adaptive control for bioreactors with unknown kinetics. *Automatica*, **40**, 1379–1385.

Marcos N., Guay M. and Dochain D. (2004). Output feedback adaptive extremum seeking control of a continuous stirred tank bioreactor with Monods kinetics. *Journal of Process Control*, **14**(7), 807–818.

Méndez-Acosta H., Steyer J. P., Femat R. and González-Álvarez V. (2007). Robust nonlinear control of a pilot-scale anaerobic digester. *Dynamics and Control of Chemical and Biological Processes*, LNCIS **361**, 165–199.

Olsson G. (1987). *Dynamic Modelling, Estimation and Control of Activated Sludge Systems*. In: Systems Analysis in Water Quality Management, M. B. Beck (ed.), Pergamon Press, Oxford, UK.

Olsson G., Aspegren H. and Nielsen M. K. (1998). Operation and control of wastewater treatment – a scandinavian perspective over 20 years. *Water Science and Technology*, **37**(12), 1–13.

Olsson G. and Newell B. (1999). Wastewater Treatment Systems. Modelling, Diagnosis and Control, IWA Publishing, London, UK.

Olsson G., Ingildsen P., Jeppsson U., Kim C. W., Lynggaard-Jensen A., Nielsen M. K., Rosen C., Spanjers H., Vanrolleghem P. A. and Yuan Z. (2004). Instrumentation, Control and Automation – Hidden Technologies in Wastewater Treatment. Keynote paper. In: *Proceedings of the 2nd IWA Leading-Edge Water and Wastewater Treatment Technologies (LET 2004)*, Prague, Czech Republic, 1–4 June.

Olsson G., Nielsen M. K., Yuan Z., Lynggaard-Jensen A. and Steyer J.-P. (2005). Instrumentation, Control and Automation in Wastewater Systems. Scientific and Technical Report No. 15, IWA Publishing, London, UK.

Olsson G. (2012). ICA and me – A subjective review. *Water Research*, **46**(6), 1585–1624.

Olsson G., Carlsson B., Comas J., Copp J., Gernaey K. V., Ingildsen P., Jeppsson U., Kim C., Rieger L., Rodríguez-Roda I., Steyer J.-P., Takács I., Vanrolleghem P. A., Vargas A., Yuan Z. and Åmand L. (2013). Instrumentation, Control and Automation in wastewater–from London 1973 to Narbonne 2013. In: *Proceedings of the 11th IWA Conference on Instrumentation Control and Automation (ICA2013)*, 18–20 September, Narbonne, France.

Orellana G., Cano-Raya C., López-Gejo J. and Santos A. R. (2011). Online monitoring sensors. In: Treatise on Water Science, *Volume 3, Aquatic Chemistry and Biology*, Peter Wilderer (ed.), Elsevier B. V., pp. 221–261.

Rieger L., Alex J., Winkler S., Boehler M., Thomann M. and Siegrist H. (2003). Progress in sensor technology – progress in process control? Part I: sensor property investigation and classification. *Water Science and Technology*, **47**(2), 103–112.

Rieger L., Jones R., Dold P. and Bott C. (2012). Myths about ammonia feedforward aeration control. In: *Proceedings of the 85th Water Environment Federation's Annual Technical Exhibition and Conference (WEFTEC2012)*, New Orleans, USA, September 29 – October 3.

Rodríguez J., Ruiz G., Molina F., Roca E., Lema J. M. (2006). A hydrogen-based variable-gain controller for anaerobic digestion processes. *Water Science and Technology*, **54**(2), 57–62.

Sedlak R. I. (1991). Phosphorus and Nitrogen Removal from Municipal Wastewater: Principles and Practice. 2nd edn, CRC Press, New York, USA.

Simeonov I., Noykova N. and Stoyanov S. (2004). Modelling and extremum seeking control of the anaerobic digestion. In: *IFAC workshop DECOM-TT*, Bansko, Bulgaria, October 3–5.

Simeonov I., Noykova N. and Gyllenberg M. (2007). Identification and extremum seeking control of the anaerobic digestion of organic wastes. *Cybernetics and Information Technologies*, **7**(2), 73–84.

Spanjers H., Vanrolleghem P. A., Olsson G. and Dold P. (1998). Respirometry in Control of Activated Sludge Processes. IAWQ Scientific and Technical Reports, London, UK.

Spanjers H. and van Lier J. B. (2006). Instrumentation in anaerobic treatment – Research and practice. *Water Science and Technology*, **53**(4–5), 63–76.

Standard methods for the examination of water and wastewater (2012). 22nd ed, American Public Health Association/American Water Works Association/Water Environment Federation, Washington DC, USA.

Steyer J. P., Buffiere P., Rolland D. and Molleta R. (1999). Advanced control of anaerobic digestion processes through disturbances monitoring. *Water Research*, **33**(9), 2059–2068.

Suescun J., Ostolaza X., García-Sanz M. and Ayesa E. (2001). Real-time control strategies for predenitrification–nitrification activated sludge plants: biodegradation control. *Water Science Technology*, **43**(1), 209–216.

Sweeney M. I. W. and Kabouris J. C. (2013). Modeling, Instrumentation, Automation, and Optimization of Wastewater Treatment Facilities. *Water Environment Research*, 1322–1338(17).

Trillo I., Jenkins T. E., Redmon D., Hilgart T. and Trillo J. (2004). Implementation of Feed Forward Aeration Control Using On-Line Off Gas Analysis: The Grafton WWTP Experience. In: *Proceedings of the Water Environment Federation* (*WEFTEC 2004*), 91–95, 27–45(19).

Vanrolleghem P. A. and Lee D. S. (2003). On-line monitoring equipment for wastewater treatment processes: state of the art. *Water Science and Technology*, **47**(2), 1–34.

Vanrolleghem P. A., Clercq B. D., Clercq J. D., Devisscher M., Kinnear D. J. and Nopens I. (2006). New measurement techniques for secondary settlers: a review. *Water Science and Technology*, **53**(4–5), 419–429.

Chapter 11

Microbial Fuel Cells for wastewater treatment

V. B. Oliveira[1], L. R. C. Marcon[2], J. Vilas Boas[1], L. A. Daniel[2] and A. M. F. R. Pinto[1]

[1]University of Oporto, Portugal – Faculty of Engineering – Department of Chemical Engineering – Centro de Estudos de Fenómenos de Transporte

[2]University of São Paulo, Brazil – São Carlos School of Engineering – Department of Hydraulics and Sanitation

11.1 INTRODUCTION

There are different problems of using fossil fuels, which meet about 80% of the world energy demand. One is that they are limited in amount and sooner will be depleted and the other is that fossil fuels are causing serious environmental problems (air pollution, ozone layer depletion and global warming). Therefore, together with strong improvements in energy conservation and efficiency, new technologies are needed to gradually replace fossil sources by renewable ones. Based on that, in the last years, the interest in fuel cells has increased dramatically, due to their high efficiencies, low or zero emissions of pollutants, simplicity and absence of moving parts. Particularly, the interest on microbial fuel cells (MFCs) has been growing, because they are able to simultaneously treat and produce electricity directly from the wastes that our society produces and can degrade toxic compounds and pollutants (phenol, sulfates and chromium). The MFCs operate at ambient temperatures, are fueled by organic matter which is neither toxic as methanol nor explosive as hydrogen and can contribute to optimize the overall efficiency of wastewater treatment facilities.

The problem with MFCs is that they are technically still very far from attaining acceptable levels of power output, since the performance of this type of fuel cells is affected by limitations based on irreversible reactions and processes occurring both on the anode and cathode side. Electricity generation in a MFC is accomplished by microbial catabolism, electron transfer from microbes to the anode, reduction of electron acceptors at the cathode and proton transfer from the anode to the cathode and all these processes affect the MFC performance. There has been a growing amount of work on MFCs in the last years, both on the microbiological issues and on the engineering ones (electrode materials, MFC configuration, membrane technology, operational conditions), which managed to increase power outputs by an order of magnitude.

This chapter discusses the key work done in order to improve MFC performance based on the limitations described regarding both fundamental and technological aspects of microbial fuel cells. The main goal is to provide a review on the challenges and developments in MFCs and on the recent work concerning the optimization of the operational conditions (such as pH, temperature, organic load, feed rate and shear stress), on empirical and fundamental modelling and scale-up studies. Towards the introduction of MFCs in the market, a cost benefit analysis where this technology is compared with the traditional ones is, also, presented.

11.2 OPERATING PRINCIPLE OF A MFC

MFCs are similar to any other fuel cell having two electrodes separated by a proton exchange membrane (PEM). However, they use organic substrates as fuels to produce electricity instead of chemical compounds.

A MFC is an electrochemical cell that generates electricity based on the oxidation of the organic compounds and reduction of oxygen. Figure 11.1 shows schematically a typical MFC, comprising an anode, a cathode and a membrane (PEM). The membrane is permeable to protons which ate the ionic charge carrier.

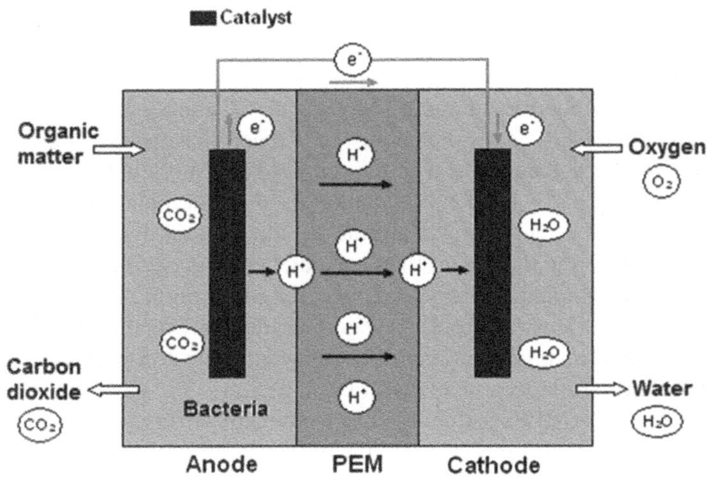

Figure 11.1 Operating principle of a Microbial Fuel Cell.

As can be seen in Figure. 11.1, the organic matter, presented in a wastewater, is fed to the anode compartment where the bacteria grow, oxidizing organic matter and releasing electrons and protons. Protons flow to the cathode through the membrane, while electrons reach the anode and move to the cathode through an external electrical wire. Simultaneously, oxygen is fed to the cathode compartment where it combines with electrons and protons to produce water.

11.3 FUNDAMENTALS AND CHALLENGES

In the last years, MFCs have been developed to generate electricity directly from complex organic wastewaters such as food, domestic, chemical, swine, marine and electroplating wastewaters (Holmes *et al.* 2004; You *et al.* 2006; Zhongjian *et al.* 2008; Yong *et al.* 2011). Although, the proof of concept for this technology was established decades ago, the first observation of electrical current generated by bacteria is generally attributed to Potter in 1991 (Potter, 1911), the low power output of MFCs remains problematic.

It is commonly accepted by the MFC scientific community that to achieve the desirable levels of energy density, a MFC system must overcome the following key challenges:

- Electrodes: there is a need for new surface materials that favour biofilm adhesion on the anode electrode surface and promote faster electron transfer to the electrode;
- Anodes with higher electron-acceptor surface properties and cathodes with higher electrochemical potentials would be welcome;
- Semi-permeable membranes or other proton exchange media: It is important to increase the proton transfer from the anode to the cathode compartment, while minimizing the risks of unwanted electron acceptor transfer from the cathode to the anode compartment;
- Microbiology issues: research is focused on the isolation of bacteria that grow on low cost substrates and are able to produce and transfer electrons at a faster rate to the anode electrode. To this end, the enrichment with suitable bacteria which form biofilms is the next step to gain enhanced power outputs. Moreover, the parameters (such as hydrodynamic stress) affecting biofilm formation, structure and composition are also important.
- Applicability in sewage treatment: is it important to verify if it is economically acceptable to develop MFCs that are able to treat wastewaters with high efficiency in order to compensate their lower electricity production and if MFCs can compete with the traditional wastewater treatments;
- Operating conditions: the exact effect of operating conditions, such as pH, temperature, aeration rate, feed rates on the MFC performance is still unknown.

The maximum performance of a MFC depends on the electrochemical reactions occurring between the organic matter and the final acceptor, oxygen. The real cell potential is always lower than its maximum value due to three irreversible loss types: the activation, the Ohmic and the concentration losses (Figure 11.2).

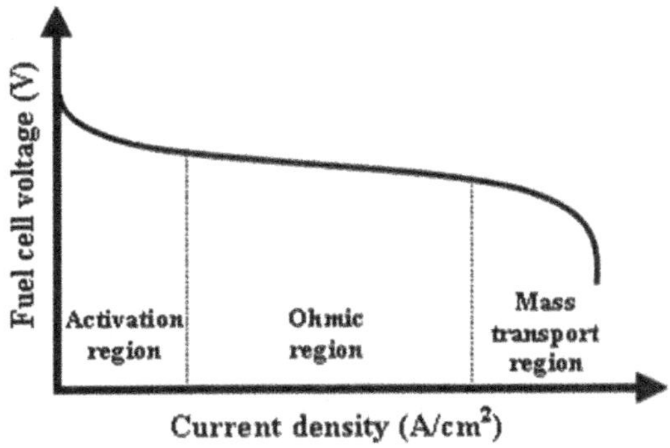

Figure 11.2 Schematic representation of a typical MFC polarization curve.

The activation loss is dominant at low current densities and is due to the activation energy that must be overcome by the reacting species. Phenomena involving adsorption or desorption of reactant species, transfer of electrons and the nature of the electrode surface contribute to the activation polarization. In MFCs where microbes do not readily release electrons directly on the anode electrode surface, activation polarization gets lower when mediators are added to the anode compartment. In the middle of the operating range, the predominant loss is the Ohmic loss and is due to ionic and electronic conduction. This loss can be reduced by shortening the distance between the two electrodes and by increasing the ionic conductivity of the electrolytes. At very high current densities, the major loss is the concentration loss and is due to the inability to maintain the initial substrate concentration in the bulk fluid, mainly due to mass transport limitations. As can be seen in Figure 11.2, the polarization curve of a MFC indicates the various loss types and their extent over the current density range, pointing to possible measures to minimize them and approach the maximum potential. These measures may include selection of new microbes, mediators, substrates, modifications in the MFC design (short spacing between electrodes) and configurations (improvement in electrode structure, new catalyst and electrolytes). Besides the three major losses described that lead to a decrease in power output, on the anode compartment of a MFC methanogens compete with electrochemically active microorganisms to convert organic material in methane, reducing the electricity generation process. Therefore it is extremely important to study the effect of the different operating conditions on methane production in order to avoid that and increase the power production.

In the earliest MFC concept, electrical energy was produced from living cultures of *Escherichia coli* and *Saccharomyces spp.* by using platinum electrodes, although, the power output was very low. This concept only generated much interest when it was discovered that current density and power output could be greatly enhanced by the addition of electron mediators (Ieropoulos *et al.* 2005; Du *et al.* 2007). Good mediators should be able to cross the cell membrane easily, possess a high electrode reaction rate, should be non-biodegradable and non-toxic to microbes and have low cost. Typical synthetic exogenous mediators include dyes and metallorganics (neutral red (NR), methylene blue (MB), Fe(III)EDTA), but their high toxicity and cost, instability and low efficiency limit their application in MFCs (Ieropoulos *et al.* 2005; Du *et al.* 2007). These facts made scientists look more closely at the microbiological features in the anode and to use bacteria that could transfer electrons directly to the electrode and increase the Coulombic efficiency, the so-called electricigens (*Geobacter* and *Rhodoferax* spp.) (Lovley & Nevin, 2008). The use of these species is of interest because closely related microorganisms have been found on the anodes of fuel cells harvesting electricity from aquatic sediments and complex wastes (Holmes *et al.* 2004; Jung & Regan, 2007). Furthermore, these species have the ability to completely oxidize organic substrates to carbon dioxide with an anode serving as the electron acceptor. Also these types of bacteria develop biofilms on the MFC electrodes, allowing considerable conversion capacity, and thus a potentially efficient microbial system to enhance the electricity (Rittmann, 2008; Lovley, 2008). Despite the efforts made in order to select the best bacterial consortia to achieve power densities needed for MFC applications, the performances of this type of fuel cells still remain lower than the ideal one. For a MFC it is also possible to enhance performance by optimization of the operating conditions, such as electrode materials, fuel type, proton transfer material, substrate concentration and feed rate, pH, temperature and hydrodynamic stress.

11.4 SCALE UP

Most MFCs studies were conducted at lab scales, however in order to make MFCs suitable for real applications, such as wastewater treatment plants, it is critical to achieve high power densities at a large scale. Also, for scaling up a MFC it is mandatory to develop low-cost, simple construction and easy to maintenance systems that can generate high power outputs. Many efforts have been made in order to achieve these goals (Tender *et al.* 2008; Ieropoulos *et al.* 2008; Dewan *et al.* 2008; Clauwaert *et al.* 2008; Dekker *et al.* 2009; Fan *et al.* 2010; Logan, 2010; Liu *et al.* 2008; Zhang *et al.* 2013; Zhu *et al.* 2013; Cheng & Logan, 2011; Cheng *et al.* 2014; Zhuang & Zhou, 2009; Liu *et al.* 2008; Zhuang *et al.* 2009, 2010b, 2012a, 2012b; Ieropoulos *et al.* 2010a, Kim *et al.* 2012; Gurung *et al.* 2012; Zuo *et al.* 2007, 2008; Lefebvre *et al.* 2009; Heijnea *et al.* 2011), but until now, no consensus was found. The scale-up process can be accomplished by connecting together single small units or by increasing the volume of a single unit (Tender *et al.* 2008; Ieropoulos *et al.* 2008; Dewan *et al.* 2008; Clauwaert *et al.* 2008;

Dekker *et al.* 2009; Fan *et al.* 2010; Logan, 2010; Liu *et al.* 2008; Zhang *et al.* 2013; Zhu *et al.* 2013; Cheng & Logan, 2011; Cheng *et al.* 2014; Zhuang & Zhou, 2009; Liu *et al.* 2008; Zhuang *et al.* 2009, 2010b, 2012a, 2012b; Ieropoulos *et al.* 2010a, Kim *et al.* 2012; Gurung *et al.* 2012; Zuo *et al.* 2007, 2008; Lefebvre *et al.* 2009; Heijnea *et al.* 2011).

The first demonstration of using a microbial fuel cell as an alternative and viable power supply for a low power consuming application was reported by Tender *et al.* (2008). The specific application reported was a meteorological buoy that measures different parameters such as air temperature, pressure and relative humidity, and water temperature. The MFC employed in this demonstration was the benthic microbial fuel cell (BMFC), which operates at the bottom of marine environments. It is maintenance free, does not deplete and is sufficiently powerful to operate a wide range of low power marine deployed scientific instruments normally powered by batteries.

Despite this first attempt, bringing the MFC technology out of laboratory appears as a challenge in the MFC developments, due to the high power densities and low cost materials required. Also, the reactor configuration, the operation at large scale, the electrode performance and the longevity are key factors in MFC scale-up. In the last years, some efforts have been made in order to scale-up MFCs (Tender *et al.* 2008; Ieropoulos *et al.* 2008; Dewan *et al.* 2008; Clauwaert *et al.* 2008; Dekker *et al.* 2009; Fan *et al.* 2010; Logan, 2010; Liu *et al.* 2008; Zhang *et al.* 2013; Zhu *et al.* 2013; Cheng & Logan, 2011; Cheng *et al.* 2014; Zhuang & Zhou, 2009; Liu *et al.* 2008; Zhuang *et al.* 2009, 2010b, 2012a, 2012b; Ieropoulos *et al.* 2010a, Kim *et al.* 2012; Gurung *et al.* 2012; Zuo *et al.* 2007, 2008; Lefebvre *et al.* 2009; Heijnea *et al.* 2011), but there are still many challenges that must be overcome before that.

One of the challenges of scaling up MFCs is to maintain the power outputs at levels needed for real applications, since previous studies concerning this issue revealed that the power density decrease in the scale-up process (Ieropoulos *et al.* 2008; Dewan *et al.* 2008; Clauwaert *et al.* 2008; Dekker *et al.* 2009; Fan *et al.* 2010). As mentioned before, one possible way of scale-up MFCs is to increase its volume. However, large MFCs units can alter electrode spacing which will affect the power density through changes in the internal resistance and non-uniform current distribution (Logan, 2010; Liu *et al.* 2008; Zhang *et al.* 2013; Zhu *et al.* 2013; Cheng & Logan, 2011; Cheng *et al.* 2014). Questions such as whether the energy generated by MFCs increases linearly or not with their size and how power is related to the surface area of electrodes and electrode spacing should be therefore answered.

Dewan *et al.* (2008) studied the relation between the limiting electrode surface area (anode compartment) and the cell power output and found that the maximum power density is not directly proportional to the surface area of the anode, but is proportional to the logarithm of the surface area. Cheng and Logan (2011) studied the effect of the anode and cathode electrode surface area on the MFC performance and found that doubling the anode and cathode size can increase the power output to 12% and 62% respectively. Based on these findings, the three-dimensional electrodes having the largest surface area for the same reactor volume (compared to the other alternatives, such as carbon paper or carbon cloth), are expected to be the best option for full-scale MFC (Cheng & Logan, 2011).

The electrode spacing is another important issue on MFC scale-up, since it has significant influence on the cell internal resistance. Liu *et al.* (2008) studied the effect of the electrode spacing on MFC performance. The electrode spacing was found to be a key factor affecting the specific area, internal resistance and power density. Also, in MFCs scale-up the large electrode array will affect the biofilm growth on the electrode surface and will lead to a non-uniform internal distribution in the cell. This will cause low energy generation, substrate utilization and electrode utilization and less biomass production (Zhang *et al.* 2013; Zhu *et al.* 2013). Therefore, to maintain the power density during scale-up, the larger reactor architecture must maintain or even reduce the electrode spacing by applying a parallel electrode orientation (Liu *et al.* 2008; Zhang *et al.* 2013).

In the MFC scale-up process, as the reactor dimensions become larger, the distance between the points where the electrons are generated and the leading-out terminal where current flows out increases and consequently the ohmic losses increase. Cheng *et al.* (2014) studied the impact of the anode ohmic resistance on large-scale MFCs by changing the leading-out configurations and found that a major part of the power loss from small to large scale was due to a bad leading-out terminal. They found that an effective way to reduce the power loss is changing the connecting configuration and the anode material since it reduces the anode resistance, which is one of the limitations for high performances in large-scale MFCs.

To avoid some of the problems of increasing the volume of a single MFC unit, some effort has been puton the other commonly accepted configuration for scale-up; the stack configuration. This consists of connecting single small units together in series or parallel (Ieropoulos *et al.* 2008; Dekker *et al.* 2009; Zhuang & Zhou, 2009; Liu *et al.* 2008; Zhuang *et al.* 2012a, 2012b; Ieropoulos *et al.* 2010a, Kim *et al.* 2012; Gurung *et al.* 2012). Ieropoulos *et al.* (2008) compared the performance of three different sizes of microbial fuel cell (MFC) (large, medium and small), when operated under continuous flow conditions using acetate as the fuel substrate, by means of polarization curves. The authors, also, investigated the connection of multiple small-scale MFCs, in series, parallel and series–parallel configurations. Among the three combinations tested, the series–parallel proved to be the more efficient one, stepping up both the voltage and current of the system. The results from this study suggested that MFC scale-up may be better achieved by connecting multiple small-sized units together rather than increasing the size of an individual unit. When the single units are connected in stacks, voltage, current or both can be increased depending on the stack size and configuration. This is very useful since the power output needed can be met by adjusting the size of the stack through the removal or addition of single units. It should be mentioned, that when cells are connected in series the voltage of the MFC stack is the sum of the individual cell unit voltages. When the stack presents a parallel configuration the stack current is the sum of the individual currents (Ieropoulos *et al.* 2008, 2010a). However, connecting multiple units together in series and/or parallel present some problems, such as voltage reversal, voltage losses and unpredictable operation (Zhuang & Zhou, 2009; Liu *et al.* 2008). Moreover, the connection of multiple MFCs may be complicated if the units are running under continuous flow conditions, which involves electrical and hydraulic connections. Connecting multiple MFCs in continuous flow requires the units to be joined to an inflow and outflow stream (Ieropoulos *et al.* 2008). However, a study regarding this issue shows that there is a voltage loss when the cells are hydraulically and electrically connected (Zhuang & Zhou, 2009). To overcome that, different studies regarding the stack configuration have been performed (Liu *et al.* 2008; Zhuang *et al.* 2012a, 2012b). Another problem of the stack configuration is the fact that the materials used for a single operating cell may be not the best choice for a stack ofcells. In their work, Ieropoulos *et al.* (2010a) found that the PEM that allows the highest energy output level in lab-scale MFCs was not the best option for stacks.

One major challenge to MFCs became suitable for real applications, is developing low-cost, simple constructions and easy to maintenance systems that can generate high power outputs. The noble metals used as catalysts (like Platinum) are very expensive and therefore unsuitable for large-scale applications. Using different cathode catalyst materials or other electron acceptors instead of oxygen to increase current densities is desirable (Zuo *et al.* 2007, 2008; Lefebvre *et al.* 2009; Heijnea *et al.* 2011; Zhuang *et al.* 2009, 2010b).

11.5 OPERATIONAL CONDITIONS

Operating conditions like pH, organic load, feed rate, shear stress and temperature are key parameters for MFC optimization. The comprehension of their interdependence is also important.

11.5.1 Effect of pH

The pH is crucial to the MFCs power output. Anodic pH microenvironment influences substrate metabolic activity and in turn affects the electron and proton generation mechanism. Generally, bacteria require a pH close to neutral for their optimal growth and respond to the changes of internal and external pH by adjusting their activity. Depending on the bacteria and growth conditions, variations in pH can cause modifications in several primary physiological parameters, such as, ion concentration, membrane potential, proton-motive force and biofilm formation (Zhang *et al.* 2011; Yaun *et al.* 2011). Most MFCs are operated at neutral pH in order to optimize bacterial growth conditions (Biffinger *et al.* 2008; Jadhav & Ghangrekar, 2009; Behera and Ghangrekar, 2009; He *et al.* 2008; Puig *et al.*

2010, Li *et al.* 2013b; Guerrini *et al.* 2013; Vologni *et al.* 2013). However, the low concentration of protons at this pH makes the internal resistance of the cell relatively high compared to chemical fuel cells that use acidic electrolytes.

The anode reactions produce protons and the cathode reaction consumes protons. Accumulation of protons due to slow and incomplete proton diffusion and migration through the membrane will cause a pH decrease in the anode. This will lead to a decrease in the bacterial activity and the electron transfer in the anode compartment. On the other hand, the continuous proton consumption by the oxygen reduction reaction results in a pH increase in the cathode compartment which, according to the Nernst equation, results in a decrease in current generation (the oxygen reduction reaction rate decreases with an increase of pH) (Biffinger *et al.* 2008; Jadhav & Ghangrekar, 2009; Behera & Ghangrekar, 2009; Erable *et al.* 2009; Zhuang *et al.* 2010a). The difference between the pH values in the anode and cathode compartments causes a pH gradient, and consequently, a decrease in the voltage efficiency and power generation especially at high current densities. Therefore, neutral pH at the anode side and lower pH values at the cathode side are desired. This can be achieved in a traditional dual-chamber MFC since two different pH conditions can be maintained to optimize, respectively, the anodic and cathodic reactions (Zhang *et al.* 2011; Yaun *et al.* 2011; Erable *et al.* 2009; Zhuang *et al.* 2010a; Nimje *et al.* 2011). However, this is impossible in the case of air-cathode MFCs, because only one electrolyte is present and the pH of this electrolyte affects the reactions in both compartments (He *et al.* 2008; Raghavulu *et al.* 2009). Since the air-cathode MFC configuration is more advantageous due to higher power outputs and simplified reactor configuration, the research is focused on maintaining a favorable pH value for both anode and cathode. According to He *et al.* (2008), the air-cathode MFC can tolerate an electrolyte pH as high as 10 with the optimal values ranging between pH 8 and 10. The anodic bacterial activity is optimal at a neutral pH, while the cathodic reaction was improved at a higher pH. Moreover, it was shown that the polarization resistance of the cathode was the dominant factor limiting power output. Contrarily, in a continuous flow air-cathode MFC, the higher power output was observed at acidophilic conditions (pH 6.3) (Martin *et al.* 2010). Raghavulu *et al.* (2009), also, observed higher current densities at acidophilic conditions (pH 6) when compared to neutral (pH 7) and alkaline (pH 8). The results showed that a better substrate degradation was achieved at neutral conditions. The second best was for alkaline conditions and last one for acidophilic conditions. Due to the poor cathodic oxygen reduction and the negative buildup of a pH gradient between anode and cathode, Cheng *et al.* (2010) proposed a solution to overcome these limitations. By inverting the polarity of the MFC successively in the same half-cell they could neutralize the pH effects. They found that a mixed culture forming an acidophilic biofilm can also catalyze the cathodic reaction of oxygen in a single bioelectrochemical system. However, further research is needed to explore the application of such bidirectional microbial catalytic properties for sustainable MFC processes.

The use of buffers is another way to maintain suitable values of pH in both compartments (Li *et al.* 2013b; Vologni *et al.* 2013; Fan *et al.* 2007a, 2007b; Min *et al.* 2008; Qiang *et al.* 2011; Nam *et al.* 2010; Torres *et al.* 2008; Fornero *et al.* 2010a; Ahn & Logan, 2013). The buffers affect MFC performance due to their chemical composition and interaction with electrodes, bacteria, and membrane. Moreover, an ideal buffer should be able to maintain constant pH without interfering with chemical reactions or microbial physiology, facilitate proton transfer to the electrode and increase the solution conductivity. Different buffers, such as phosphate, bicarbonate, zwitterionic and borax buffers, have been used for this purpose (Vologni *et al.* 2013; Fan *et al.* 2007a, 2007b; Min *et al.* 2008; Qiang *et al.* 2011; Nam *et al.* 2010; Torres *et al.* 2008; Fornero *et al.* 2010a).

Phosphate buffers are the most commonly used in MFC and it was found that increasing the buffer concentration the MFC power output is improved (Vologni *et al.* 2013; Fan *et al.* 2007a; Min *et al.* 2008). However the cost of a high concentration phosphate buffer as well as the depletion of phosphate in the environment make their use problematic and efforts are being made to recover phosphate from the effluent before discharge. Bicarbonate buffer sare considered to be low cost and effective alternatives, although they can enhance the growth of methanogens (Fan *et al.* 2007b). The zwitterionic buffers have a pKa slightly higher than the one needed for the anode compartment, which is advantageous for maintaining the pH value as the substrate is consumed. They are chemically stable and do not interfere with biochemical reactions, but they can be toxic to bacteria (Nam *et al.* 2010).

Despite the different types of buffers mentioned and their advantages, they all present some problems and their use is not practical for full scale applications since they entail an extra cost. Moreover, the type of buffer used has a hig himpact on power production, since the various buffers produce different solution conductivities, resulting in different ohmic resistances and maximum power outputs (Nam *et al.* 2010). In order to avoid some of these problems and to ensure a continuous feed of a buffer at a low cost, the addition of acid buffers to the cathode compartment in the form of carbon dioxide (Torres *et al.* 2008; Fornero *et al.* 2010a) and the use of saline solutions in both compartments have been studied (Ahn & Logan, 2013; Li *et al.* 2013b). Carbon dioxide is an acid that combines with the hydroxide ions in the cathode compartment producing bicarbonate and carbonate. These species can migrate to the anode compartment as hydroxide ion carriers at a rate much higher than the hydroxide ions themselves and act as buffers on this compartment. Therefore, carbon dioxide/carbonate or bicarbonate buffered catholyte systems are produced (Torres *et al.* 2008; Fornero *et al.* 2010a). The carbon dioxide is available as a waste gas in industrial settings, such as cement, chemical and petrochemical industries, making it a low cost buffer. The use of saline solutions, such as sodium chloride, was found to be an effective way to control pH and lead to an increase of the solution conductivity, a decrease of the internal resistance and an increase of the power density (Ahn & Logan, 2013; Li *et al.* 2013b). However, caution should be taken when using these highly saline solutions at the anode side, since the anodic microbial communities can be affected (Lefebvre *et al.* 2012).

11.5.2 Effect of temperature

In conventional fuel cells, most of the experimental studies are performed at high temperatures to favor the electrochemical kinetics in the anode and cathode, increase the open-circuit voltage, reduce the activation over voltage (according to the Arrhenius relation) and, thus, increase the performance. Moreover, an increase in temperature leads to an increase in solution conductivity with a consequent decrease in the ohmic resistance. However, the high cell temperature decreases the membrane stability and the oxygen partial pressure. Both positive and negative effects of temperature are similar to all types of fuel cells. In the case of biological fuel cells, such as MFCs, the high temperature levels of the conventional fuel cells are detrimental to the microorganism. In MFCs, temperature is one of the most important parameters that affects directly the growth and the metabolism of microbial populations, affecting not only the intracellular biochemical processes but also, the extracellular chemical and biochemical ones (Jadhav *et al.* 2009; Martin *et al.* 2010; Min *et al.* 2008; Guerrero *et al.* 2010; Campo *et al.* 2013; Michie *et al.* 2011; Liu *et al.* 2012; Li *et al.* 2013a; Wei *et al.* 2013). An increase in temperature, leads to an increase of the intracellular biochemical reaction rate, an increase on the microorganisms metabolism rate, which results in an increase of the microorganisms growth rate and an increase inpower outputs. But in the long run, when the microorganisms are under high temperatures, the other important compounds of the cell such as proteins, nucleic acids or other temperature sensitive parameters, may suffer an irreversible damage, which will lead to a decline in cell function. Therefore, the microorganisms growth rate and concentration decrease and consequently the power output decreases. The growth and reproduction rates of microorganism will be fastest and the cell function will not be affected only if they live in the optimum temperature (Min *et al.* 2008; Wei *et al.* 2013).

Generally, different types of microorganism grow optimally at different temperature ranges. The microorganisms used in MFCs are categorized in mesophilic (32–42°C) and thermophilic (48–55°C), while at the transitory range of 40–45°C, both mesophilic and thermophilic microorganisms function under suboptimal conditions. The methanogens are more sensitive to rapid changes of temperature. To some extent, this is an exciting field for research since the results would have potential application in areas with large temperature variations. In the case of MFCs, it was found that the more advantageous operating temperature range is between 30°C and 40°C, leading to higher power outputs and microbial activity (Jadhav *et al.* 2009; Martin *et al.* 2010; Min *et al.* 2008; Guerrero *et al.* 2010; Campo *et al.* 2013; Michie *et al.* 2011; Liu *et al.* 2012; Li *et al.* 2013a; Wei *et al.* 2013). However, the startup time for a MFC operating at higher temperatures (above 40°C) is longer than for low temperatures (below 40°C), despite the fact that the voltages increase with an increase in temperature. Moreover, under temperatures higher than 40°C, it was found that the power output was initially high, but decreased during operation and remained at low values (Min *et al.* 2008; Wei *et al.* 2013). In the sequel, it was found that decreasing the operation temperature of a MFC from 40°C during start-up to 30°C during normal operation, is beneficial. Once the biofilm is formed during startup, the microorganisms are more robust to functioning at lower temperature levels without significant lost in their performance (Michie *et al.* 2011; Liu *et al.* 2012; Li *et al.* 2013a) and saving some heating costs. Martin *et al.* (2010) studied the influence of anode and cathode temperature on power and methane productionin a MFC. They found that the methane production on the anode compartment was more temperature sensitive than power generation, since an increase in the anode temperature increased the methane production more than the power production. The cathode temperature appeared to have a more significant impact on the power production, than the anode temperature, indicating that a significant charge transfer limitation prevails in the cathode at low temperatures. Cathode heating may increase the catalytic activity, while the effect on anodophilic populations was less pronounced. They concluded that temperature rising may be useful if performed at the cathode, to increase cathodic kinetics without increasing activity of methanogenic populations in the anodic chamber. Also, increasing the cathode temperature instead of the anode temperature could avoid some of the problems mentioned above regarding the effect of the temperature on the microorganisms growth and metabolism.

11.5.3 Organic load

In the anodic chamber of a MFC, microorganisms produce electrons and protons due to the oxidation of organic compounds. Therefore, the organic load presented in the effluent will affect both the performance and the microbial community of a MFC. Some studies have been conducted in order to study the effect of the organic load on the MFC performance through the parameter, organic loading rate (OLR), which is related to the capacity of conversion the organic substrates per reactor volume (Jadhav *et al.* 2009; Behera *et al.* 2009; Martin *et al.* 2010; Guerrero *et al.* 2010; Campo *et al.* 2013; Aelterman *et al.* 2008; Mohan *et al.* 2009; Lorenzo *et al.* 2010; Reddy *et al.* 2010; Juang *et al.* 2011; Velvizhi *et al.* 2012; Ozkaya *et al.* 2013; Jia *et al.* 2013). These studies allowed to conclude that there is an optimum range of values for OLR which has a direct influence on the power output, COD (chemical oxygen demand) removal and coloumbic efficiency (CE). Generally, increasing the OLR, leads to an increase in the cell voltage, power output, as well as, the COD removal rate. This is due to the fact that under higher organic loads, more substrate is available to sustain the metabolic activity and more organic matter is used for power generation. Moreover, increasing the OLR, the ionic strength of the anodic solution is increased and the activity and concentration of the anodophilic microorganisms are also favored (Martin *et al.* 2010; Velvizhi *et al.* 2012). However, higher organic loading rate lead to a decrease on power generation, even though the increase on the substrate degradation occurs (Jadhav *et al.* 2009; Martin *et al.* 2010; Campo *et al.* 2013; Aelterman *et al.* 2008; Mohan *et al.* 2009; Lorenzo *et al.* 2010; Reddy *et al.* 2010; Juang *et al.* 2011; Velvizhi *et al.* 2012; Jia *et al.* 2013). This can be explained by the fact that a further increase on the organic load leads to an excessive nutrient supply. This additional supply will be metabolized by the non-electricity generation microorganisms and/or will hindered the performance of the biocatalyst, generating less current. The increase in the substrate removal rate observed at high loading rates may be attributed to the direct anodic oxidation (DAO) mechanism. This helps the further oxidation of the substrate leading to enhanced removal rates (Mohan *et al.* 2009). Despite the fact that an increase on the organic load leads to an increase on the power output and COD removal, the Coulombic efficiency decreases (Jadhav *et al.* 2009; Martin *et al.* 2010; Campo *et al.* 2013; Aelterman *et al.* 2008; Mohan *et al.* 2009; Lorenzo *et al.* 2010; Reddy *et al.* 2010; Juang*et al.* 2011; Velvizhi *et al.* 2012; Ozkaya *et al.* 2013). This means that most of the electrons produced from the oxidation of organic compounds are diverted to non-electricity generating processes, such as methane production. High organic loading rates cause saturated conditions which will lead to competition between microorganisms involved in electricity production and other processes (such as methane production), leading to a greater organic matter removal that may not be directly related to current generation. Therefore, the Coulombic efficiency decreases, despite the fact that the substrate degradation increases.

Studies regarding the effect of the OLR on methane production, show that the volumetric methane production rate increases with an increase of the organic loading rate (Martin *et al.* 2010). Therefore to avoid that, low organic loading rates should be used (Martin *et al.* 2010; Lorenzo *et al.* 2010; Velvizhi *et al.* 2012), but working in such conditions will lead to lower power outputs, higher internal resistances and lower COD removal rates. So, it is necessary to find a way to simultaneously increase the power and Coulombic efficiency and decrease the methane production. Studies regarding that suggested that an optimized OLR can be used to avoid in some extent the methane production and to shift the electricity-to-methane production rate towards the electricity production (Martin *et al.* 2010; Lorenzo *et al.* 2010; Velvizhi *et al.* 2012).

11.5.4 Feed rate and shear stress

Hydrodynamic conditions is one of the key parameters that affects the microbial adhesion and biofilm formation (Pham *et al.* 2008; Rochex *et al.* 2008). Moreover, biofilm development and electrochemical activity affect the MFC performance on treating wastewater (Pham *et al.* 2008; Rochex *et al.* 2008). Pham *et al.* (2008) studied the influence of shear rates on the establishment and structure of the biofilm and on the MFC performance. They found that an increase on the shear rate leads to a power output two to three times higher, the biofilm formed showed a double average thickness and the biomass concentration increased by a factor of 5. The development of a thicker biofilm may be due to an increase of the biofilm cohesion as aresponse to the high detachment forces induced by the high shear stress. Moreover it could be due to an increase of the biomass production resulting from higher mass transfer. In these conditions, a higher electron concentration is present on the anode electrode surface resulting in a higher power production. However, there is a maximum value of shear stress, beyond which, cell detachment prevails and reduced power production is observed (Pham *et al.* 2008). Rochex *et al.* (2008) studied the effect of shear stress on the composition of biofilm bacterial communities and found that higher shear rates decreased the biofilm diversity. As a result, one species dominated the bacterial community making it more uniform in terms of age diversion. The "young" microbial cells exhibiting high metabolic activity expelled the mature in age cells and formed a more active biofilm as a whole. A biofilm with these characteristics improves the electron conduction via direct contact and enhances transfer via electron shuttles as more cells are involved in electron transfer and more shuttles can be produced.

The ability of a microorganism to function under different flow rates would be an important factor in the case of continuous MFCs installed in wastewater treatments plants. Also, in order to achieve the best performance with a MFC, the time needed for the formation of an efficient biofilmis crucial. Studies regarding the effect of the flow rate on the MFC performance showed that the power output of a MFC increases with the flow rate until a maximum value where an increase on flow rate leads to a decrease in power output (Aelterman *et al.* 2008; Lorenzo *et al.* 2010; Aaron *et al.* 2010; Juang *et al.* 2012; Ieropoulos *et al.* 2010b). Higher flow rates may wash out the microbial community decreasing the microbial concentration in the reaction media and consequently the power production. These severe conditions also affect the biofilm formation, maturity and stabilization leading to thinner biofilms attached to the electrode surface and consequently lower power densities. Moreover, an increase in the flow rate limits the time available for the microorganisms to degrade the organic compounds and/or form the biofilm and as a result, both the Coulombic efficiency and COD removal are decreased (Aelterman *et al.* 2008; Lorenzo *et al.* 2010; Aaron *et al.* 2010; Juang *et al.* 2012; Ieropoulos et el., 2010b).

11.6 MODELLING STUDIES

Fuel cell modelling has received much attention over the last decade in an attempt to understand the phenomena occurring within the cell. Different modeling approaches have been developed and have led to analytical, semi-empirical and mechanistic models (Oliveira *et al.* 2007). In Figure 11.3 the models developed for MFCs are categorized according to the features studied, such as dimension, mass transport, pH effects, multispecies, biofilm, biochemical, electrochemical and dynamic (Zeng *et al.* 2010; Picioreanu *et al.* 2007, 2008, 2010a, 2010b;

Pinto *et al.* 2010, 2012; Oliveira *et al.* 2013, Zhang *et al.* 2014; Wen *et al.* 2009; Marcus *et al.* 2007; Merkey & Chopp, 2012).

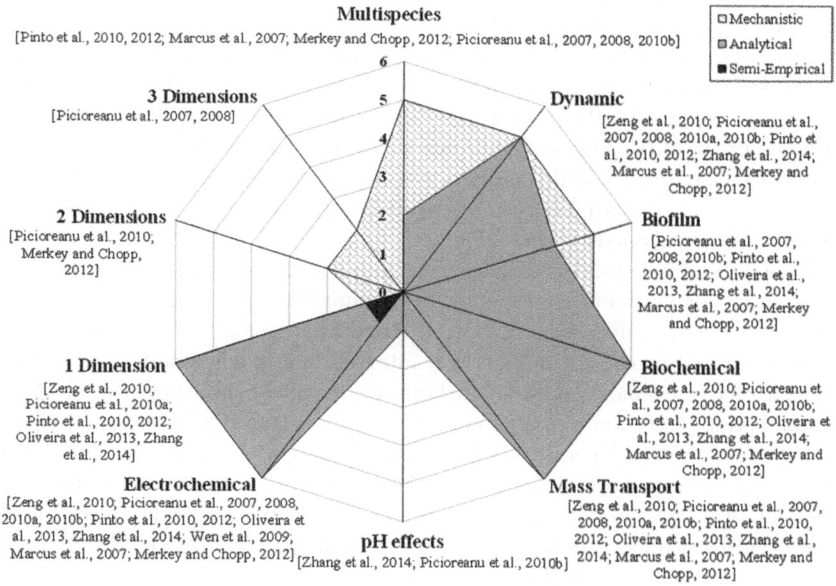

Figure 11.3 MFC model categorization based on areas of investigation.

Analytical models are an adequate tool to understand the effect of basic variables on fuel cell performance. They can be used to predict voltage losses for simple designs and can be useful for rapid calculations in these systems (Zeng *et al.* 2010; Picioreanu *et al.* 2010a; Pinto *et al.* 2010, 2012; Oliveira *et al.* 2013, Zhang *et al.* 2014). However, they neglect some of the crucial features as the transient performance and the dynamic behavior. Semi-empirical models allow designers and engineers to predict the fuel cell performance as a function of different operating conditions using simple empirical equations. They give quick predictions for existing designs but fail to predict innovative ones, since their validity is limited to a narrow window of operating conditions and cannot predict accurately the performance outside this range (Wen *et al.* 2009). Mechanistic models are complex models using differential and algebraic equations whose derivation is based on the electrochemistry, biologic and physics governing the phenomena taking place in the cell. These models involve extensive calculations, but predict accurately all the phenomena occurring in the cell (Marcus *et al.* 2007; Merkey & Chopp, 2012; Picioreanu *et al.* 2007, 2008, 2010b). It should be noted that most of the models developed only consider the phenomena occurring in the anode compartment assuming a constant over potential in the cathode compartment, neglecting the processes taking place there (Picioreanu *et al.* 2007, 2008, 2010a, 2010b; Pinto *et al.* 2010, 2012; Wen *et al.* 2009; Marcus *et al.* 2007; Merkey & Chopp, 2012). There are only a few models extended to both anode and cathode chambers (Zeng *et al.* 2010; Oliveira *et al.* 2013, Zhang *et al.* 2014).

There are, also, some models based on partial differential equations to simulate the multi species biofilm formation in the anode compartment (Marcus *et al.* 2007; Merkey & Chopp, 2012; Picioreanu *et al.* 2007, 2008, 2010b). However, these models are complex, require long computational times and may be difficult to implement widely. To overcome these difficulties, simplified models that adequately describe MFCs at various operating conditions while being suitable for process design and optimization are needed.

Among the simple models that describe both anode and cathode chambers, only few account for the biofilm formation in the anode compartment (Oliveira *et al.* 2013, Zhang *et al.* 2014). Zeng *et al.* (2010) and Zhang *et al.* (2014) developed one dimensional and analytical models that integrate the biochemical reactions with mass and charge balances and simulate both steady and dynamic behavior of a MFC, including voltage, power density, fuel concentration, and the influence of various parameters on power generation. The proposed models are easy to implement and can serve as a framework for modelling other types of MFCs. However, the major drawback of the model developed by Zeng *et al.* (2010) is the fact that the biofilm formation in the anode compartment is not considered. The model developed by Oliveira *et al.* (2013), is also a one dimensional and analytical model considering the effects of coupled charge, mass transfer and biofilm formation, along with the bio-electrochemical reactions occurring in the cell, but includes the heat transfer effects.Therefore, this model allows the assessment of the effect of operating and the design parameters on the biofilm thickness and on the temperature and concentration profiles along the cell and consequently on the cell performance.

11.7 ECONOMIC EVALUATION

The world demand for adequate sanitation and access to portable water leads to the need to treat the wastewaters. As example in Unites States (US) 4% to 5% of the energy produced is used in water infrastructures, such as water collection, treatment and distribution. The costs of maintaining such infrastructures are significant and is expected that will increase over the next years to maintain and improve this infrastructures (Logan, 2008). Wastewaters contain energy, in the form of organic matters, that we expend energy to remove rather than trying to recover it. At a conventional wastewater treatment plant it was estimated that there was 9.3 times as much energy in the wastewater than used to treat it (Logan, 2008). Industrial, agricultural and domestic wastewaters are estimated to contain a total of 17 Gigawatts, which is, for example, the same amount of energy that is used for the whole water structure in US. Therefore, if this energy can be recovered a water treatment plant can be self-sufficient. The most common technology that is used to extract this energy on a commercial scale is the anaerobic digestion. However, anaerobic digestion requires meso-to thermophilic temperatures to achieve sufficient turnover and limited methane solubility. The major operating costs for wastewater treatment are water aeration, sludge treatment and wastewater pumping. The aerobic treatment consumes large amounts of electrical energy for aeration accounting for half of the operation costs at a typical wastewater treatment plant.Also, the sludge treatment cost to wastewater treatment plants can reach up to 500€/ton dry matter. Therefore, eliminating or reducing this two costs can save appreciable amount of energy and can decrease the economic balance of the process. Based on that, the MFCs appear as an interesting technology for the production of energy from wastewaters. MFCs have many advantages over the conventional technologies used for generating energy from organic matter, such as: direct conversion of organic matter into electricity, which allows higher conversion efficiencies; can operate at ambient or low temperatures reducing the heat costs; do not require gas treatment since the gas produced in MFCs is carbon dioxide which as no useful energy content; do not need energy input for aeration since the cathode can be passively aerated; can be used in remote locations (locations lacking electrical infrastructures); can use a diversity of fuels allowing to satisfy the energy requirements for each application; and can reduce considerably the solids production at a wastewater treatment plant, reducing the operating costs for solids handling. However, the MFC technology has to compete with the well establish anaerobic digestion technology since both can use the same biomass to produce energy and have advantages and disadvantages. To implement this kind of system, the cost which is one of the MFC challenges, needs to be faced and exceeded. In order to become an advantageous technology, the cost benefits of the MFC system (energy production and wastewater treatment) should be higher than the total costs (implementation an operational cost). To achieve this, the benefits should be maximized and the costs minimized. In general, high conversion rates, whose evaluation is based on the power output, are a major condition for low costs. However, this rate is limited by the energy losses in the systems, already explained in the fundamentals and challenges section. Therefore, it is extremely important to reduce these losses and as presented above different studies have been done in order to achieve this. Moreover, it is fundamental to identify the major costs related to the MFCs technology and find ways to reduce them (Sleutels *et al.* 2012; Fornero *et al.* 2010b; Rozendal *et al.* 2008; Pham *et al.* 2006; Rabaey & Verstraete, 2005).

In the work developed by Rozendal *et al.* (2008) an overview of the estimated implementation costs of a MFC based on the materials available commercially and commonly used on these systems, as well as, a comparison of estimated costs among the different wastewater treatments available is presented. They compared the costs of MFCs technology with the anaerobic digestion (AD) treatment. The authors concluded that with the materials available, the MFC capital costs are higher than those of conventional treatments, 8€/kg COD for MFC and 0.01€/kg COD for AD. Among the different costs associated to the MFC technology, they showed that the major percentage for the total cost is the cathode side material (47%) due to the need of using platinum as cathode electrode to promote the oxygen reduction reaction, with an estimate cost of 500€/m^2. The second major contribution for the overall cost (38%) is related to the membrane used to separate both compartments with an estimated cost of 400€/m^2. Therefore, MFC systems can only became economically interesting if these larger capital costs are reduced, based on changing the materials commonly used for less expensive ones. The authors estimated a total costs of 0.4€/kg COD, which was acceptable and near the costs of the AD. Also, it should be emphasized that 1 kg of COD can be converted to 1kWh using anaerobic digestion while the same amount of COD can be converted in 4 kWh when a MFC is used (Pham *et al.* 2006; Rabaey & Verstraete, 2005). This shows a higher energy conversion rate when MFC are used instead of the AD. Therefore, the slightly higher costs of the MFC technology may be compensated by its higher conversion rate. The MFCs have, also, other benefits over the AD such as operation on a small-scale, with low COD concentrations and temperatures. Moreover, economic value can be added to this technology, if besides the COD removal, the system is used for the production of valuable chemical products instead of electricity. The production of hydrogen gas, caustic or hydrogen peroxide that can be directly used in the treatment process, or the removal of persistent and toxic wastewater compounds may provide an additional cost benefit for the wastewater treatment plants when a MFC is used.

11.8 SUMMARY

The fundamentals and challenges on MFCs have been summarized and the recent modeling and experimental studies have been reviewed. An economic outlook of this technology was also presented. As was mentioned for practical applications of this technology cost effectiveness is essential. Therefore to a MFC be cost effective the value of the products and of wastewater treatment (revenues) need to be higher than the capital and operational costs. The production of the valuable chemicals with MFCs as well as the reduction of the material costs are expected to offset the higher capital investments of this technology. Apart from the economic aspects, the MFC are a sustainable technology that can be adaptable to a wide variety of application. They produce electrical energy from different array of electron donors with different concentrations at low and moderate temperatures and no other existing technology can achieve that.

Many studies have been focused on analyzing and improving single parts of MFCs and different materials, designs, microbes, mediators and operating conditions have been suggested, which manage to increase the power outputs. However, power generation have not yet reached the levels needed for commercial use. Also, the best operating and design conditions for one type of MFC may be not necessary the same to another type. Therefore, when developing a MFC, the materials and conditions need to be carefully considered and chosen in order to achieve the best performance which means the highest power output. Since significant challenges still exist before a MFC can be ready to commercialization, further studies on more cost-efficient materials and optimization of configurations are needed in order incorporate large-scale MFCs in conventional wastewater treatment systems.

The development of mathematical models is essential to a better understanding and prediction of the main processes occurring in a MFC. These models can be used to predict the effect of various operating and structural parameters on cell performance and are a fundamental tool for the design and optimization of fuel cell systems. In spite of the modelling work on MFCs developed in the past few years, new models are still needed to describe the main biological, mass transfer, energy and charge effects in the anode and cathode compartment of a MFC, as well as the biofilm formation.

The developments both on experimental and modeling issues have led to an increase on the MFC power output at lab-scale. Therefore, the scaling-up and durability need to emerge as future development areas in the MFC research.

11.9 ACKNOWLEDGEMENTS

V. B. Oliveira acknowledges the post-doctoral fellowship (SFRH/BDP/91993/2012) supported by the Portuguese "Fundação para a Ciência e Tecnologia" (FCT), POPH/QREN and European Social Fund (ESF). POCI (FEDER) also supported this work via CEFT.

11.10 REFERENCES

Aaron D., Tsouris C., Hamilton C. Y. and Borole A. P. (2010). *Energies*, **3**, 592–606.

Aelterman P., Versichele M., Marzorati M., Boon N. and Verstraete W. (2008). *Bioresource Technology*, **99**, 8895–8902.

Ahn Y. and Logan B. E. (2013). *Bioresource Technology*, **132**, 436–439.

Behera M. and Ghangrekar M. M. (2009). *Bioresource Technology*, **100**, 5114–5121.

Biffinger J. C., Pietron J., Bretschger O., Nadeau L. J., Johnson G. R., Williams C. C., Nealson K. H. and Ringeisen B. R. (2008). *Biosensors and Bioelectronics*, **24**, 900–905.

Campo A. G., Lobato J., Cañizares P., Rodrigo M. A. and Morales F. J. F. (2013). *Applied Energy*, **101**, 213–217.

Cheng K. Y., Ho G. and Ruwisch R. C. (2010). *Environmental Science & Technology*, **44**, 518–525.

Cheng S. and Logan B. E. (2011). *Bioresource Technology*, **102**, 4468–4473.

Cheng S., Ye Y., Ding W. and Pan B. (2014). *Journal of Power Sources*, **248**, 931–938.

Clauwaert P., Aelterman P., Pham T. H., Schamphelaire L. De, Carballa M., Rabaey K. and Verstraete W. (2008). *Applied Microbiology and Biotechnology*, **79**, 901–913.

Dekker A., Heijne A. T., Saakes M., Hamelers H. V. M. and Buisman C. J. N. (2009). *Environmental Science & Technology*, **43**, 9038–9042.

Dewan A., Beyenal H. and Lewandowski Z. (2008). *Environmental Science & Technology*, **42**, 7643–7648.

Du Z., Li H. and Gu T. (2007). *Biotechnology Advances*, **25**, 464–82.

Erable B., Etcheverry L. and Bergel A. (2009). *Electrochemistry Communications*, **11**, 619–622.

Fan Y., Han S. K. and Liu H. (2012). *Energy & Environmental Science*, **5**, 8273–8280.

Fan Y., Hu H. and Liu H. (2007a). *Journal of Power Sources*, **171**, 348–354.

Fan Y., Hu H. Q. and Liu H. (2007b). *Environmental Science & Technology*, **41**, 8154–8158.

Fornero J. J., Rosenbaum M., Cotta M. A. and Angenet L. T. (2010a). *Environmental Science & Technology*, **44**, 2728–2734.

Fornero J. J., Rosenbaum M. and Angenet L. T. (2010b). *Electroanalysis*, **22**, 832–843.

Guerrero A. L., Scott K., Head I. M., Mateo F., Ginesta A. and Godinez C. (2010). *Fuel*, **89**, 3985–3994.

Guerrini E., Cristiani P. and Trasatti S. P. M. (2013). *International Journal of Hydrogen Energy*, **38**, 345–353.

Gurung A. and Oh S. E. (2012). *Energy Sources*, Part A **34**, 1591–1598.

He Z., Huang Y., Manohar A. K. and Mansfeld F. (2008). *Bioelectrochemistry*, **74**, 78–82.

Heijnea A. T., Liu F., van Rijnsoever L. S., Saakes M., Hamelers H. V. M. and Buisman C. J. N. (2011). *Journal of Power Sources*, **196**, 7572–7577.

Holmes D. E., Bond D. R., O'Neil R. A., Reimers C. E., Tender L. R. and Lovley D. (2004). *Microbial Ecology*, **48**, 178–190.

Ieropoulos I., Greenman J. and Melhuish C. (2008). *International Journal of Energy Research*, **32**, 1228–1240.

Ieropoulos I., Greenman J. and Melhuish C. (2010a). *Bioelectrochemistry*, **78**, 44–50.

Ieropoulos I., Winfield J. and Greenman J. (2010b). *Bioresource Technology*, **101**, 3520–3525.

Ieropoulos I. A., Greenman J., Melhuish C. and Hart J. (2005). *Enzyme and Microbial Technology*, **37**, 238–245.

Jadhav G. S. and Ghangrekar M. M. (2009). *Bioresource Technology*, **100**, 717–723.

Jia J., Tang Y., Liu B., Wu D., Ren N. and Xing D. (2013). *Bioresource Technology*, **144**, 94–99.

Juang D. F., Yang P. C., Chou H. Y. and Chiu L. J. (2011). *Biotechnology Letters*, **33**, 2147–2160.

Juang D. F., Yang P. C. and Kuo T. H. (2012). *International Journal of Environmental Science and Technology*, **9**, 267–280.

Jung S. and Regan J. M. (2007). *Applied Microbiology and Biotechnology*, **77**, 393–402.

Kim D., An J., Kim B., Jang J. K., Kim B. H. and Chang I. S. (2012). *ChemSusChem*, **5**, 1086–1091.

Lefebvre O., Ooi W. K., Tang Z., Abdullah-Al-Mamun Md., Chua D. H. C. and Ng H. Y. (2009). *Bioresource Technology*, **100**, 4907–4910.

Lefebvre O., Tan Z., Kharkwal S. and Ng H. Y. (2012). *Bioresource Technology*, **112**, 336–340.

Li L. H., Sun Y. M., Yuan Z. H., Kong X. Y. and Li Y. (2013a). *Environmental Technology*, **34**, 1929–1934.

Li X. M., Cheng K. Y. and Wong J. W. C. (2013b). *Bioresource Technology*, **149**, 452–458.

Liu H., Cheng S., Huang L. and Logan B. E. (2008). *Journal of Power Sources*, **179**, 274–279.

Liu L., Tsyganova O., Lee D. J., Suc A., Chang J. S., Wang A. and Ren N. (2012). *International Journal of Hydrogen Energy*, **37**, 15792–15800.
Liu Z., Liu J., Zhang S. and Su Z. (2008). *Biotechnology Letters*, **30**, 1017–1023.
Logan B. E. (2008). Microbial Fuel Cells. Wiley Bicentennial, United States of America.
Logan B. E. (2010). *Applied Microbiology and Biotechnology*, **85**, 1665–1671.
Lorenzo M. D., Scott K., Curtis T. P. and Head I. M. (2010). *Chemical Engineering Journal*, **156**, 40–48.
Lovley D. R. (2008). *Geobiology*, **6**, 225–231.
Lovley D. R. and Nevin K. P. (2008). Electricity production with electricigens. In: Bioenergy, J. D. Wall, C. S. Harwood and A. Demain (eds), Chap. 23, ASM Press, Washington, DC, pp. 295–306.
Marcus A. K., Torres C. I. and Rittman B. E. (2007). *Biotechnology and Bioengineering*, **98**, 1171–1182.
Martin E., Savadogo O., Guiot S. R. and Tartakovsky B. (2010). *Biochemical Engineering Journal*, **51**, 132–139.
Merkey B. V. and Chopp D. L. (2012). *Bulletin of Mathematical Biology*, **74**, 834–857.
Michie I. S., Kim J. R., Dinsdale R. M., Guwy A. J. and Premier G. C. (2011). *Environmental Science & Technology*, **4**, 1011–1019.
Min B., Román O. B. and Angelidaki I. (2008). *Biotechnology Letters*, **30**, 1213–1218.
Mohan S. V., Raghavulu S. V., Peri D. and Sarma P. N. (2009). *Biosensors and Bioelectronics*, **24**, 2021–2027.
Nam J. Y., Kim H. W., Lim K. H., Shin H. S. and Logan B. E. (2010). *Biosensors and Bioelectronics*, **25**, 1155–1159.
Nimje V. R., Chen C. Y., Chen C. C., Tsai J. Y., Chen H. R., Huang Y. M., Jean J. S., Chang Y. F. and Shih R. C. (2011). *International Journal of Hydrogen Energy*, **36**, 11093–11101.
Oliveira V. B., Falcão D. S., Rangel C. M. and Pinto A. M. F. R. (2007). *International Journal of Hydrogen Energy*, **32**, 415–424.
Oliveira V. B., Simões M., Melo L. F. and Pinto A. M. F. R. (2013). *Energy*, **61**, 463–471.
Ozkaya B., Cetinkaya A. Y., Cakmakci M., Karadag D. and Sahinkaya E. (2010). *Bioprocess and Biosystems Engineering*, **36**, 399–405.
Pham H. T., Rabaey K., Aelterman P., Clauwaert P., Schamphelaire L. D., Boon N. and Verstraete W. (2006). *Engineering in Life Sciences*, **6**, 285–292.
Pham H. T., Boon N., Aelterman P., Clauwaert P., Schamphelaire L. D., Oostveldt P., Verbeken K., Rabaey K. and Verstraete W. (2008). *Microbial Biotechnology*, **6**, 487–496.
Picioreanu C., Head I. M., Katuri K. P., van Loosdrecht M. C. M. and Scott K. (2007). *Water Research*, **41**, 2921–2940.
Picioreanu C., Head I. M., Katuri K. P., van Loosdrecht M. C. M. and Scott K. (2008). *Water Science & Technology*, **57**, 965–971.
Picioreanu C., Head I. M., Katuri K. P., van Loosdrecht M. C. M. and Scott K. (2010a). *Journal of Applied Electrochemistry*, **40**, 151–162.
Picioreanu C., van Loosdrecht M. C. M., Curtis T. P. and Scott K. (2010b). *Bioelectrochemistry*, **78**, 8–24.
Pinto R. P., Srinivasan B., Manuel M. F. and Tartakovsky B. (2010). *Bioresource Technology*, **101**, 5256–5265.
Pinto R. P., Tartakovsky B. and Srinivasan B. (2012). *Journal of Process Control*, **22**, 1079–1086.

Potter M. C. (1911). *Proceedings of the Royal Society of London. Series B*, **84**(571), 260–276.

Puig S., Serra M., Coma M., Cabré M., Balaguer M. D. and Colprim J. (2010). *Bioresource Technology*, **101**, 9594–9599.

Qiang L., Yuan L. J. and Ding Q. (2011). *HuanjingKexue/Environmental Science*, **32**, 1524–1528.

Rabaey K. and Verstraete W. (2005). *Trends in Biotechnology*, **23**, 290–298.

Raghavulu S. V., Mohan S. V., Reddy M. V., Mohanakrishna G. and Sarma P. N. (2009). *International Journal of Hydrogen Energy*, **34**, 7547–7554.

Reddy M. V., Srikanth S., Mohan S. V. and Sarma P. N. (2010). *Bioelectrochemistry*, **77**, 125–132.

Rittmann B. E. (2008). *Biotechnology and Bioengineering*, **100**, 203–212.

Rochex A., Godon J.-J., Bernet N. and Escudie R. (2008). *Water Research*, **42**, 4915–4922.

Rozendal R. A., Mahelers H. V. M., Rabaey K., Keller J. and Buisman C. J. N. (2008). *Trends in Biotechnology*, **26**, 450–459.

Sleutels T. H. J. A., Heijne A. T., Buisman C. J. N. and Hamelers H. V. M. (2012). *ChemSusChem*, **5**, 1012–1019.

Tender L. M., Gray S. A., Groveman E., Lowy D. A., Kauffman P., Melhado J., Tyce R. C., Flynn D., Petrecca R. and Dobarro J. (2008). *Journal of Power Sources*, **179**, 571–575.

Torres C. I., Lee H. S. and Rittmann B. E. (2008). *Environmental Science & Technology*, **42**, 8773–8777.

Velvizhi G. and Mohan S. V. (2012). *International Journal of Hydrogen Energy*, **37**, 5969–5978.

Vologni V., Kakarla R., Angelidaki I. and Min B. (2013). *Bioprocess and Biosystems Engineering*, **36**, 635–642.

Wei L., Han H. and Shen J. (2013). *International Journal of Hydrogen Energy*, **38**, 11110–11116.

Wen Q., Wu Y., Cao D., Zhao L. and Sun Q. (2009). *Bioresource Technology*, **100**, 4171–4175.

Yong L., Renduo Z., Guangli L., Jie L., Bangyu Q., Mingchen L. and Shanshan C. (2011). *Bioresource Technology*, **102**, 3827–3832.

You S. J., Zhao Q. L., Jiang J. Q. and Zhang J. N. (2006). *Chemical and Biochemical Engineering Quarterly*, **20**, 407–412.

Yuan Y., Zhao B., Zhou S., Zhong S. and Zhuang L. (2011). *Bioresource Technology*, **102**, 6887–6891.

Zeng Y., Choo Y. F., Kim B. H. and Wu P. (2010). *Journal of Power Sources*, **195**, 79–89.

Zhang L. and Deshusses M. (2014). *Environmental Technology*, **35**, 1064–1076.

Zhang L., Li C., Ding L., Xu K. and Ren H. (2011). *Journal of Chemical Technology & Biotechnology*, **86**, 1226–1232.

Zhang L., Li J., Zhu X., Ye D. and Liao Q. (2013). *Chemical Engineering Journal*, **223**, 623–631.

Zhongjian L., Xingwang Z. and Lecheng L. (2008). *Process Biochemistry*, **43**, 1352–1358.

Zhu X., Zhang L., Li J., Liao Q. and Ye D. (2013). *International Journal of Hydrogen Energy*, **38**, 15716–15722.

Zhuang L., Zhoua S., Wang Y., Liu C. and Geng S. (2009). *Biosensors and Bioelectronics*, **24**, 3652–3656.

Zhuang L., Zhou S., Li Y. and Yuan Y. (2010a). *Bioresource Technology*, **101**, 3514–3519.

Zhuang L., Feng C., Zhou S., Li Y. and Wang Y. (2010b). *Process Biochemistry*, **45**, 929–934.

Zhuang L., Yuan Y., Wang Y. and Zhou S. (2012a). *Bioresource Technology*, **123**, 406–412.

Zhuang L., Zheng Y., Zhou S., Yuan Y., Yuan H. and Chen Y. (2012b). *Bioresource Technology*, **106**, 82–88.

Zhuang L. and Zhou S. (2009). *Electrochemistry Communications*, **11**, 937–940.

Zuo Y., Cheng S., Call D. and Logan B. (2007). *Environmental Science & Technology*, **41**, 3347–3353.

Zuo Y., Cheng S. and Logan B. (2008). *Environmental Science & Technology*, **42**, 6967–6972.

Part II

Innovative technologies and economics in sewage treatment plants – case studies

Chapter 12

Management optimisation and technologies application: a right approach to balance energy saving needs and process goals

P. Ragazzo[1], L. Falletti[2], N. Chiucchini[1] and G. Serra[3]

[1]ASI Azienda Servizi Integrati S.p.A., Process Control and Technological Development, Venice, Italy

[2]PH.D. Dep. Industrial Engineering University, Padua, Italy

[3]Azienda Servizi Pubblici Sile-Piave S.p.A., Process Control, Treviso, Italy

12.1 INTRODUCTION

Energy saving is a widespread goal in any country but dealing with problems of consumption rationalization is complicated. In fact even though it is clear that in each energy sector, saving should be obtained by multilevel strategies, the general trend is to pursue the goal without considering integrated approaches, sometimes even adopting conflicting actions.

Integrated water cycle services play an important role in energy consumption – *about 6.3 billion kWh per year in Italy, in 2009–2011 (Terna, 2012)* – and the contribution of the WWTP (Waste Water Treatment Plant) – *about 3 billion kWh per year in Italy, in the same period* – can be really significant.

Therefore as the energy consumption for wastewater treatment plants is a significant fraction of total energy consumption – *about 0.9% in Italy in 2009–2011 (DPS, 2010; Co.N.Vi.RI, 2011)* – and above all it represents a significant part of the WWTP managements costs, applying energy saving strategies to wastewater treatments is important; but it must be done by considering the overall factors and avoiding negative impacts on depuration process.

Often this objective is only dealt with in terms of improvement of sludge treatment, however other important aspects of the WWTPs – *such as the lack of a sufficient flexibility in the treatments, the oversizing and inflexibility of important sections as the aeration, the pumping and mixing systems, as well as the adoption of operative procedures not well adjusted on process real needs* – can undoubtedly lead to severe limitations on management and energy issues. Therefore, the correction of structures and management practices, represents the first important step in terms of energy saving purposes, also because these factors often have an impact on wastewater treatment processes, undermining the quality of discharges.

Some published data (Terna, 2012; DPS, 2010) indicate that, in the case of Italy, the average energy consumption is about 37 kWh per equivalent inhabitant (e.i.) per year, however in the case of absence of suitable flexibility, the effective values can be much higher. This is the case for example of WWTPs supplied by combined sewage systems where the designs are based on reference data provided by technical handbook, for which the consumptions can be 4 times higher than the average value above referred.

But also adopting very rigorous criteria of designing – such as circumstanced data for the aeration dimensioning; pipelines with minimal head losses; flexibility of treatment lines and devices – significant energy wastes could be caused by wrong operating practices and/or lack of proper controls; for this reason what is essential is adopting multi-level monitoring systems for the main process variables and the influent characteristics.

Another problem is the tendency to adopt high energy-consuming technologies without real needs, as it often occurs with membrane bioreactors (MBR) or ultraviolet (UV) disinfection technologies, both inconsistent with energy saving purposes but more and more encouraged in the last few years.

In this chapter, on the basis of the experience of two companies managing integrated water service in northeastern Italy, some significant energy saving experiences, obtained in municipal WWTPs in the years 2010–2013, are reported.

12.2 ENERGY SAVING WITH MAINTENANCE AND CONTROL OPERATIONS

12.2.1 Initial situation of plants

All the WWTPs reported in this chapter treat municipal wastewater collected by combined sewage systems; they operate with a conventional treatments sequence, with the exception of WWTP7 that uses tubular membranes for final sludge separation; some of them are located in the hinterland of the Veneto region, while others serve tourist resorts near Venice and are subject to high seasonal fluctuations in flow and organic loads. Only two WWTPs are made of multiple lines (Quarto d'Altino, Jesolo), in this way they can work flexibly during the year. Table 12.1 reports operational conditions in terms of design e.i., actual flow rate and organic load during year 2010 for the plants hereafter discussed.

The operation of the plants and the technologies applied were evaluated to highlight the critical aspects in relation to both process and energy efficiency. This was possible over a significant period of time using different multilevel control instruments for the main process variables – *such as sludge waste flow and recirculation ratio, oxygen and total suspended solids in biological tank* – and energy consuming sections.

Table 12.1 Operational conditions of WWTPs – average values 2010.

Plant	Location	Lines	Design e.i.	Flow rate m^3/d		Organic load $kgBOD_5/d$	
				Year	H.S.	Year	H.S.
$WWTP_1$	Quarto d'Altino	2	50,000	11,860		1,330	
$WWTP_2$	Musile Piave	1	10,000	3,190		140	
$WWTP_3$	Eraclea(1)	1	32,000	3,200	3,555	310	600
$WWTP_4$	Caorle(1)	1	120,000	11,980	15,197	1,440	3,330
$WWTP_5$	Fossalta	1	3,600	830		65	
$WWTP_6$	Jesolo(1)	3	160,000	26,840	32,115	1,810	3,270
$WWTP_7$	Torre di Mosto	1	3,000	390		45	

WWTP: wastewater treatment plant; e.i.: equivalent inhabitants; H.S: high season (summer), (1) seasonal plant.

In most reported cases, the efficacy of the treatment was the primary goal of the interventions and once the conditions able to guarantee the process reliability were identified and set out, the energy saving goals were pursued. In other cases instead the goal was the efficiency and the energy savings but when no the whole wastewater treatment has been assessed, the effectiveness of some steps was sometimes compromised. In Table 12.2 the typical energy consumptions of the plants before the interventions are reported.

Table 12.2 Energy consumptions – average values 2010.

Plant	Total Consumption		Specific Consumption	
	kWh/year	kWh/m^3	kWh/e.i.f.	kWh/e.i.b.
$WWTP_1$	1,934,630	0.45	39	87
$WWTP_2$	309,250	0.27	27	144
$WWTP_3$	620,070	0.54	55	151
$WWTP_4$	1,468,880	0.34	34	117
$WWTP_5$	184,190	0.62	64	194
$WWTP_6$	2,740,370	0.30	30	160
$WWTP_7$	636,930	4,5	462	969

e.i.: equivalent inhabitants; WWTP: wastewater treatment plant; kWh/e.i.f. and kWh/e.i.b.: calculated on the basis of flows and BOD_5 values respectively.

Regarding WWTP1 instead, where the reduction of energy consumption was the main purpose, the areas requiring corrections were identified by monitoring all the utilities in terms of energy consumption and operation hours. Each section was analysed by means of field surveys and spreadsheets, specifically designed to compare actual and designed conditions and as a result the actions were conceived only for pump stations, aeration systems and mechanical devices.

In all cases the interventions were developed on 3 subsequent levels: a first level of immediate correction through management and maintenance procedures; a second level with the implementation and minimal structural upgrades; and finally a third long-term level, not dealt with in this chapter, concerning the up-grade of entire treatment sections and/or the restoring of sewage systems. In Table 12.3 the main critical situations founded in all the WWTPs are listed.

Table 12.3 Plants main critical aspects related to energy wasting.

Section	Pumps and pipings	Mixers and motors	Air systems
Critical aspects	Pumps size	Devices size	Devices size
	Pumping station configuration	Aging-wearing out	Flexibility
	Regulation devices		Regulation devices
	Pipes configuration		Balanced distribution
	Hydraulic overloads		Aging-wearing out
	Aging-wearing out		Diffusers fouling

12.2.2 Interventions on pumps and piping system

Verifying the adequacy of the use of the pumps was the first phase of the work in this sector. In each station the pumps were monitored by measuring the power consumption, the pressures and flows and their operating conditions were assessed in relation to their performance curves. This enabled, where possible, the optimisation of the hydraulic levels and/or the reduction of the head losses, or the replacement of the inappropriate pumps.

In particular components and devices – *such as impellers, shim plates, coupling foot, valves etc...* – were maintained to solve and avoid wearing-out, damage or obstruction; level control devices were installed to maintain optimal working conditions or to allow dischargings by gravity; inverters were installed where possible, to let working conditions closest to the best efficiency point.

The second crucial evaluated point was the location of pumps' installation in the station, because excessive closeness could imply interferences among the suction flows, causing even serious power loss; as well as when working in multiple lines, discharging into the same pipe frequently leads to a significant reduction in the overall system flow capacity. The flow capacity of the pumping station at the $WWTP_1$ for example – *where 6 pumps were connected in parallel to one single pipe* – is reduced up to 40% with pumps working simultaneously. Considering the trend of the hydraulic load to the plant, this could imply an energy waste of 40.000 kWh/year, therefore justifying the redesigning of the pipe system.

Finally another important aspect evaluated in this area was the configuration of the pipe systems, particularly in relation to their shape and connections, because the presence of elbow curves, restrictions or T-connections can cause serious increases in head losses.

Some examples of the interventions carried out in this area with the respective achieved energy savings are shown in Table 12.4.

Table 12.4 Pumps and pipings – interventions and savings at $WWTP_1$.

List of operations	Total saving	Specific savings		
	kWh/year	kWh/m³	kWh/e.i.f.	kWh/e.i.b.
Pump and valve maintenance	20,000	0.005	0.5	0.5
Level for gravity discharge	60,000	0.02	1.5	1.3

e.i.: equivalent inhabitants; WWTP: wastewater treatment plant; kWh/e.i.f. and kWh/e.i.b.: calculated on the basis of flows and BOD_5 values respectively.

12.2.3 Interventions on mixers and engines

Using the same above-mentioned criteria, namely improving the energy efficiency without compromising the process requirements, the mixing systems were studied in terms of actual performance and power absorptions, under operating conditions.

On the basis of the analysis even in this area, different interventions – *such as management procedures changes and/or devices maintenance or replacement* – were performed.

In this way for example at the $WWTP_1$, ten old mixers in the denitrification compartment – *total power 24 kW* – were replaced by 6 units with higher specific axial trust – *bigger and more performing propeller, gear between motor and propeller- total power 13 kW* – thus obtaining an annual energy saving around 94.000 kWh.

With reference to the engines and the other electromechanical utilities similarly, as the need was to renew some part of the equipment, high efficient new units with all their regulation devices, were immediately adopted.

With this approach at $WWTP_1$, four new high efficient blower engines replaced the old ones, working for oxidation, digestion and waste treatment, and this led to 10% efficiency increase.

At the same time some engines in sludge dewatering, pumping and air compartments were equipped with inverters to adjust power consumption to real needs. Table 12.5 summarizes the savings obtained in this sector.

Table 12.5 Mixers and engines – interventions and savings at $WWTP_1$.

List of operations	Total saving	Specific savings		
	kWh/year	kWh/m³	kWh/e.i.f.	kWh/e.i.b.
Mixers with gearbox	94,000	0.04	3.9	3.6
Speed regulators	79,000	0.02	2	1.8
High efficiency engines				
Oxidation (*)	130,000	0.09	8.2	7.5
Digestion and waste treatment (*)	60,000	0.02	1.5	1.4

e.i.: equivalent inhabitants; kWh/e.i.f. and kWh/e.i.b.: calculated on the basis of flows and BOD_5 values respectively; (*): estimated values.

12.2.4 Interventions on air compression and distribution

This constitutes a high energy consuming sector where it is important to focus the efforts on the rationalization of energy use: significant margins for energy savings in fact are often possible in the oxidation compartments and they could be pursued even using simple management interventions. Therefore the management of the aeration systems and relative devices in each plant was carefully analysed to adapt their use and the energy consumption involved to the different process needs; where possible and useful of course, the blowers were also equipped with inverters to modulate the oxygen supply. Some procedures were given out to the operators to both detect the best working configuration and to provide guidelines to maintain it. Simple interactive spreadsheets for the management moreover, allowed to rapidly highlight and correct any deviation from the established operating conditions.

Hereafter some examples of the main interventions in this area are reported:

- In the oxidation compartment at the $WWTP_2$, since a sufficient sludge stabilization was achieved already in this tank, it was possible to replace the two over-dimensioned blowers (28.5 kW each) – *causing, although regulated, oxygen concentrations higher than the process needs (from 4.5 up to 9 mg/L)* – with the digestion one of 13 kW. This change resulted in an energy saving higher than 90,000 kWh/year.
- In two other plants – *$WWTP_3$ $WWTP_4$* – both subjected to significant seasonal fluctuations in hydraulic and organic loads, the rationalisation of blowers, aerators and turbines use, and their modulation during the summer, allowed a total energy saving of 55,000 kWh as compared to the same period and conditions in the previous year; Figure 12.1 shows an example of manual modulation implemented at the $WWTP_4$ from April to December 2012.
- At $WWTP_1$ the air excess in oxidation process was deviated towards the sand and grit removal compartment, allowing to switch off the dedicated blower: this led to 36,000 kWh energy savings per year.

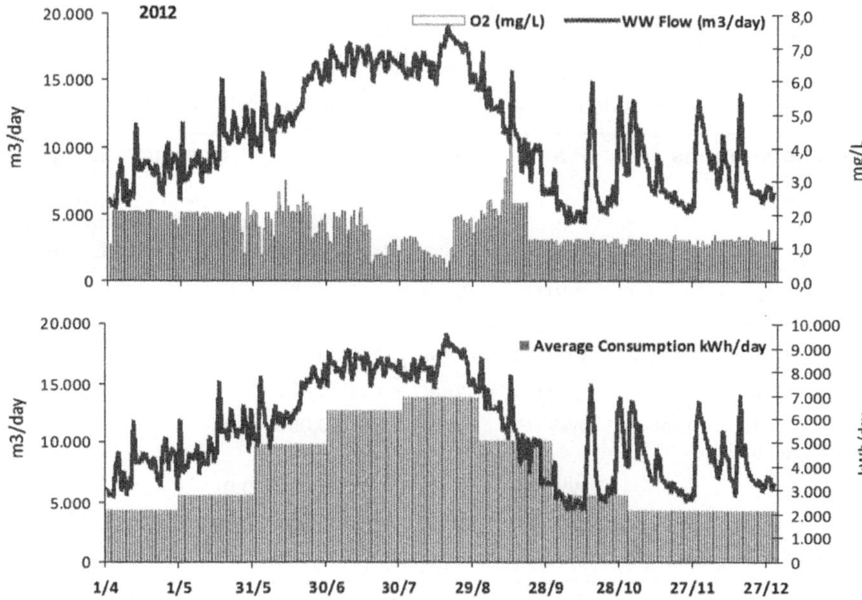

Figure 12.1 Incoming wastewater load and oxygen concentration in oxidation tank of $WWTP_4$ with different blower combinations and power consumption.

Another crucial point was to improve the efficiency of the air distribution network and diffusers, introducing some periodical maintenance procedures such as: the removing of the material interfering with the diffusers on the bottom of the aeration tanks; the cleaning of the air supply networks by means of an online washing system; the replacing of the damaged or obstructed membranes of the diffusers.

All the different interventions – *enabling lower working pressures of membranes and networks* – led to a general improvement in the oxygen transfer efficiency.

Lastly the aeration devices were analyzed in terms of the installed powers and energy yields in relation to the requirements, to identify those systems no longer able to guarantee the required performance with suitable power consumption and renew them. In this way, for example at the $WWTP_1$, two lobe blowers were replaced by a single turbo-blower equipped with inverter, this resulted in 200,000 kWh/year estimated savings; a similar intervention has been scheduled for 2014 at the $WWTP_4$ – *a turbo-blower of 75 kW in replacement of a 90 kW lobe blower*- with estimated savings of 66.000 kWh/year. In some other plants moreover old air systems were replaced by fine bubbles diffusing devices.

A good choice, as it was for the $WWTP_1$, could be to install pressure control devices in the air networks, with the purpose of monitoring the diffusers fouling trend and correct it at the right time.

Table 12.6 refers to some examples of the interventions and the relative energy savings obtained.

Table 12.6 Air systems and distributions – interventions and savings

List of operations	Plant	Period	Total saving	Specific savings		
			kWh/period	kWh/m³	kWh/e.i.f.	kWh/e.i.b.
Diffusers replacement and tanks cleaning	$WWTP_1$	Year	150,000	0.07	6.3	7.2
Oxidation air excess reuse	$WWTP_1$	Year	36,000	0.01	0.9	1
Blower changes	$WWTP_1$(*)	Year	200,000	0.09	8.3	7.5
Air systems rationalization and modulation	$WWTP_2$	Year	90,000	0.07	6.7	41
	$WWTP_3$	H.S.	25,000	0.07	1.7	12
	$WWTP_4$	H.S.	30,000	0.08	1.8	1.3
Air systems up-grade	$WWTP_5$	6 months	37,000	0.14	6.4	48

e.i.: equivalent inhabitants; kWh/e.i.f. and kWh/e.i.b.: calculated on the basis of flows and BOD_5 values respectively; WWTP: wastewater treatment plant; (*): estimated values; H.S.: high season (summer).

12.2.5 When energy and process efficiency do not agree

When the energy savings purposes are pursued without firstly considering all the treatment requirements, an energy improvement may lead to some negative impact on process. This has been encountered at $WWTP_5$ where the inefficient submerged aeration system was renewed with the fine bubble membrane diffusers (Table 12.6). In this WWTP in fact, where no denitrification treatment was included, it was just the inefficiency of the aeration system to let the creation of an appropriate oxygen gradient into the flocks of activated sludge; thus permitting a balance between the ammonia oxidation and nitrate reduction in oxidation tank. When the fine bubble membrane diffusers were installed, the oxygen penetration into the flocks did not permit anymore the suitable redox conditions for the co-denitrification that before led to a 50% of total nitrogen removal. Neither the subsequent replacement of membranes with other ones having a lower number of holes, nor the partial switch-off of the air networks led to a solution to the problem (Figure 12.2).

The same effect was observed in the $WWTP_6$ – *working with multiple lines* – where, despite the presence of a denitrification treatment, the replacement of turbines in one oxidation tank, significantly reduced the total nitrogen removal. Laboratory respirometric tests confirmed such worsening, with NUR changing (Nitrogen Uptake Rate) from 0.08 to 0.01 gN/g VSS d (Figure 12.3).

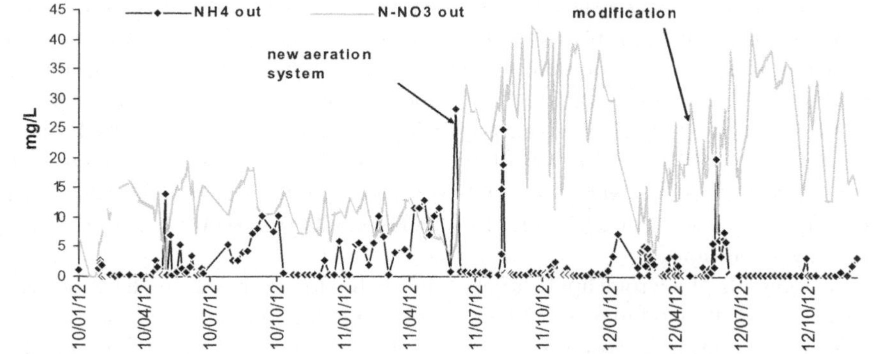

Figure 12.2 Concentration of NH4+ and NO3-N in the effluent of $WWTP_5$ in 2010–2012 (Ragazzo *et al.* 2013).

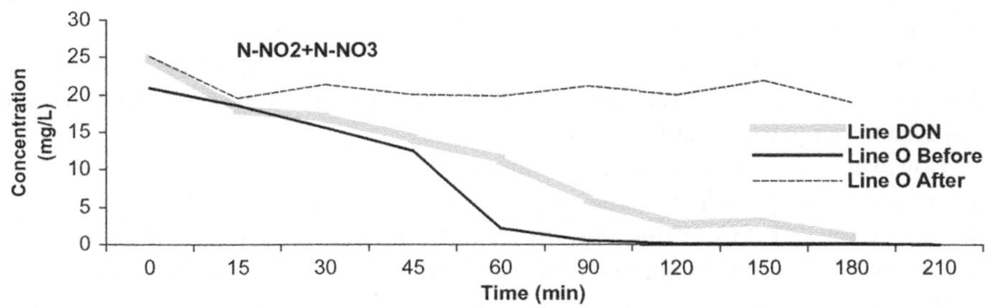

Figure 12.3 Respirometric tests (NUR) of sludge of $WWTP_6$ (Ragazzo *et al.* 2013).

12.3 ENERGY SAVING CHOOSING THE RIGHT TECHNOLOGY

Choosing the correct technology should be based on a complete and comprehensive evaluation of real goals and needs, taking into account the actual available alternatives and their benefits. Thus difficult balances among different contrasting aspects should often be found: management costs and energy consumptions are among these, substantial issues that have to be considered. This kind of approach however is difficult and choices frequently are not supported by any circumstantial knowledge and are directly influenced by market trends.

Therefore even if advanced treatments are not required, they are adopted resulting in high energy waste.

This is the case for example with membrane bioreactors (MBR) systems that have been continuously proposed in the last few years as the best solution for all the problems in municipal wastewater treatment. They are certainly a good choice for WWTPs having space problems and reuse purposes, but extending their application anywhere is not a good solution: this in fact could entail energy consumption up to 10 times higher as compared to conventional systems. This is evident from the management experiences of advanced MBR technologies documented by Brepols and Schäfer in (2010) where, for a large scale WWTP (80,000 e.i.) characterized by high flow variations, energy consumptions were from 1.7 up to 9.3 times higher than those registered in conventional WWTPs (30,000–160,000 e.i.) working in the same conditions.

An example of a wrong application of this technology is represented by the case where a cross flow tubular membrane system, suitable to treat homogeneous and constant concentrated wastewaters, was adopted at a municipal wastewater plant – *$WWTP_7$ 3000 e.i.*– receiving sewage from a combined system. In this plant, the energy consumptions ranged between 4 and 6.4 kWh/m^3 and, even though many efforts were done to reduce the wastings – *the optimization of washing procedures, the addition of a pre-filtration stage and the redesigning of the pumping system* – consumptions are still from 3 up to 15 times higher than those of similar conventional activated sludge WWTP.

Another system clearly in contrast with energy saving goals is the UV disinfection technology, often proposed as the most credible alternative to chemical disinfection in wastewater treatment.

However considering energy consumptions issues only, at doses expected to be effective (40–80 mJ/cm^2), the only UV radiation entails an additional impact between 0.05 and 0.15 kWh/m^3 and this could be even higher according to some literature data stating the actual UV doses reducing pathogens are higher.

In any case, in contrast with the reasons supporting the adoption of such technology, the chemical dosage is always required either to pursue the bacterial reductions for reuse purposes and/or to support the UV inefficiency to reduce pathogens or to maintain the cleaning of the mandatory filtration steps.

Moreover considering that available chemical alternatives are supposed to be safe for the environment (also chlorine could be used without a significant by-products production if dosed at a specific ratio with ammonium nitrogen), adopting UV technology that implies such high energy consumptions, often does not make sense.

12.4 CONCLUSIONS

The energy saving in relation to wastewater treatment is an important goal, but stable results can be achieved only using integrated approaches and acting at different levels: on the design criteria; on technology choice; on plant monitoring and management practices.

Even though the energy saving should start from restoring the sewage systems itself, the fact that the WWTPs are rigidly conceived on the basis of the only designing reference data and are provided with inflexible structures and machines, implies significant energy wasting. Therefore a fundamental step to pursue energy saving goals should be the cooperation among the different subjects – *plant managers, designers and so on* – to meet the actual treatment needs, realizing the most appropriate and flexible solutions.

This work, reporting the energy savings achieved in northern Italy through some general interventions on the management and controls as well as through minimal up-grades on machines and structures, shows how the multilevel approach could lead to significant changes in energy consumption (Table 12.7).

Table 12.7 Energy savings for each WWTP, deriving from all interventions 2010–2013.

Plants	Period	Saving	Specific saving		
		kWh/period	KWh/m³	kWh/e.i.f.	kWh/e.i.b.
$WWTP_1$	Year	829,000	0.30	21	19
$WWTP_2$	Year	90,000	0.07	6.7	41
$WWTP_3$	H.S.	25,000	0.07	1.7	12
$WWTP_4$	H.S.	30,000	0.08	1.8	1.3
$WWTP_5$	6 months	37,000	0.14	6.4	48
$WWTP_7$	Year	113,000	0.69	63	137

e.i.: equivalent inhabitants; kWh/e.i.f and kWh/e.i.b.: calculated on the basis of flows and BOD_5 values respectively; WWTP: wastewater treatment plant; H.S.: high season (summer).

Lastly a careful evaluation between advantages and disadvantages is recommended when we assess the opportunity to adopt advanced technology implying very high energy costs: the authors' opinion in fact is that a conventional wastewater treatment, if properly dimensioned and provided by flexible structures, still represents the best way to constantly maintain effluent quality at even very restrictive requirements.

12.5 REFERENCES

Brepols C. and Schäfer H. (2010). Operational Cost. In: Operating Large Scale Membrane Bioreactors for Municipal Wastewater Treatment, C. Brepols (ed.), IWA Publishing, Chapter 7, London, UK.

Co.N.Vi.R.I. (National Committee for surveillance of water resources) (2011). Rapporto sullo stato dei servizi idrici (Report on the state of water services). Annual Report to the Parliament, Roma.

DPS (Development Department-Ministry of Economic Development) (2010). Obiettivi di servizio – Servizio idrico integrato- S.11 (Service objectives – Water Service – S11). http://www.dps.tesoro.it/obiettivi_servizio/servizio_idrico.asp (accessed 13 December 2013).

Terna S. p. A. (2012) Dati Statistici sull'energia elettrica in Italia (Statistical Data on Electricity in Italy) Available at: http://www.terna.it/default/Home/SISTEMA_ELETTRICO/statistiche/dati_statistici.aspx (accessed 10 December 2013).

Ragazzo P., Falletti L., Chiucchini N. and Serra G. (2013). Management optimisation and technologies application: a right approach to balance energy saving needs and process goals. *Water Practice and Technology*, **8**(2), 263–271.

Chapter 13

Energy factory: the Dutch approach on wastewater as a source of raw materials and energy

Ruud M. W. Schemen[1], Rutger Dijsselhof[2], Ferdinand D. G. Kiestra[3], Ad. W. A. de Man[4], Coert P. Petri[5], Jan Evert van Veldhoven[6], Erwin de Valk[7] and Henry M. van Veldhuizen[7]

[1]*Dutch Water Authorities Den Haag, The Netherlands*

[2]*Regional Water Authority Reest en Wieden, Meppel, Drenthe, The Netherlands*

[3]*Regional Water Authority Aa en Maas, 's-Hertogenbosch, Noord Brabant, The Netherlands*

[4]*Regional Water Authority Implementing Body Limburg, Roermond, Limburg, The Netherlands*

[5]*Regional Water Authority Rijn en IJssel, Doetinchem, Gelderland, The Netherlands*

[6]*Regional Water Authority de Dommel, Boxtel, Noord Brabant, The Netherlands*

[7]*Regional Water Authority Vallei en Veluwe, Apeldoorn, Gelderland, The Netherlands*

ABSTRACT

The energy factory is a sewage treatment plant (STP) that produces all the required energy itself (Energy factory). The concept has given a strong boost to sustainable innovation within the regional water authorities. This concept supports the philosophy behind the 'Wastewater management roadmap towards 2030', which gives a new perspective on wastewater as a source of raw materials, instead of a waste product.

It also contributes to a more sustainable, circular economy.[1] The energy factory concept enables the Dutch waterboards to achieve their sustainable targets as agreed with the national government. These regional authorities came together to create a vision of how to develop energy-producing sewage treatment plants. Most of them succeeded relatively quickly in defining a project of their own. Where there were uncertainties or gaps in knowledge, joint investigation provided the answers. This research was supported and facilitated by the Dutch Foundation for Applied Water Research (STOWA).[2] This practical approach is appealing, and was the keystone of the 'Green Deal' that the Dutch Water Authorities signed with the national government in 2011. In exchange, the department of economic affairs provided the sector with advice on legislative matters and economic support for enhancing biogas recovery from wastewater in the future. It has been shown that small modifications of an existing wastewater treatment installation can render it non energy consuming. By applying sludge pretreatment technologies (e.g., thermal pressure hydrolysis (TPH)), there is an energy surplus. The best opportunities are found at the local level in partnership with municipalities and industry. This vision is presented in more detail in the Wastewater management roadmap towards 2030. A number of energy factories have already been completed or are being constructed. Several others are still in the preparation phase, with attention being focussed on the techniques to be used, effect on sludge processing (dewaterability and dry solid concentration) and the period for return on investments. This chapter is about the development of the concept of the energy factory in the Netherlands and its opportunities for the Dutch waterboards. The described case studies show the first examples how this is brought into practise and gives global information about the cost savings.

Keywords: Energy Factory, Sustainability, Water authorities, Wastewater management roadmap, struvite, sidestream treatment, phosphate removal, Nereda, Anammox, TPH, biogas, CNG and LNG.

13.1 ENERGY FACTORY

13.1.1 The concept

Wastewater contains potentially 8 times the energy needed to purify it (Energy Factory). This means that an STP can generate energy, at least in theory. From this aspect, an energy-producing STP can considered to be an energy factory. This involves two major aspects: minimizing energy demand and maximizing energy production.

There are two pathways to *minimizing energy demand*: (1) Use of energy-efficient equipment (high-efficiency motors, plate aerators, etc.); (2) use of process

[1]A Resource-efficient Europe – flagship initiative of the Europe 2020 Strategy cas.

[2]STOWA is the knowledge centre for regional water managers in the Netherlands. STOWA develops, collects and shares the knowledge water managers need to properly perform their tasks.

configurations with minimal energy consumption (A-stage[3], sidestream treatment, pre-sedimentation, aerobic granular sludge (Nereda), etc.) (Nereda®) (Specialists day pre-treatment). Naturally, an important requirement is ensuring that effluent discharge limits are never exceeded.

On the other side of energy productive processes, digestion of sludge is currently the most important step, along with the reclamation/utilization of heat coming from the wastewater treatment. There are several different options for *maximizing this energy yield*: (1) Maximizing the volume of primary sludge (addition of chemicals in the pre-sedimentation and/or A-stage), (2) pretreatment of the sludge (particularly the secondary sludge which is non-readily biodegradable) to allow it to be further broken down in the digesters (mechanical or with high pressure/temperature, (3) increasing the yield of the conversion of biogas into electricity in Combined Heat and Power (CHP) systems with Organic Rankine Cycle (ORC) or fuel cells.

One of the advantages of thermal pressure hydrolysis (TPH) is that it almost doubles the capacity of anaerobic digesters in dry solid load. The application of TPH requires sidestream treatment in order to efficiently remove the extra nitrogen load. These systems commonly also reclaim phosphate, which is present in high concentrations in the digestate.

13.1.2 The history

In light of the increasing worldwide demand for raw materials and other issues, the international Global Water Research Coalition (GWRC) has called for new concepts and visions for the future of the STP. The Dutch contribution to this discussion was the NEWater concept (NEWater). Nutrients, Energy and Water are the major products addressed in this concept.

In early 2008, there was very little experience in the Netherlands on sludge pretreatment for increasing biogas yield and/or faster sludge decomposition. WBL (the regional water Board implementing body for Limburg) ascertained early on that there were real energy gains to be obtained from secondary sludge, but the available hydrolysis and other techniques were only economically viable for very large installations. In 2008, together with Dutch firm Sustec, they began investigating the potential of treating the sludge in a continuous process, ultimately applying the process at pilot scale in 2009. The results were encouraging enough for WBL to move to full scale application of thermal pressure hydrolysis (TPH) at the STP in Venlo. The first TPH in the Netherlands was completed in 2012. In the sequel, two more TPH pilots were constructed in 2010, one at the STP in Amersfoort and another at the STP Hengelo. Both were successful, and they ultimately led multiple regional water authorities to adopt the TPH technology. The regional water authority Reest & Wieden applied a different technology, a two-stage digestion process (thermophilic followed by mesophilic).

[3]A-stage as part of the AB-process.

The development of the energy factory concept is the result of a unique approach. Four regional water authorities (Aa & Maas, Hoogheemraadschap Hollands Noorderkwartier, Veluwe and Rivierenland) developed the energy factory concept as a response to a competitive project called by the Dutch Water Authorities. Together these four authorities investigated the potential to convert an STP into an energy factory. This joint approach generated so much enthusiasm that it did not remain limited to these four pioneers; it very quickly became apparent that over half of Dutch regional water authorities were interested in joining this network organisation. By becoming a member, every water authority undertook the task to transform its own STP into a concrete business case, setting out first the economic target, and then actually implementing the conversion. The projects described in the second part of this chapter are the tangible results.

The water authorities have come to agreement with the national government on several issues.

- Multiyear Arrangement (MYA): in the period of 2005 to 2020, the annual energy efficiency increases with 2% so in 2020 the specific energy consumption is 30% less compared to 2005.
- Climate Agreement/Energy Agreement: in 2020, 40% of the energy consumed by regional water authorities should be self-generated.

When national and international policies get connected on climate and energy, the political support on a concept promoting these goals is increasing. This gave an extra boost to this concept. Likewise, the national government is enthusiastic about the water authorities' approach. This led to the 2011 'Green Deal' between the Dutch Water Authorities and the ministries of Economic Affairs, Agriculture & Innovation and Environment & Infrastructure. In this deal, apart from making commitments to eliminate a wide range of impediments, the government also allocated a research budget of half a million euro. This research budget has been used to make further technological development in the field of gasification and supercritical gasification possible.

13.1.3 The present state

In several cases (Kampen, Nijmegen) the STPs were converted to energy-neutral systems with minor modifications such as fine bubble aeration, high-efficiency CHPs and centralised sludge treatment. Furthermore, in order to create an energy-producing system, greater investment and preparation time are required. In Venlo the secondary sludge is treated to produce extra biogas. This biogas is needed to produce electricity (CHP) and for the production of steam for the TPH proces. In Echten they build an thermophilic digester to produce extra biogas. Both examples are discussed in the second part of this chapter.

In 2014 a number of energy factories are currently being built in the Netherlands using a combination of various technologies. It is interesting to note that in practice, these generally involve energy production as well as recovery of nutrients, due to the fact that the digestion process releases a large amount of nitrogen and phosphate into the digestate.

Although the regional water authorities were initially limited in improving their own systems, at present they are seeking for new options in collaboration with the local community. For example:

- Heat from the local environment being used by the STP or, vice versa, heat from the STP being used to heat local homes (STP Apeldoorn).
- Biogas produced by the STP is being used as a green fuel for vehicles or being supplied directly to industrial consumers (STP Den Bosch).

Local conditions lead to different opportunities and solutions and a blueprint for the best Energy factory doesn't exist.

The ultimate choice for technology depends on the results of the tendering process. In order to take maximum advantage of the knowledge in the market, the general practice is not to prescribe a specific technology but rather to define results (i.e., biogas and phosphate production, dewatering efficiency and breakdown of organic dry solids (ODS)). This allows the regional water authority to be the launching customer for the application of innovative techniques.

At the time of this writing (May 2014), the energy factories in Venlo and Echten are operating, Amersfoort, Apeldoorn[4] and Tilburg are under construction, and tendering for Hengelo and Den Bosch has started. Additionally, with modification to the aeration, changes in operations and replacement of CHPs, many STPs have already begun producing more biogas and electricity. Table 13.2 shows the key figures for these energy factories. The projects are described in more detail in the second part of this chapter.

13.1.4 Economic aspects

The treatment of wastewater costs energy and generates sludge alongside purified wastewater. In the Netherlands, sewage sludge can no longer be used in agriculture, due to the high content of heavy metals such as copper and zinc (Disposal of sewage sludge by destination). As a result, most sludge is incinerated in mono-incinerators. Generally speaking, the costs of sludge incineration are considerable (€40–€60 per tonne of sludge cake). Converting this sludge into biogas at this point serves two purposes: (1) The extra biogas can be converted into electricity, reducing the amount of electricity that needs to be purchased elsewhere. (2) It reduces the amount of sludge remaining, which reduces the costs of sludge

[4]The STP Apeldoorn is being implemented in phases. This refers to the last section (see also case description).

processing. In addition, it has been shown that some measures have a positive effect on sludge dewatering, meaning that less water need be transported and evaporated. This combination makes the conversion of STPs into energy factories financially attractive as well. The Investments for techniques like TPH are high and are only attractive for larger STP. Therefore in practise they are mostly combined with centralised sludge treatment. The same applies for the recovery of phosphate. Struvite deposits can lead to extra maintenance costs, especially when a higher concentration of dry matter is treated in the digestion process. By removing the phosphate in a controlled manner (in the form of struvite), this product can be sold separately, and it also means reduced maintenance costs and less sludge.

13.1.5 The future (Wastewater management roadmap towards 2030)

In the forthcoming years, the focus within the wastewater management sector will expand into the additional utilisation of other sustainable energy sources and the recovery of raw materials. The following advances are envisioned:

- Further development of technologies to reduce energy consumption and increase energy yield (Cold Anammox, fuel cells, gasification, supercritical gasification).
- Conversion of biogas to transport fuel (Compressed Natural Gas (CNG) and Liquefied Natural Gas (LNG)) to expand sustainability gains into the transportation sector.
- Energy supply from renewable sources (wind turbines and solar panels) to STPs.
- Further monetization of raw materials from wastewater such as biopolymers and bioplastics.
- Use of microfine filters for the separation of cellulose.

Additionally to these technological advances, the individual local characteristics will become more important. This concerns the physical environment, the administrative structure and the priorities for spatial planning. The local need and availability of raw materials, energy and water are relevant as well. The roadmap describes a set of local perspectives for future sustainable solutions (Wastewater management roadmap towards 2030). These are not ready-made blueprints, but should be seen as a useful starting point for identifying the options and targets. The cooperation with the local community on jointly reclaiming energy and raw materials is a component of the closed-circle economy.

The water authorities were asked to define the impact of the energy factory on energy production in future. The results are shown in Table 13.1. At present, the annual total energy consumption for wastewater purification by all regional water authorities is approximately 8 PJ per year (primary energy consumption).

Table 13.1 Impact of the energy factory on the energy production in future.

Year	2005	2012	2016	2020
Energy generated by STP (PJ/year)	1.7	2.3	3.1	4
Percentage of self-consumption (%)	20	30	40	50

13.2 CASES

In this section different cases are described. A general description of the plants is given in Table 13.2 and Table 13.3 gives an overview of the measures taken to transform the STP into an energy factory.

13.2.1 LNG production at 's-hertogenbosch

The major part of the 's-Hertogenbosch STP dates from 1973 and will not achieve the stricter discharge requirements nowadays (Ntot 12 → 7 mg/l; Ptot 2 → 0,7 mg/l). The new system slated for construction will be designed for 320,000 p.e. In addition, the secondary sludge from three other STPs (Aarle-rixtel, Asten and Dinther) will be centrally digested in 's-Hertogenbosch STP.

Tendering for the installation has yet to take place and the exact treatment technologies have therefore not been finalised. The following measures are under consideration:

- Improved pre-sedimentation by means of injection of polyelectrolyte (PE).
- Augmentation of the biology by means of Nereda technology (Nereda®).
- Improvement of the digestion process by application of TPH or thermophilic digestion.
- The STP will no longer generate its own electricity, and the heat required will be obtained from the municipal biomass power station.
- The biogas will be partially converted into bio-LNG, which will be supplied to the municipal waste service as vehicle fuel. The rest will be supplied to industrial consumers, which are eager to increase their green credentials.
- Sidestream handling of the resulting digestate, after dewatering.

Calculations show that the direct supply of the biogas to industrial consumers is financially neutral as compared to the STP generating its own electricity in the CHP. The refinement of biogas to bio-LNG is financially interesting if the product can be sold directly to an end consumer as can be seen from Table 13.4 At the same time, the municipality saves on diesel costs (Table 13.5). The moment that the heat from the municipal biomass power station (fired by brush/cuttings) is used for heating the digestion, the capacity of this installation will be used optimally.

With these mutual deliveries, there are local opportunities for all parties to seize some sustainability (energy and emissions) and financial (total savings across the value chain of up to €800,000 or more per year) advantages.

Table 13.2 Description of the WWTP *before* transformation to an energy factory.

	Plant Echten	**Plant Apeldoorn**	**Plant Olburgen**	**Plant Tilburg**	**Plant Venlo**	**Plant Den Bosch**	**Plant Amersfoort**
Design capacity in p.e. based on 150 g TOD (total oxygen demand)	169.000	317.000	137.500	350.000	300.000	310.000	307.000
Waterline							
Primary treatment	Screens (2 × 3 mm), grit- and sand removal	Screens, primary clarifiers, grit- and sand removal	Screens, primary clarifiers, grit- and sand removal	Screens, grit and sand removal, primary clarifiers	Screens, grit and sand removal	Screens, primary clarifiers, grit- and sand removal	Screens, primary clarifiers, grit- and sand removal
Secondary treatment	Activated sludge (50% with biological P-removal and pre-denitrification)	Activated sludge (with pre-denitrification) m-UCT Biological and chemical P-removal	Activated sludge (with biological and chemical P-removal and pre-denitrification)	Activated sludge (with pre-denitrification and chemical P-removal)	Activated sludge (with biological P-removal)	Activated sludge (with pre-denitrification)	Activated sludge (with pre-denitrification) chemical P removal
Tertiary treatment	–	–	–	–	–	–	Sand filtration (enhanced phosphorus removal)
Sidestream proces (after transformation to an Energy Factory)	Demon Anammox and Struvite precipitation (Airprex)	Demon (since 2010) (must be increased, not planned yet) After Nuresys voor P-removal	Struvite precipitation and Anammox	Anammox	–	? (project is in the tendering phase)	Demon since 2013 After Enhanced Demon and Pearl-reactor voor P-recovery

Sludge line							
Primary sludge	–	Gravity thickener	Gravity thickener	Gravity thickener	–	Gravity thickener	Gravity thickener and After transformation Gravity followed by belt thickener
Secondary Sludge	Gravity thickener followed by Belt thickener (new)	Belt thickener pre thickening existing BT, after buffer end thickening with ZBP to 12%	Gravity thickener	Belt thickener	Gravity thickener (2×350 m^3)	Stirred up with the primary sludge and treated with gravity thickeners	Belt thickener (4,55) and after Wasstrip thickening by ZBP to 13%
Sludge digestion Before/after	None/1 mesophilic & 1 thermophilic tank	4[1] mesophilic tanks parallel (before and after)	2 mesophilic tanks (before and after)	2 mesophilic tanks 3 mesophilic tanks	None/2 mesophilic tanks	2 mesophilic tanks (before and after	3 mesophilic tanks (before and after)
Dewatering	Plate and frame filter press	3 centrifuges	2 centrifuges	2 centrifuges	2 centrifuges	1 centrifuge	2 centrifuges
Energy							
Using the biogas	CHP	CHP (future CNG *and LNG), heating* households	Heating purpose *at potato* processing industry	CHP (future CNG?)	–	CHP (future biogas *and LNG)*	CHP
Energy consumption before transformation to an Energy factory	3,5 MWh/year Incl bellen	9,5 MWh/year	2,4 MWh/year	7,7 MWh/year	10 MWh/year	5,3 MWh/year	6,7 MWh/year
Energy production before transformation to an Energy factory	0	12,3 MWh/year 22.4 GJ/year Heat	1,0 MWh	5,0 MWh/year	0 MWh	4,5 MWh/year	3,1 MWh/year

(Continued)

Table 13.2 Description of the WWTP *before* transformation to an energy factory (*Continued*).

	Plant Echten	**Plant Apeldoorn**	**Plant Olburgen**	**Plant Tilburg**	**Plant Venlo**	**Plant Den Bosch**	**Plant Amersfoort**
Energy consumption after transformation to an Energy Factory	4,5 MWh/year	10 MWh/year	1,8 MWh		7–8 MWh/year	7,5 MWh/year (mainly because of the double sludge amount to be treated)	7,1 MWh/year
Energy production after transformation to an Energy Factory	4,7 MWh/year	18.7 MWh/year 30 GJ/year Heat	Extra biogas 300.000 m^3/year (replacing natural gas)		2–3 MWh/year	0 MWh	8.2 MWh/year
Energy costs before transformation to an Energy Factory	420.000 euro	–160.000 €/year	290.00 euro/year	2.688.000 euro/year	1.000.000 euro/year	100.000 euro/year	370.000 €/year
Energy costs after transformation to an Energy Factory	60.000 euro	–520.000 €/year	150.000 euro/year		700.00 – 800.000 euro/year	750.000 – 1.200.000 = 450.000 euro benefit	– 140.000 €/year
Sludge production before and after transformation	6.000 tds/year 4.400 tds/year	11.100 tds/year 46.500 t/year 8.700 tds/year 26.750 t/year	Equal	6.797 tds/year 14.866 tds/year	7.000 tds/year 30.000 t/year 4.200 tds/year 15.000 t/year	3.650 tds/year 6.570 tds/year	23.000 t/year 19.000 t/year
Sludge disposal costs before and after transformation	2.700.000 euro/year 2.200.000 euro/year	5.900.000 euro/year 3.400.000 euro/year	Equal	6.775.000 euro/year	1.530.000 euro/year 765.000 euro/year	785.000 euro/year 1.410.000 euro/year	2.900.000 euro/year 2.400.000 euro/year
Biogas production before and after transformation	0 2.000.000 Nm^3/year	6.000.000 Nm^3/year 9.000.000 Nm^3/year	700.000 Nm^3/year 1.000.000 Nm^3/year	2.480.000 Nm^3/year 9.850.000 Nm^3/year	0 m^3/year 2.240.000 Nm^3/year	2.0000.000 NM^3/year 4.0000.000 NM^3/year	2.200.000 Nm^3/year 4.200.000 Nm^3/year
Investment costs	12.500.000 euro	10.000.000 euro	900.000 euro		6.000.000 euro	13.000.000 euro	10.500.000 euro

[1]2 mesophilic tanks for the digestion of fluid waste.

Table 13.3 Overview of the measures taken to transform the STP into an energy factory.

STP	Aeration measures	Digestion measures	Return on investment time	Capacity of sludge processing installation (tonnes DS/year) & reduction of slib quantity (%)	Nutrient reclamation	Biogas production (Nm^3/ years)	Remarks
Tilburg	Anammox	TPH	<10 years	25,900 45% DS 59% ODS	Phosphate 438 tonnes struvite	8,000,000	In the future, potential partnership with fruit/ veg waste and manure digestion
's-Hertogenbosh	Reference design based on a Nereda system for the extra aeration	Depending on tender. Reference design based on thermophilic digestion	+/– 8 years	19,900 29%	Phosphate. Is a desirable, but not an obligation	4,000,000	Supply of bio-LNG to the municipal cleaning service, and supply of biogas to industry
Amersfoort	DEMON, surface aerators replaced with bubble aerators	TPH hydrolysis of digested sludge (internal cycle over digester) Lysotherm	6.7 years	11,555 49% DS 57% ODS	Phosphate Wasstrip & Pearl	4,200,000	In the future, potential drying of sludge with residual heat
Apeldoorn	None	TPH of secondary sludge (Sustec)	+/– 8 years	11,000 50%	Phosphate Nuresys	6,500,000	Co-digestion and supply of heat to residential area
Echten	surface aerators replaced with bubble aerators, and a DEMON	2-stage digestion thermophilic/ mesophilic	approx. 9 years	6800 27%	Phosphate 2000 tonnes struvite (MAP reactor)	1,800,000	First thermophilic digester for household wastewater
Venlo	The STP has sufficient capacity.	TPH and agro-digesters, both new CHP	+/– 8 years	7000 40%	Potential phosphate	2,240,000	First continuous TPH installation in the Netherlands
Olburgen	Plate aeration	None, extra PE dosage in pre-sedimentation tank	7 years	1300 37%	Phosphate	1,000,000	Supply of biogas to industry

Table 13.4 Comparison in costs (euro) between de different alternatives for the STP 's-Hertogenbosch.

Alternative	Investments	Annual costs	Annual revenues	Netto profit
CHP	3.100.000	360.000	250.000	–110.000[5]
LNG	7.000.000	670.000	1.350.000	725.000

Table 13.5 Comparison in costs (euro) between fuels for the municipality of 's-Hertogenbosch.

Alternative	Annual costs for transition to gas	Annual costs	Totall	Netto savings
Diesel		1.100.000	1.100.000	
LNG	120.000	880.000	1.000.000	100.000

13.2.2 Thermophilic digestion at STP Echten

Until 4 years ago, the regional water Board Reest & Wieden had three digestion installations (Beilen, Steenwijk and Meppel) operational at its seven STPs. Two (Beilen and Steenwijk) of the three would, however, require significant investments to be upgraded to cater for the future demand. A study has shown that the construction of a single central digester is more viable than renovating the existing systems, and that the STP in Echten is a suitable location for the reason that this is the central location for sludge dewatering. For this reason there is a plate and frame filter press present at the STP. To increase the gas yield, a choice has been made for digestion at two temperature levels.

The following steps have been taken:

- Successful creation of thermophilic (10 days at 55°C) and mesophilic (14 days at 35°C) digestion
- Sidestream treatment (DEMON), 360 kg Kj-N in 450 m^3 per day
- Installation of CHPs (600 kW(e)) allowing the STP to fully provide for its own energy needs (approx 4 million kWh per year)
- MAP reactor for the reclamation of 200 tonnes of struvite per year (Airprex)

In times with lower biogas production there is an external dosage of glycerine. Additionally, the use of thermophilic digestion has had a positive effect on sludge dewatering.

The installation in Echten is the first in the Netherlands to operate using this process. It became fully operational in mid-April 2013. Figure 13.1 gives an overview of the STP Echten after the transformation to an Energyfactory.

[5]Delivery route to industry is classified but neutral compared to CHP.

Figure 13.1 Overview STP Echten of the waterboard Reest & Wieden.

13.2.3 Delivering biogas from STP Olburgen to potato industry

The Olburgen STP treats primarily domestic wastewater. For the treatment of the industrial wastewater, the regional water authority established a separate entity ('Waterstromen'). Waterstromen has a system onsite at the Olburgen STP that treats the wastewater from "Aviko" (a food sector company that processes potatoes). For the purification of the industrial wastewater, Waterstromen uses anaerobic treatment followed by a struvite reactor and an Anammox reactor. New disc aerators were installed at the Olburgen STP in 2004. These experienced significant fouling, and must be replaced. In the old situation, the biogas generated at the Olburgen installation was converted into electricity in a CHP. A portion of the heat was used for heating the digestion, and the remaining portion was dissipated as waste heat.

The following steps have been taken:

- A biogas line from the STP Olburgen to Waterstromen for the supply of biogas
- A hot water line from Waterstromen to the STP Olburgen for the supply of heat/residual heat to the STP for heating of the digestion
- Installation of a chemical dosage unit on the pre-sedimentation tank in order to increase the primary sludge production
- Modification of the operation of the pre-aeration (turns off 75% of the time) for energy savings
- Replacement of the bubble aerator discs by more efficient plate aeration, leading to energy savings

With these measures, the biogas produced at the STP Olburgen can now be supplied to Waterstromen, where it is combined with a portion of its own biogas production, before being supplied to Aviko, some 5 km away from the STP.

Supplying the industrial customer directly means that the gas does not have to be transported to the natural gas network, so it does not need to be refined to natural gas quality. Aviko uses the natural gas for heating purposes, which means the energy content of the biogas is utilized virtually completely. Waterstromen, which at present also has a surplus of residual heat, supplies heat to the STP to heat the digestion process. This heat, which is released from the generation of power in the CHP, was up to now dissipated as waste heat. The earn-back period of the first four measures is 1 to 2 years. The earn-back period of the last measure, replacement of the bubble aerators, is longer (15 years). On balance, the earn-back period is 7 years.

Additionally, all the concentrated waste water from the digestion is now routed from the STP to the struvite reactor, so this phosphate is also reclaimed, creating a 'phosphate factory.' The partnership that has now been initiated is seen as a first step toward further cooperation. Further advantages of synergy that could be obtained from scaling up biogas production from the digestion of waste, cuttings, and so on are now being considered.

13.2.4 Centralised sludge treatment at STP Tilburg

The current Tilburg STP processes its own sludge as well as the sludge from two smaller installations. In addition to this location, the bulk of the sludge from regional water authority De Dommel, is dewatered at the Mierlo site (17,000 tonnes DS/year). At present, only the sludge at Tilburg is digested. The rest (primary and secondary) is, after dewatering at Mierlo, transported by road to the SNB mono-incinerator in Moerdijk. A scenario study has shown that central processing of all sludge from De Dommel at this location is the optimal scenario, in terms of energy, robustness and sustainability. This also maximally utilises the advantages of TPH.

The following steps have been taken:

- Receiving station for mixing external dewatered sludge with the liquid sludge from the Tilburg location
- Completion of a TPH installation (Cambi)
- Construction of a sidestream treatment facility operating under the Anammox principle.
- Expansion of the existing digestion volume (from $2 \times 4{,}440\ m^3$ to $3 \times 4440\ m^3$)
- Expansion of the CHP capacity from 2×265 KW to 1500 + 1200 KW.
- Expansion of the sludge dewatering.
- Phosphate removal through the use of Phospaq (438 tonnes struvite per year).

Tendering

Because the water authorities' expertise in the area of optimisation of biogas production is limited, the choice was made to adjust the tendering process accordingly. The black box approach was used, in which the request is defined in functional terms and it is then up to the market to find the best solution. This puts the responsibility and process guarantee completely in the hands of the market. The market is also responsible for adequate reduction of the extra nitrogen load released with the digestate.

Figure 13.2 gives an overview of the STP Tilburg before the transformation to an Energyfactory.

Figure 13.2 Overview of the STP Tilburg.

13.2.5 Hydrolizing secondary sludge with TPH at STP Venlo

The Venlo STP has no pre-sedimentation tanks, and so has no digestion capacity. In the past, WBL elected to dry the sludge and sell it to ENCI, where it was burned in the cement kilns along with other fuel. Drying sludge is relatively expensive and energy-intensive. Now that these installations are reaching the end of their useful life, alternatives are being considered. Based on laboratory research and pilot studies, the regional water authority opted for continuous thermal pressure hydrolysis. No additional sidestream treatment is required for nitrogen or phosphorous removal, since the STP has sufficient residual capacity.

The following steps have been taken:

- Construction of two agro digesters (each 2200 m^3)
- Construction of a filter in the excess sludge feed to the installation (approx. 1 mm mesh size)

- Construction of a TurboTec® installation with peripherals, retention time 1 hour at 4 bar and 140°C
- Operation and maintenance contracted for a period of 6 years

The installation in Venlo processes sludge from Venlo STP, Venray STP and Gennep STP. The hydrolysed secondary sludge is converted into biogas in the digester. Dry solids breakdown is 40%, and the breakdown in ODS is 55–60%. There is a guarantee for 28.8% DS for the sludge dewatering. The total investment (filter, mechanical concentration, TPH, digestion, CHP and dewatering) is €6 million. The annual revenues are approximately €1 million, 25% of which is reduced electricity costs and 75% of which is attributable to lower sludge disposal costs. The investments will be earned back in approximately 8 years. The regional water authority has another sludge dryer in operation, at the Susteren STP. The sludge dryer at Hoensbroek is already closed and the potential for a TPH on a scale of 8500 tonnes DS/year at this site is still being investigated.

13.2.6 Digestion of external biomass at STP Apeldoorn

The transition of the Apeldoorn STP into an energy factory is carried out in phases. The first step to achieve complete energy self-sufficiency, the regional water authority Vallei & Veluwe commissioned a separate digester in June 2009 to process biomass from third parties. The biogas generated from this digester (3,500,000 m^3 in 2013) is combined with the biogas from the STP (3,000,000 m^3 in 2013) and converted into electricity in a CHP (8,750,000 kWh from external biomass and 12,000,000[6] kWh for the STP). The combustion of the biogas in the CHP releases more heat than the STP itself requires. In cooperation with the municipality of Apeldoorn and power company Essent, a useful application for this heat was found: supplying heat to Essent, which distributes it through the heating network to the residential neighbourhood Zuidbroek. In addition to the heat from the STP (3000 GJ per month), Essent generates heat by burning biomass. The heat from the STP is enough to heat 1170 homes. This arrangement has helped all three parties to meet their own sustainability goals.

For the STP, this is the first step towards increasing energy production. At this point, there has been some research into refining the biogas to natural gas quality (CNG) for potential future sale of the extra biogas that will be generated after implementation of TPH.

The following steps have been implemented since 2009.

- Construction of extra digester for the processing of external biomass (volume 2600 m^3)
- Pasteurisation step for the processing of category 3 waste (min. 1 h at 70°C).
- Replacement of existing CHP with high-efficiency CHPs

[6]Normally, this figure is 14,000,000 kWh, but due to maintenance this was less in 2013.

- Use of ORC at the CHP
- Refining the biogas into green gas (trial)

The following steps will be implemented in the next 3 years

- Improvement of the biogas production from secondary sludge through the use of TPH, thereby increasing the available digestion volume.
- Trial with refining biogas into LNG, including filling station.

Based on the experiences at Apeldoorn, Vallei & Veluwe has identified similar opportunities for Ede (residential heating) and Harderwijk (green gas).

13.2.7 Reclamation of energy and resources at STP Amersfoort

To attain sustainable purification, the Amersfoort STP pursued a process starting with a focus on energy, which was then expanded to cover raw materials. The next step was a regional approach (including the neighbouring purification facilities of Nijkerk, Soest and Woudenberg). This led to a project consisting of the following phases: (1) central digestion with TPH (2013), (2) Bio-P and P reclamation (2014) and (3) sludge drying with residual heat (2015). In parallel a study was conducted which showed that PE dosage for better separation in the pre-sedimentation tank could have a negative effect on the effluent quality, and that use of ORC in the CHP is not economically viable on this scale. Phases 1 and 2 have been tendered simultaneously in accordance with UAV-GC (Dutch Form of Contract), with a maximum investment amount. Table 13.6 gives the result of the tendering phase.

Table 13.6 Guarantees that have been received for Amersfoort STP.

Result	Unit	Guarantee
STP energy-neutral	%	130
Region[a] partially self-sufficient	%	70
Sludge dewatering	% DS	31
Return on investments period	years	6.7
Phosphate reclamation regional	%	42
Reduction in chemical use	%	50

[a]STP's Amersfoort, Nijkerk, Woudenberg and Soest.

The drying of the sludge with residual heat at STP Amersfoort has proven to be impossible within the project period. The choice has now been made for a pilot at STP Ede in the 2014–2016 period. The results of this pilot will be reworked for STP Amersfoort in 2015.

13.3 CONCLUSION(S)

It has been shown that the innovative approach of the Dutch regional water authorities has led to a broadly supported concept with a new perspective on wastewater (Waste as a Resource for Energy and Raw materials). The results of this new concept have been proven by applying it in practice directly and initially on a small scale. The coordinated, sector-wide approach has further increased the support base for it. Additionally, this has also given the industry the freedom to put their innovative ideas into practice, with the regional water authority taking on the role of launching customer. The projects were not limited to the field of wastewater purification, but instead the STPs were considered as a component of the environment. As a result broad-spectrum solutions with greater benefit to the environment and financial profits arose. The more sustainable way need not always to be the more expensive way.

13.4 REFERENCES

Energy Factory – Water Boards Inside Out, February 2009, www.energiefabriek.com. (accessed 18 February 2014).

Wastewater management roadmap towards 2030 – A sustainable approach to the collection and treatment of wastewater in the Netherlands, November 2013, www.uvw.nl. (accessed 18 February 2014).

Verslag specialistendag voorbehandelingstechnieken 21 juni 2011 (STOWA, Energiefabriek). (Report Specialits day pre-treatment techniques 21 juni 2011). www.stowa.nl. (accessed 18 February 2014).

Afzet van zuiveringsslib naar bestemming, 1981–2011. (Disposal of sewage sludge by destination 1981 – 2011), CBS, 2013 http://www.compendiumvoordeleefomgeving.nl/indicatoren/nl0154-Afzet-van-zuiveringsslib-naar-bestemming.html?i=1-4. (accessed 18 February 2014).

Nereda® praktijk onderzoek 2010–2012 (Nereda® practical research 2010–2012), STOWA, 2013, Report 29.

NEWater, Op weg naar de RWZI 2030 (The STP of the future 2030), STOWA, 2010, Report 11.

Chapter 14

A new perspective on energy-efficiency and cost-effectiveness of sewage treatment plants

Helmut Rechberger[1] *and Alicja Sobańtka*[2]

[1]*Vienna University of Technology, Institute for Water Quality, Resource and Waste Management*

[2]*Vienna University of Technology, Centre for Water Resource Systems*

14.1 INTRODUCTION

The cleaning performance of sewage treatment plants (STPs) is typically expressed by removal rates (e.g., of nitrogen (N) or phosphorus (P)). There are also mandatory emission limits, such as a N-removal rate in excess of 70% at temperatures >12°C for all STPs greater than 5,000 population equivalents (PE) as well as an NH_4^+ concentration of less than 5 mgN/L (Phillippitsch & Grath, 2006), (Thaler, 2009). The advantage of using the N-removal rate as a measure of cleaning performance is that it can be easily calculated. The N-compounds are quantified by the N-content of the influent and effluent. However, in this way, N is not differentiated in its various chemical species (e.g., NH_4^+, NO_3^-, N_2 etc.); thus, NH_4^+ and NO_3^- have equal weighting factors in the total N-content, although the environmental impact of these two forms of N are different. The Austrian Water Act and Life Cycle Impact Assessment (LCIA) assign different emission limits and impact factors to individual N-compounds (BGBl II Nr, 98/2010), (BGBl II Nr, 99/2010), (BGBl II Nr, 461/2010), (Guinée *et al.* 2002)). For example, NH_4^+ has a higher eco-toxicological impact on water bodies than NO_3^- and therefore a lower emission limit. Other studies have concluded that the role and fate of aquatic N_{org} is still unknown, which implies that N_{org} in water bodies cannot be regarded as NH_4^+ (Westgate & Park, 2010). N_2O plays a major role in global warming (Kampschreur *et al.* 2009). The omission of gaseous N-compounds is a drawback in the use of the N-removal rate, because wastewater purification can result in atmospheric pollution (e.g., the emission of N_2O and NH_3). Wang and colleagues have stressed the problem of limited integration of N_2O emissions from wastewater treatment (WWT) into the overall N pollution reduction, and propose creation of a greenhouse gas crediting system (Wang *et al.* 2011). In Austria, a performance indicator ('Leistungskennwert': LW) has been developed to describe the cleaning capacity of STPs that considers the effluent concentrations of NH_4^+ and NO_3^-. Specific weighting factors are assigned to each species that reflect their different harming potential on water bodies (ÖWAV 2000). Other N-compounds in the effluent, such as N_{org}, gaseous N-emissions such as N_2O, and N transfer to sludge, are ignored.

To overcome the disregard of the speciation of N in the evaluation process of wastewater treatment statistical entropy analysis (SEA) has been applied (Sobańtka *et al.* 2012). This allows to quantify the entropy reduction achieved by a STP expressing the benefit (cleaning performance) of the facility. Specifically, SEA quantifies the distribution of a substance (e.g., a heavy metal) among different material flows (e.g., waste, fly ash, and wastewater) before and after a process (e.g., waste incineration). The change in the distribution of the substance then indicates the concentrating power relative to the extent of dilution (dispersion) of the particular process (Rechberger & Brunner, 2002). To date, SEA has been primarily applied to the field of waste and resource management to assess the efficacy of different processes in recovering substances such as heavy metals (Kaufmann *et al.* 2008; Rechberger ,2001a, 2001b, 2012; Rechberger & Graedel, 2002; Yue *et al.* 2009). SEA has subsequently been extended (eSEA) to enable its application to processes in which the specification of chemical compounds is highly relevant, as is the case for N. Such a system can, for example, specify the N budget of a farming region. Statistical entropy, applied as a measure of concentration and dilution thus serves as an agri-environmental indicator (Sobańtka *et al.* 2013). In a separate study, the advantages of eSEA over the traditional N-removal rate for the evaluation of the N-removal performance of WWT systems have been demonstrated (Sobańtka & Rechberger, 2013).

The cleaning performance of STPs requires expenses, which can be expressed as energy consumption and costs. The energy budgets and costs of STPs have been studied extensively. Both metrics are usually referred to the pollution load of the wastewater, expressed by population equivalents (PE). According to state-of-the-art literature, large STPs (>100,000 PE) are able to operate more energy-efficiently and cost-effective than small plants (Agis, 2002), (Nowak, 2002), (Lindtner *et al.* 2008), (Kroiss & Svardal, 2009), (Hernandez-Sancho & Sala-Garrido, 2009), (Lindtner, 2010). In this chapter, both energy consumption and costs are related to the N-removal performance of numerous Austrian STPs, quantified by eSEA, and the influence of the size of STPs for energy-efficiency and cost-effectiveness of the N-removal performance is investigated.

14.2 METHODS AND DATA

14.2.1 Application of eSEA for the assessment of the N-removal performance of STPs

The eSEA is applied to evaluate the N-removal performance of Austrian STPs. First, a N-mass balance must be established: assuming that there is no storage of N in the STP, the incoming N must equal the N that leaves the STP. Figure 14.1a illustrates the flow of N-compounds before and after WWT assuming the state-of-the-art biological treatment including both nitrification and denitrification. Figure 14.1b illustrates the effect of WWT according to statistical entropy.

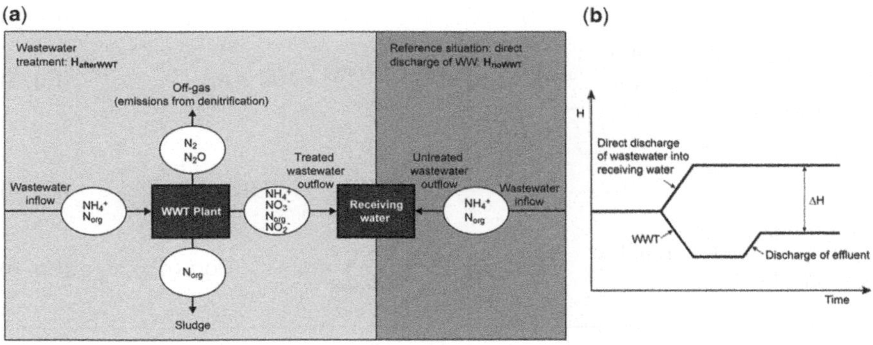

Figure 14.1 Description of the N-treatment by means of statistical entropy (a) Left: Schematic illustration of N-compounds through the WWT process; Right: Reference situation: the wastewater is discharged without treatment into the receiving water. (b) The direct discharge of wastewater to the hydrosphere results in an entropy increase. WWT reduces the entropy, and the discharge of the effluent to the receiving water results in a (small) increase. ΔH quantifies the cleaning performance of the WWT process (taken from Sobańtka & Rechberger, 2013).

The emission of every single N-compound as described in Figure 14.1a will result in a dilution process in both the receiving water and the atmosphere (i.e., the environmental compartments). A later dilution of the N in the sludge due to further sludge utilization is not considered. The water or air mass that dilutes a N emission depends on both the emission concentration and the concentration of the respective N-compound in the receiving environmental compartment. For example, surface waters with a good ecological status contain approximately 1 mgN/L of NO_3^- (BGBl II Nr, 461/2010). Thus, an emission of 1 mgN-NO_3^-/L would not contribute to the statistical entropy, but an emission of 10 mgN-NO_3^-/L would produce entropy because of dilution. A detailed mathematical description of eSEA can be found in

(Sobańtka *et al.* 2012). The calculation of the statistical entropy of N-compounds is additionally outlined in Table 14.1.

Table 14.1 Brief overview of the computation of the statistical entropy of N-compounds in STPs.

Step 1: measured data	$\dot{M}_i, c_{im}, c_{im,geog}$	
Step 2: normalization of the mass flow	$m_{im} = \frac{\dot{M}_i}{\sum_i \sum_m \dot{X}_{im}}$	(Eq. 1)
	where	
	$\dot{X}_{im} = \dot{M}_i * c_{im}$	(Eq. 2)
Step 3: calculation of the diluting masses for each N-compound	$m'_{im} = m_i * \frac{c_{im} - c_{im,geog}}{c_{im,geog}} * 100 + m_i$	(Eq. 3)
	and	
	$c'_{im} = \frac{c_{im} * c_{im,geog}}{c_{im} - 0.99 * c_{im,geog}} * 0.01$	(Eq. 4)
Step 4: computation of the statistical entropy	$H(m'_{im}, c'_{im}) = -\sum_i \sum_m m_{im} * c'_{im} * \log_2(c'_{im})$	(Eq. 5)

$\dot{M}_i$ is the measured mass flow in kg per day. The index i refers to the wastewater, effluent, off-gas, sludge, and the environmental compartments atmosphere and hydrosphere. The variable c_{im} (in kgN/kg) corresponds to the measured concentration of a N-compound (m) in the particular mass flow i. The background concentration of a compound m in an environmental compartment i is indicated by $c_{im,geog}$. The mass flows are normalized according to Equation 14.1. The denominator of Equation 1 equals the total flow of N through the STP such that the masses m_i are related to one mass-unit of N (e.g., kg in the effluent per kg of the total N throughput). Normalization is required to make processes of different size comparable. In the next step, the diluting masses m'$_{im}$ and the corresponding concentration terms c'$_{im}$ are calculated according to Equation 3 and Equation 4, respectively. The mass-function calculates how much water or air is needed to dilute the emitted concentration to its corresponding background concentration. The dimensionless mass-function for the N in the sludge is computed according to Equation 1. The statistical entropy H is then calculated according to Equation 5 for every N-compound in the effluent, the atmosphere and the sludge. Thus, the statistical entropy is a function of the mass flows, such as the wastewater, effluent, air and sludge, the emission concentrations and the corresponding background concentrations of all N-compounds. More diluted mass flows and larger differences between the emitted and the background concentrations result in increased dilution in the environment and, consequently, in higher entropy values. Given the assumption that dilution should be avoided whenever possible for sustainable resource management, low entropy values are desired. The entropy values of all emitted N-compounds are added to obtain the total statistical entropy of N after WWT ($H_{afterWWT}$). This value is compared with the statistical entropy determined for a hypothetical scenario in which the untreated wastewater is directly discharged into the receiving waters (H_{noWWT}). The benefit of a STP for N-treatment is then expressed as the reduction in the statistical entropy (ΔH) relative to the direct discharge of wastewater into receiving waters (see Figure 14.1b). A higher ΔH value indicates a more favorable performance of a particular STP because it results in less dissipation of N-compounds into the environment. The main advantage of eSEA for the assessment of the N-performance of STPs is that all N-compounds, as well as the flows of wastewater, effluent and sludge, can be considered both qualitatively and quantitatively. The environmental impact is reflected in the quantification of the dilution. Thus, a comprehensive evaluation of the N-removal performance of a STP can be provided by this analysis. The disadvantages of eSEA are the relatively large data requirement, the inability to perform the calculation with a single equation, in contrast to the estimation of the N-removal rate, and the lack of other studies that can be used to compare the results.

14.2.2 Data of Austrian STPs

The annual average values of the concentrations and respective fractions of NH_4^+, NO_3^- and N_{total} in the wastewater and the effluent, the wastewater inflow, the sludge volumes and the amount of N in the sludge are available for 56 Austrian STPs. All of these STPs use both nitrification and denitrification processes and meet mandatory emission standards (Phillippitsch & Grath, 2006). The data for each STP are measured in their respective laboratories according to German standards. Because gaseous emissions are not directly measured, the N_2O emissions of each STP are estimated to be 0.5% of the total N-input (based on an Austrian study), and the N_2 amount is estimated from the N mass balance (Kroiss *et al.* 2007). For each plant, the energy consumptions (in kWh) and the total operating costs (in EUR) of the mechanical-biological treatment process are available. The operating costs include the costs for labor, energy, external services, and additional materials, as well as other costs. The mean atmospheric background concentrations used in this paper were obtained from the ESPERE Climate Encyclopedia (Uherek, 2004). Both the NO_3^- and NH_4^+ background concentrations in the surface and ground water are set in accordance with a very good ecological status of the water (BGBl II Nr, 98/2010), (BGBl II Nr, 99/2010), (BGBl II Nr, 461/2010). N_{org} is hardly present in the hydrosphere; thus its background concentration is assumed to be approximately four orders of magnitude smaller than the background concentration of NH_4^+.

14.3 RESULTS AND DISCUSSION

14.3.1 Assessment of the N-removal performance of STPs: eSEA vs N-removal rate

In Figure 14.2 the N-removal performance of 5 selected Austrian STPs according to both the reduction in statistical entropy (ΔH) and the N-removal rate is shown. The scaling for both indicators is between 0% and 100%, where 0% refers to the absence of N-removal, for example, the discharge of untreated wastewater (thus the complete load of N in the wastewater) into the receiving water. A 100% removal rate describes a hypothetical situation in which all of the N-compounds in the wastewater are removed. A 100% eSEA performance result indicates that the WWT process transforms all of the N-compounds in the wastewater to either harmless N species, such as N_2, which has a high natural background concentration in the atmosphere or concentrates the N in the sludge. Both situations (0% and 100%) are not realistic, but serve as a reference.

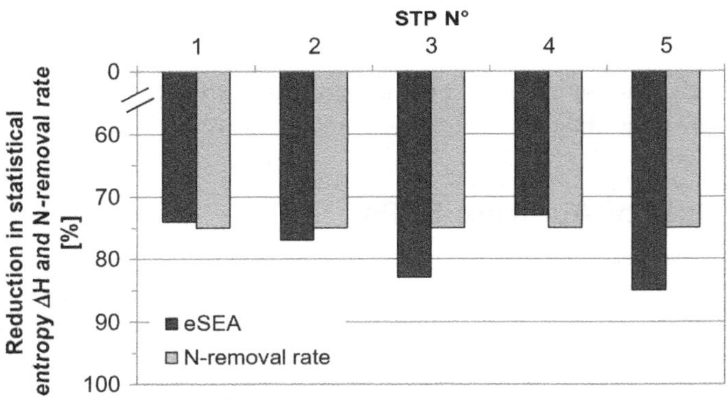

Figure 14.2 Assessment of the N-removal performance of 5 selected Austrian STPs based on eSEA and the N-removal rate.

All of the STPs report N-removal rates of ca. 75%. Consequently, these plants would be considered to exhibit equally good cleaning performances. The eSEA results, however, reveal that there are differences in the N-removal performance of the individual STPs. STP N°5 achieves the highest reduction in statistical entropy (ΔH = 85%) and is thus the most favorable one. STP N°4 is the least favorable, with a ΔH of 73%. STP N°1 achieves a ΔH of 74%, STP N°2 has a ΔH of 77%, and STP N°3 attains a ΔH of 83%. In Table 14.2 the proportion of each N-compound in the total N amount and its contribution to the statistical entropy after WWT (H_{afterWWT}), which has a direct influence on the reduction in the statistical entropy ΔH, are summarized.

Table 14.2 Proportion of N-compounds to the total N after WWT and contribution of the individual N-compounds to the statistical entropy after WWT for 5 Austrian STPs.

	STP N°1	STP N°2	STP N°3	STP N°4	STP N°5
Proportion of N-compounds to total N after WWT [%]					
NH_4^+, aq.	3	10	2	2	1
NO_3^-, aq.	16	15	6	5	9
N_{org}, aq.	6	0	6	15	5
N_2, gas.	53	63	65	56	74
N_2O, gas.	0.5	0.5	0.5	0.5	0.5
N_{org}, sludge	21	12	21	21	11
Contribution of the N-compounds to the statistical entropy after WWT ($H_{afterWWT}$) [%]					
NH_4^+, aq.	11	39	14	9	5
NO_3^-, aq.	45	49	23	14	46
N_{org}, aq.	28	0	40	63	39
N_2, gas.	0.2	0.3	0.2	0.3	0.3
N_2O, gas.	3	3	4	3	4
N_{org}, sludge	13	9	19	11	6

STP N°4 emits more N in the form of N_{org} in the effluent to the river (15%) than does STP N°5 (5%). N_{org} represents a composition of various organic compounds that are naturally present in scarce quantities in rivers with a good ecological status. N_{org} is therefore the main contributor to the statistical entropy after WWT by STP N°4. STP N°5 achieves a higher denitrification rate because most of the N is emitted as N_2 (74%). By contrast, only 56% of the N leaves STP N°4 as N_2. The emission of N_2 into the atmosphere does not generate entropy because of the high concentration of N_2 that is already present in the atmosphere (75% mass fraction). Similar considerations can be made for the other STPs. These examples demonstrate that the N-removal performance of a STP can appear quite different if the different N-compounds, their distribution in the individual mass flows and their dilution in the environment are considered.

14.3.2 Determination of the best practice STP: energy-efficiency and cost-effectiveness

N-treatment requires expenses in the form of energy and costs. Energy is mostly used for aeration processes during nitrification. A certain concentration of dissolved oxygen ($O_{2,diss.}$) is important in the nitrification process and, consequently, for long-term water quality. However, a very high concentration of $O_{2,diss.}$ can lead to undesirable effects due to the incomplete denitrification caused by the recirculation of O_2 from the aerobic to the anoxic reactor thus inhibiting the denitrification process (Flores-Alsina *et al.* 2011). To compare STPs of different sizes, the values of both the energy-consumption and the costs are divided by the individual PE. The energy-efficiency is defined as the energy consumption that is required for every PE to achieve a reduction in statistical entropy. The cost-effectiveness is calculated as the ΔH per PE-specific costs, which defines the cost-effectiveness of the N-treatment as the reduction in the statistical entropy that is achieved for every EUR and PE. In Table 14.3 the N-removal performance of the 5 analyzed Austrian STPs according to eSEA and the energy-efficiency and cost-effectiveness are presented. The best performances are indicated in bold.

Table 14.3 Comparison of the N-removal performance of five Austrian STPs as assessed by eSEA, energy-efficiency and cost-effectiveness.

	STP N°1	**STP N°2**	**STP N°3**	**STP N°4**	**STP N°5**
ΔH [%]	74	77	**83**	73	**85**
Energy-efficiency [kWh/PE/ΔH%]	0.15	**0.11**	**0.12**	0.19	[a]
Cost-effectiveness [ΔH%/€/PE]	**62**	[a]	23	32	8

[a]No data available.

According to the eSEA results, STPs N°3 and N°5 exhibit the best N-removal performance. However, STPs N°2 and N°3 exhibit the most energy-efficient N-treatment, and STP N°1 achieves the highest cost-effectiveness. The results indicate that an energy-efficient N-removal performance does not necessarily imply cost-effectiveness. Because all five STPs comply with Austrian mandatory emission standards, it is reasonable to nominate the most energy-efficient or the most cost-effective STP as the best practice STP, which in this case would be STP N°1, N°2 or N°3. A different approach would be to propose the best practice STP as the plant with good results in all 3 categories, which, according to the analysis presented in this work, would be STP N°3. In Austrian benchmarking, cost-effectiveness plays the decisive role, which is reasonable because the costs for energy consumption are included and the relationship between the energy consumption and the energy costs is usually proportional (Lindtner, *et al.* 2002), (Lindtner, 2009). The low cost-effectiveness of STP N°3 compared to STP N°1, for example, originates from higher expenses for labor and external services.

14.3.3 The influence of plant size

Large STPs tend to operate more energy-efficiently and cost-effective, according to the classical definition of both terms. In this work, both the energy-efficiency and cost-effectiveness are related to the N-removal performance as assessed by eSEA for 56 Austrian STPs. Table 14.4 first gives an overview over the number of STPs among the different size groups.

Table 14.4 Overview over the number of STPs among the different size groups.

Size group/PE									
10,000 – 20,000	20,000 – 30,000	30,000 – 40,000	40,000 – 50,000	50,000 – 60,000	60,000 – 70,000	70,000 – 80,000	80,000 – 90,000	90,000 – 100,000	>100,000
Number of STPs among the different size groups									
10	10	8	10	1	4	0	2	1	10
STPs < 50,000 PE				STPs > 50,000 PE					
Total number of small and large STPs									
38				18					

Data on the cleaning performance of 38 small STPs (<50,000 PE) and 18 large STPs (>50,000 PE) are available. Among the small plants all size groups are equally represented while among the large STPs those that are designed for more than 100,000 PE are overly represented. Note, however, that data on energy consumption and costs are not available for some of these STPs.

Figures 14.3 and 14.4 reveal the newly defined energy-efficiencies and cost-effectiveness of all the STPs for which data were available.

The average energy-efficiency at 0.16 kWh/(ΔH% * PE) is better for the large STPs (>50,000 PE) than for the small plants (<50,000 PE), at 0.21 kWh/(ΔH% * PE). However, many small STPs can operate as energy-efficiently as larger plants (cf. Figure 14.3). For example, a STP, which serves 167,000 PE reports the same energy-efficient N-removal performance (0.13 kWh/(PE * ΔH)) as a STP, which is responsible for 42,000 PE, and another STP, which serves only 22,000 PE. In total, 14 out of 15 large STPs and 16 out of 26 small STPs achieve energy-efficiencies between 0.1 and 0.2 kWh/(ΔH% * PE) (numerical results are not shown). These findings contradict the results of the state-of-the-art literature, which state that large STPs operate more energy-efficiently (Nowak, 2002; Lindtner *et al.* 2008, Kroiss & Svardal, 2009; Lindtner, 2010).

None of the smaller STPs reaches the high cost-effectiveness of the largest STPs (79 ΔH%/(€ * PE)) (cf. Figure 14.4). The average value for the large STPs (>50,000 PE) is at 46 ΔH%/(€ * PE) more than double the average value for the cost-effectiveness of the small plants (<50,000 PE), at 21 ΔH%/(€ * PE). This result is in agreement with literature findings (Lindtner *et al.* 2008), (Hernandez-Sancho & Sala-Garrido, 2009; Lindtner 2010). However, some STPs seem capable of achieving cost-effectiveness in the range of those of the large STPs. For example, 3 of the large STPs that serve 100,000 PE, 150,000 PE, and 200,000 PE, and 5 of the small STPs one responsible for 16,800 PE and four others designed between 40,000 and 48,000 PE report cost-effectiveness of approximately 23 ΔH%/(€ * PE). Another 3 large STPs designed for 70,000 PE, 180,000 PE, and 400,000 PE achieve cost-effectiveness in the range of 32 and 37 ΔH%/(€ * PE), comparable to the cost-effectiveness of several small STPs that serve between 39,000 and 45,000 PE. A STP designed for 167,000 PE also reports higher cost-effectiveness (79 ΔH%/(€ * PE)) than a STP, which serves 950,000 PE (62 ΔH%/(€ * PE)). Another STP constructed for only 35,000 PE achieves cost-effectiveness of 75 ΔH%/(€ * PE), a value in the range of the best cost-effectiveness among the large STPs (>50,000 PE) (numerical results are not shown). These results can be of particular importance for the design and operation of small, decentralized STPs.

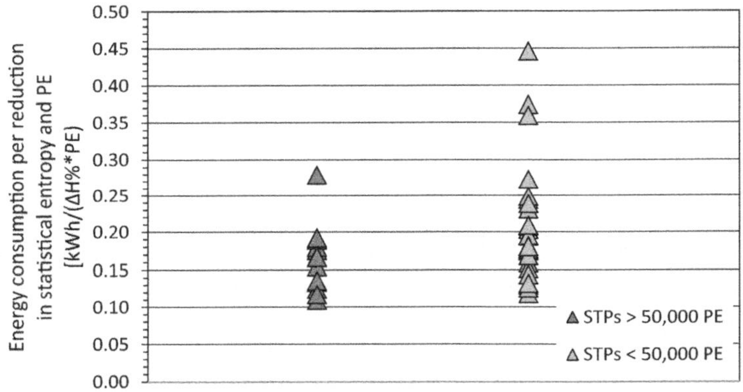

Figure 14.3 Energy-efficiency of 15 large STPs > 50,000 PE and 26 small STPs < 50,000 PE.

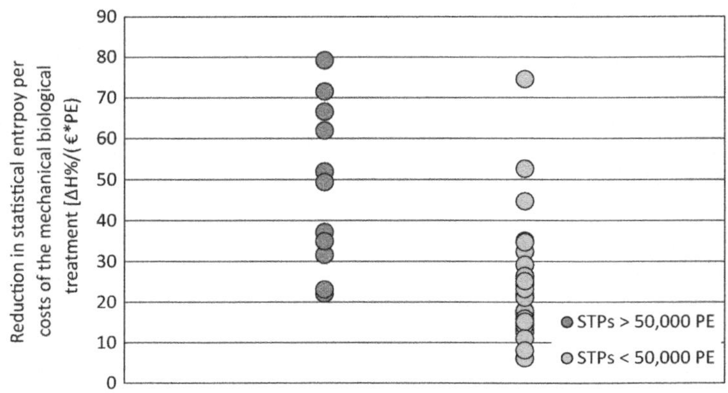

Figure 14.4 Cost-effectiveness of 12 large STPs > 50,000 PE and 38 small STPs < 50,000 PE.

14.4 CONCLUSIONS

The use of eSEA offers a more comprehensive assessment of the N-removal performance of STPs than the N-removal rate because it considers different N-compounds, including gaseous emissions; the distribution of the N in the wastewater, effluent, and sludge; and the dilution of the emissions in the environment. The application of eSEA rewards STPs that transform and transfer N-compounds from the wastewater into harmless (or less harmful) species, such as N_2 or NO_3^-, instead of into NH_4^+ or N_{org}, which would be discharged into water bodies. The eSEA results can be related to the energy-consumption and costs of the N-treatment. Thus, the evaluation can be extended to economic factors. The results of the analysis of 5 Austrian STPs demonstrate that an energy-efficient plant is not necessarily cost-effective. The N-removal performances, the energy-efficiencies and cost-effectiveness of 56 different size STPs are compared, revealing that individual, small STPs (10,000–50,000 PE) are able to compete with the larger plants (50,000–950,000 PE). These results can contribute to the discussion about the advantages and disadvantages of different size STPs offering a new perspective on the efficiency of small, decentralized STPs.

14.5 REFERENCES

Agis H. (2002). *Energieoptimierung von Kläranlagen* (Energy optimization of waste water treatment plants). Wiener Mitteilungen, Band 176, pp. 133–177.

BGBl II Nr, 9., 99/2010. *Verordnung des Bundesministers für Land- und Forstwirtschaft, Umwelt und Wasserwirtschaft über die Festlegung des ökologischen Zustandes für Oberflächengewässer Teil II* (Quality Objective Ordinance – Ecological Status of Surface Waters), Vienna, Austria; Bundesministerium für Land- und Forstwirtschaft, Umwelt und Wasserwirtschaft (Austrian Ministry of Agriculture, Forestry, Environment and Water Resources).

BGBl II Nr, 4., 461/2010. *Verordnung des Bundesministers für Land- und Forstwirtschaft, Umwelt und Wasserwirtschaft über die Festlegung des chemischen Zustandes für Oberflächengewässer Teil II* (Quality Objective Ordinance – Chemical Status of Surface Waters), Vienna, Austria; Bundesministerium für Land- und Forstwirtschaft, Umwelt und Wasserwirtschaft (Austrian Ministry of Agriculture, Forestry, Environment and Water Resources).

Flores-Alsina X., Corominas L., Snip L. and Vanrolleghem P. (2011). Including greenhouse gas emissions during benchmarking of wastewater treatment plant control strategies. *Water Research*, **45**, 4700–4710.

Guinée J. B., Gorrée M., Heijungs R., Huppes G., Kleijn R., Koning A. de, Oers L. van, Wegener Sleeswijk A., Suh S., Udo de Haes H. A., Bruijn H. de, Duin R. van and Huijbregts M. A. J. (2002). Handbook on life cycle assessment. Operational guide to the ISO Standards. I: LCA in perspective. IIa: Guide. IIb: Opeartional annex. III: Scientific background. Kluwer Academic Publishers, Dordrecht, The Netherlands.

Hernandez-Sancho F. and Sala-Garrido R. (2009). Technical efficiency and cost analysis in wastewater treatment processes: a DEA approach. *Desalination*, **249**, 230–234.

Kampschreur M. J., Temmink H., Kleerebezem R., Jetten M. S. M. and van Loosdrecht M. C. M. (2009). Nitrous oxide emission during wastewater treatment. *Water Research*, **43**, 4093–4103.

Kaufman S., Kwon E., Krishnan N., Castaldi M. and Themelis N. (2008). Use of Statistical Entropy and Life Cycle Analysis to Evalaute Global Warming Potential of Waste Management Systems. Proceedings of NAWTEC16, 16th Annual North American Waste-to-Energy Conference May 19–21–2008, Philadelphia, Pennsylvania, USA.

Lindtner S. (2009). *Benchmarking: Grundlagen/Praxiserfahrungen* (Benchmarking: Basics and experiences), k2W Enginnering Kaltesklareswasser, Vienna, Austria.

Lindtner S. (2010). *Benchmarking für Kläranlagenfür das Geschäftsjahr 2009* (Benchmarking of WWTPs for 2009), k2W Engineering Kaltesklareswasser, Vienna, Austria.

Lindtner S., Nowak O. and Kroiss H. (2002). *Benchmarking für Abwasserreinigungsanlagen* (Benchmarking of WWTPs). Institute for Water Quality, Resource and Waste Management, Vienna University of Technology, Vienna, Austria.

Lindtner S., Schaar H. and Kroiss H. (2008). Benchmarking of large municipal waste water treatment plants >100,000 PE in Austria. *Water Science and Technology*, **57**(19), 1487–1493.

Kroiss H. and Svardal K. (2009). *Energiebedarf von Abwasserreinigungsanlagen* (Energy consumption of wastewater treatment plants). Institute for Water Quality, Resource and Waste Management, Vienna University of Technology, Vienna, Austria.

Kroiss H., Paraviccini V. and Svardal K. (2007). *Abschätzung der Veränderung der klimarelevanten Gase CO_2, CH_4 und N_2O am Standort der Regionalkläranlage Linz-Astendurch den Betrieb der Schlammbelüftungsanlage* (Estimation of the climate relevant gases CO_2, CH_4, and N_2O from an Austrian WWTP), Institute for Water Quality, Resource and Waste Management, Vienna University of Technology, Vienna, Austria.

Nowak O. (2002). *Energie-Benchmarking von Kläranlagen – Überlegungen aus abwassertechnischer Sicht* (Energy-benchmarking of WWTPs). Institute for Water Quality, Resource and Waste Management, Vienna University of Technology, Vienna, Austria.

Phillippitsch R. and Grath J. (2006). *Erhebung der Wassergüte in Österreich – Jahresbericht 2006 A/3 Fliessgewässer* (Evaluation of the water quality 2006), Vienna, Austria: Bundesministerium für Land- und Forstwirtschaft, Umwelt und Wasserwirtschaft (Ministry for Land, Forestry, Environment and Water Resources).

Rechberger H. (2001a). An entropy based method to evaluate hazardous inorganic substance balances of waste treatment systems. *Waste Management and Research*, **19**, 186–192.

Rechberger H. (2001b). The use of the statistical entropy to evaluate the utilisation of incinerator ashes for the production of cement. *Waste Management and Research*, **19**, 262–268.

Rechberger H. and Brunner P. H. (2002). A new, entropy based method to support waste and resource management decisions. *Environmental Science and Technology*, **34**(4), 809–816.

Rechberger H. (2012). Waste-to-Energy (WTE): Decreasing the Entropy of Solid Wastes and Increasing Metal Recovery. In: Encyclopedia of Sustainability Science and Technology, Robert A. Meyers (ed.), Springer, New York.

Österreichischer Wasser- und Abfallwirtschaftsverband (ÖWAV) (2000). *Kennzahlen für Abwasserreinigungsanlagen Teil 1: Ablaufqualität – Bewertung und Beurteilung* (Indicators for WWTPs Part 1: assessment of the effluent quality), Vienna, Austria.

Sobańtka A., Zessner M. and Rechberger H. (2012). Extension of statistical entropy analysis to chemical compounds. *Entropy*, **14**, 2413–2426.

Sobańtka A., Thaler S., Zessner M. and Rechberger H. (2013). Extended statistical entropy analysis for the evaluation of nitrogen budgets in Austria. *International Journal of Environmental Science and Technology*, doi: 10.1007/s13762-013-0401-2.

Sobańtka A. and Rechberger H. (2013). Extended statistical entropy analysis (eSEA) for improving evaluation of wastewater treatment plants (WWTPs). *Water Science and Technology*, **67**(5), 1051–1057.

Thaler S. (2009). European Wastewater Benchmarking reporting – Discussion of Current Activities and Chances of Cross-national Comparison, European Water Association. Hennef, Germany.

Uherek E. (2004). Environmental Science Published for Everybody Round the Earth (ESPERE), Max Planck Institute for Chemistry. Mainz, Germany.

Wang J. S., Hamburg S. P., Pryor D. E., Chandran K. and Daigger G. T. (2011). Emissions Credits: Opportunity to Promote Integrated Nitrogen Management in the Wastewater Sector. *Environmental Science and Technology*, **45**(15), 6239–6246.

Yue Q., Lu Z. W. and Zhi S. K. (2009). Copper cycle in China and its entropy analysis. *Resource, Conservation and Recycling*, **53**, 680–687.

Chapter 15

Techno-economic assessment of sludge dewatering devices: A practical tool

Matteo Papa and Giorgio Bertanza

DICATAM-Department of Civil, Environmental, Architectural Engineering and Mathematics, University of Brescia, via Branze 43, I-25123 Brescia, Italy

15.1 INTRODUCTION

Sewage sludge management is a well-known key point in the operation of biological WasteWater Treatment Plants (WWTPs): in effect, sludge treatment and disposal often account for one half of the plant's operating cost (Neyens *et al.* 2004; Saveyn *et al.* 2008; Ruiz-Hernando *et al.* 2010; Ozdemir & Yenigun, 2013), so that wastewater treatment processes may convert a water pollution control problem into a solid waste disposal problem (Weemaes & Verstraete, 1998). Several EU research funding programs have been issued in this field during the last decade: among the latest, ROUTES (Novel processing routes for effective sewage sludge management) and END-O-SLUDG (Wastewater transformed for good) projects can be mentioned within the Seventh Framework Programme.

Moreover, sludge production has been unceasingly increasing, as a consequence of both the growing quantity of collected and treated wastewaters, and the application of progressively stricter standards for WWTPs effluent quality (Neyens *et al.* 2004; Kouloumbos *et al.* 2008). In EU, the implementation of the European Urban Waste Water Treatment Directive (UWWTD – 91/271/EEC) led to a strong rise in sludge generation, up to 50% (Fytili & Zabaniotou, 2008): a total yearly amount around 10 million tons of dry solids (referred to 2007 for EU27 + Norway) has been reached (EEA European Environment Agency; Eurostat), and the projection for the year 2020 is over 13 million (Kelessidis & Stasinakis, 2012). Similarly, in China approximately 3 million tons of dry solids were generated during the year 2007 (Wang *et al.* 2010).

Activated sludge is a poorly dewaterable matrix (Dursun *et al.* 2006; Yu *et al.* 2008), nevertheless a strong reduction of sludge volume can be achieved by means of mechanical dewatering (Qi *et al.* 2011), that, on the other hand, represents a costly operation. High efficiency and reliability of dewatering systems must therefore be pursued.

A crucial issue for mechanical dewatering lies in the choice of the proper conditioning agent: chemical additives, such as high molecular weight polymer flocculants (Boran *et al.* 2010), are commonly used to control the inter-particle interactions in the suspension (Aziz *et al.* 2000). Capillary Suction Time (CST) and Specific Resistance to Filtration (SRF) are established laboratory-scale tests used to understand sludge behavior under different conditioning patterns (Scholz, 2005; Peng *et al.* 2011): nevertheless, obtained information cannot be used to predict the performance of full-scale facilities. Other critical factors, such as the pollution load of the rejected water stream (i.e., the supernatant, which is recirculated back to the plant and can induce an increase up to 25% for nitrogen: Mulder *et al.* 2001; Volcke *et al.* 2006; Yang *et al.* 2007), device reliability, energy consumption, and so on, can be assessed only by means of field experimentations.

From the other hand, full-scale investigations usually are merely focused on dry solids concentration in dewatered sludge (Emir & Erdincler, 2006; Chen *et al.* 2010), based on the assumption that this is the leading parameter describing system performance. On the contrary, an integrated approach is fundamental to properly assess dewatering devices (Mamais *et al.* 2009; Uggetti *et al.* 2011): the global mass balance of solids (also considering the supernatant) and all process parameters should be accounted for (Gratziou *et al.* 2005; Hong *et al.* 2009).

In this chapter, a practical tool for the evaluation of sludge dewatering devices is presented: it can be adopted by WWTP managers as a Decision Support System (DSS) to compare different machines or modes of operation. The whole procedure, together with the decision making process, is illustrated through the application to a case study.

15.2 DESCRIPTION OF THE METHODOLOGY

The innovative approach we propose is aimed at determining both the performance and the cost of dewatering systems, based on experimental data. First of all, the description, step by step, on how to carry out the experimentation is reported, and then the data processing is in detail explained.

15.2.1 Operating procedure for test execution

The experimental phase consists of running the machine(s) to be evaluated under real conditions. Two different scenarios can be separately assessed: the first one is aimed at determining the best performance that the machine can achieve under the working conditions (namely 'optimal') being necessarily defined by skilled operators; optionally, also a second condition, namely 'reference', may be tested, through the *a priori* setting of one/more working parameters: in this way, the machine flexibility to external constraints can be evaluated. In effect, this situation can take place in WWTPs, if, for example, the type of conditioning agent has been already chosen according to previously established commercial agreements with suppliers.

The guidelines for procedure application are reported in Figure 15.1, as a practical sheet to be followed.

INSTRUMENTS / HUMAN RESOURCES / TIME

The instruments necessary for the tests are:

- the devices (industrial-scale) to be tested, either installed in the plant or on mobile units;
- scales for conditioning weighing;
- counters to detect water consumption;
- boxes for sludge collection;
- flow-meters (for inlet sludge and conditioning, if not already detectable from the control panel of the machine);
- current clamp for energy consumption (if not already detectable from the control panel of the machine);
- trucks for sludge weighing;
- graduated tank for supernatant measurement.

The necessary staff is estimated in one person for each test, plus another operator (with laboratory skills) for analytical measurements.

Time to be spent depends on the number of tests (at least one day for each test).

PRELIMINARY OPERATIONS

The device and all the instrumentations must be correctly installed at the WWTP, and the sampling points defined.

In order to obtain as much significant results as possible, minimizing the number of variables that can influence the performance, tests on more than one machine must be executed in parallel, i.e. feeding the same sludge.

TEST EXECUTION

The number / duration of each run has to ensure the significance of the results: for example, continuous systems (i.e. centrifuges and belt-presses) should work for 3 hours (the first one dedicated to set-up); for batch systems (i.e. filter-presses), at least two upload/download cycles must be performed. During the tests, sampling and monitoring phases are executed (one per hour, and in the correspondence of any changes in device operating conditions), together with experimental and analytical measurements.

In particular, the parameters to be measured are:

- total dry and volatile solids, on sludge samples (both inlet and dewatered);
- total solids, COD (Chemical Oxygen Demand) and total nitrogen, on supernatant samples (raw and, optionally, filtered).
- flow-rate of fed sludge;
- production of dewatered sludge and supernatant: if the experimental measurement of these flows is not possible (e.g. unavailability of trucks), they can be calculated by means of mass balances (both global and on solids);
- power consumption;
- conditioning agent consumption;
- water consumption;
- optionally, other machine-specific parameters (e.g., in the case of centrifuges, RPM - Revolutions per Minute -, bowl-screw differential, etc..).

Figure 15.1 Practical guidelines for test execution.

Table 15.1 Items used for the calculation of the total treatment cost for sludge dewatering (as Euro per ton of total dry solids of *fed sludge*).

Cost item (€/t_{TS})	Symbol	Calculation
Power: cost due to energy used by the machine	$\$_{E.E.}$	$= E.E. \cdot u.c._{E.E.}$ E.E.: power consumption of device [kWh/$t_{TS\ fed\ sludge}$] $u.c._{E.E.}$: power unitary cost [€/kWh]
Conditioning: cost due to conditioning agent	$\$_{cond}$	$= Cond \cdot u.c._{cond}$ Cond: conditioning dosage [kg/$t_{TS\ fed\ sludge}$] $u.c._{cond}$: conditioning agent unitary cost [€/kg]
Water: cost due to water used by the machine	$\$_{H2O}$	$= Q_{H2O} \cdot u.c._{H2O}$ Q_{H2O}: water consumption of device [$m^3/t_{TS\ fed\ sludge}$] $u.c._{H2O}$: water unitary cost [€/m^3]
Depreciation: machine depreciation cost, assuming a service life *n* and an interest rate *r*	*D*	$= \frac{I}{T \cdot M_{sludge}}$ *I*: capital cost [€] *T*: depreciation time [*y*] $= [(1+r)^n - 1]/[r(1+r)^n]$ M_{sludge}: yearly sludge load fed to the machine [$t_{TS\ fed\ sludge}/y$]
Maintenance: maintenance cost of the machine, typically provided by the company	*M*	$= \frac{k \cdot I}{M_{sludge}}$ *k*: specific maintenance cost, in the range 5–10% of investment cost per year [y^{-1}]

Labor: labor cost		*L*	$=y.c._{labor} \cdot h_{man}/M_{sludge}$ $y.c._{labor}$: yearly cost of the operators working on the machine [€/y] hman: daily man hours spent by the operators for machine run [%]
Sludge disposal: cost due to the final disposal of dewatered sludge		*S.D.*	$=u.c._{sludge\ disposal} \cdot SL$ SL: mass flow of dewatered sludge [$t_{dewat\ sludge}/t_{TS\ fed\ sludge}$]
Supernatant recirculation (treated with a conventional biological process)	*Oxygen demand:* cost due to oxygen consumption during the biological purification of supernatant recirculated in the wastewater treatment line (the hydrolyzed fraction, assumed a half of the total BOD)	O_2	$=[0.5 \cdot \alpha \cdot BOD + \gamma \cdot (N_{tot} - N_{ass})] \cdot u.c._{O2}$ α: specific oxygen consumption for BOD oxidation [kg O_2/kg BOD] BOD: BOD load in the supernatant [kg BOD/$t_{TS\ fed\ sludge}$] γ: specific oxygen consumption for ammonia oxidation [kg O_2/kg N] N_{tot}: nitrogen load in the supernatant [kgN/$t_{TS\ fed\ sludge}$] N_{ass}: nitrogen load assimilated by the activated sludge [kgN/$t_{TS\ fed\ sludge}$] $u.c._{O2}$: cost for the supply of 1 kg O_2 [€/kg]
	Supernatant sludge treatment and disposal: cost due to treatment and disposal of (i) the additional sludge obtained from supernatant purification (the hydrolyzed fraction) (ii) the supernatant un-hydrolyzed fraction	*S.S.T.D.*	$=0.5 \cdot \left(\frac{BOD \cdot Y_{obs}}{TS_{out}} + \frac{[TS]_{supern} \cdot SUP}{TS_{out}}\right) \cdot u.c._{sludge\ disposal} + 0.5 \cdot \left(\frac{BOD \cdot Y_{obs}}{TS_{in}} + \frac{[TS]_{supern} \cdot SUP}{TS_{in}}\right) \cdot (\$_{E.E.} + \$_{cond} + \$_{H2O})$ SUP: mass flow of supernatant [$t_{supern}/t_{TS\ fed\ sludge}$] TS_{in}: total dry solids concentration in fed sludge [%] TS_{out}: total dry solids concentration in dewatered sludge [%] $[TS]_{supern}$: total solid concentration in the supernatant [g/L] $u.c._{sludge\ disposal}$: unitary cost for final disposal of dewatered sludge [€/$t_{dewat\ sludge}$] Y_{obs}: observed sludge yield [kg TS/kg BOD]

Reprinted from *Journal of Environmental Management*, 132, Bertanza, G., Papa, M., Canato, M., Collivignarelli, M. and Pedrazzani, R., How can sludge dewatering devices be assessed? Development of a new DSS and its application to real case studies, 86–92, (2014), with permission from Elsevier.

15.2.2 Data processing

Experimental data processing is aimed at:

(a) determining the technical performance of system: for a synthetic evaluation, total dry solids (TS) concentration in dewatered sludge and COD concentration in the supernatant are suggested as key-factors;
(b) calculating the treatment costs: key items are listed and illustrated in Table 15.1, which also briefly reports the mathematical expressions used for the calculation.

15.3 APPLICATION TO A REAL CASE STUDY

An example of application of the DSS is described hereinafter, in order to highlight the main strengths of the tool and its ability to drive through the decision making process on the basis of both economic and technical issues.

The experiment was conducted in a WWTP consisting of 2 CAS lines and 1 MBR line (design size 400,000 p.e.; $Q \approx 70{,}000\ m^3/d$), treating domestic and industrial wastewater. The process scheme includes pre-treatments (fine screen stage, grit/oil removal and equalization/homogenization), pre-denitrification, oxidation-nitrification, secondary settling (for conventional lines) and ultrafiltration (for MBR line). The sludge treatment line consists of: dynamic thickening, anaerobic digestion and mechanical (centrifuge) dewatering.

Besides the machine currently working at the WWTP, three industrial size centrifuges (labelled as #1, #2 and #3: main devices characteristics summarized in Table 15.2), installed on mobile units, were assessed in order to rank them, determining the most suitable one. Centrifuges were fed with anaerobically digested and post-thickened secondary sludge. A pair wise comparison was performed and tests were conducted under both 'optimal' and 'reference' conditions. In the latter case, the conditioning agent and its dosage were *a priori* established, according to the commercial supply contract currently in force.

Table 15.2 Main characteristics of tested devices.

Characteristics \ Devices	CENTRIFUGE #1	CENTRIFUGE #2	CENTRIFUGE #3
Class (diameter)	480 mm	520 mm	535 mm
Max. weight	5000 kg	5300 kg	15,000 kg
Drum material	Stainless steel	Stainless steel	Stainless steel
Max. water capacity	70 m^3/h	50 m^3/h	–
Max. solid flow-rate	1800 kg TS/h	1200 kg TS/h	–
Max. drum rotational speed	3650 rpm	3650 rpm	3600 rpm
Max. acceleration of gravity	$3.6 \cdot g\ [m/s^2]$	$3.6 \cdot g\ [m/s^2]$	$3.6 \cdot g\ [m/s^2]$

15.3.1 Technical issues

Technical performances are graphically summarized in Figure 15.2 as the mean ± standard error, chosen in lieu of standard deviation to evaluate data variability, being aware of the modest number of samples collected during each test ($n = 4$). In summary:

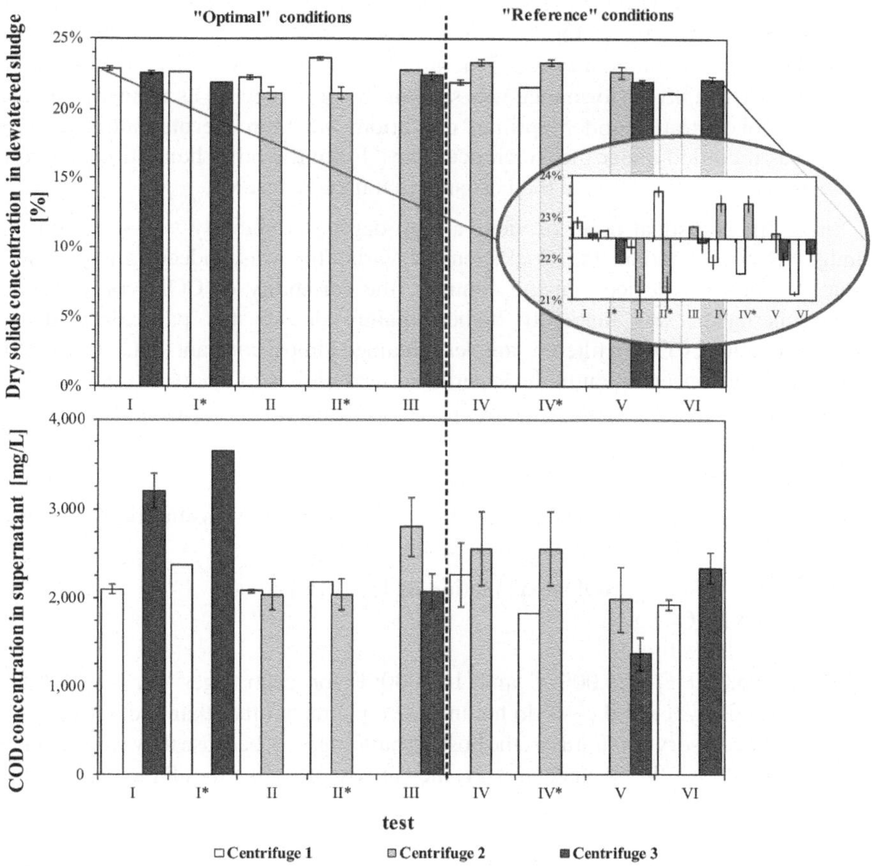

Figure 15.2 Dewatering efficiency of tested devices: total dry solids concentration in dewatered sludge (up) and COD concentration in supernatant (down). *Source*: Reprinted from *Journal of Environmental Management*, **132**, Bertanza, G., Papa, M., Canato, M., Collivignarelli, M. and Pedrazzani, R., How can sludge dewatering devices be assessed? Development of a new DSS and its application to real case studies, 86–92, (2014), with permission from Elsevier.

- Centrifuge #1 obtained the highest sludge TS concentration under 'optimal' conditions, but the lowest under the 'reference' ones (thus pointing out a modest flexibility). The best results were recorded during tests marked with "*" symbol (I*, II*, and IV*), during which the machine was operated under the "supernatant recirculation" mode, that is an advanced working system involving the return of a supernatant fraction (up to 50%) inside the centrifuge. In addition, the supernatant was slightly contaminated (low COD concentration).
- Centrifuge #2 achieved a sludge TS content lower than centrifuge #1 and similar to centrifuge #3, under 'optimal' conditions; on the contrary, it proved to be the best under the 'reference' conditions (showing a high flexibility). The supernatant quality was generally the worst (high COD concentration).
- Centrifuge #3 performance was similar to centrifuge #2 as concerns sludge TS concentration under 'optimal' conditions, while an intermediate efficiency was recorded under the 'reference' ones. From the other hand, this machine was capable to produce the best supernatant (see test V).

From the statistical point of view, a high degree of stability was recorded for sludge TS concentration: the Coefficient of Variability was, indeed, always lower than 5% for each device. On the contrary, the variability of COD concentration in the supernatant was markedly higher, mainly due to the particulate matter content: in effect, COD in filtered samples remained almost constant (data not shown). Total nitrogen concentration in the supernatant was always in a narrow range (1400–1700 mg/L) for all the centrifuges, as well, nitrogen being mainly in the dissolved form.

15.3.2 Economic issues

The total treatment cost was calculated under the following assumptions (i.e., the numerical value of coefficients appearing in Table 15.1):

- $\alpha = 0.5$ kg O_2/kg BOD (explanation in Table 15.1);
- $\gamma = 4.5$ kg O_2/kg N;
- $h_{man} = 1$ h/8 h;
- $I = 105{,}000$ €, 125,000 € and 150,000 € for centrifuges #1, #2 and #3, respectively; capital costs do not include equipment other than the device itself (e.g., accessory apparatuses, the hosting building, …), because they were already available at WWTP, and, anyway, they are exactly alike for all the machines;
- $n = 10$ y;
- $r = 8\%$;
- $u.c._{cond}$ = ranging from 1.5 to 1.7 €/kg;
- $u.c._{E.E.} = 0.10$ €/kWh;
- $u.c._{H2O} = 1$ €/m^3;
- $u.c._{O2} = 2.63$ €cent/kg;
- $u.c._{sludge\ disposal} = 100$ €/$t_{dewatered\ sludge}$;
- $y.c._{labor} = 35{,}000$ €/y;
- $Y_{obs} = 0.5$ kg TS/kg BOD.

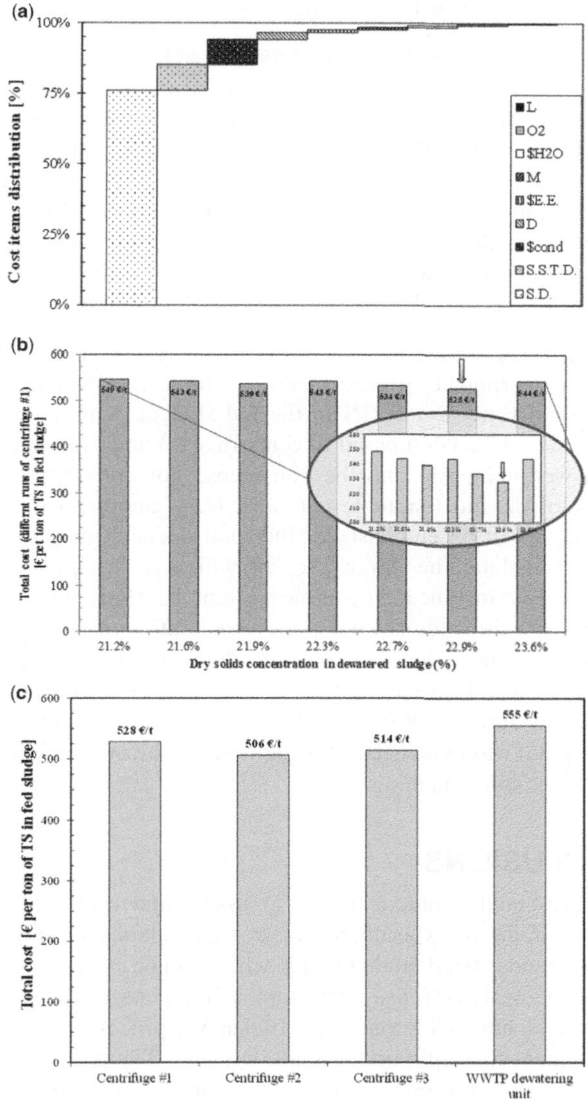

Figure 15.3 Economic assessment of tested devices: (a) weight of different cost items (the meaning of the labels is reported in Table 15.1); (b) cost as a function of dewatering efficiency; (c) overall comparison among tested devices. *Source*: Reprinted from *Journal of Environmental Management*, **132**, Bertanza, G., Papa, M., Canato, M., Collivignarelli, M. and Pedrazzani, R., How can sludge dewatering devices be assessed? Development of a new DSS and its application to real case studies, 86–92, (2014), with permission from Elsevier.

Some general outcomes arise from the analysis:

(1) the most important cost items are represented by sludge disposal (≈70–80% of total treatment cost), conditioning agent (≈5–10%) and treatment and disposal of the additional sludge generated by supernatant purification (≈5–10%), as can be clearly evinced in the average distribution reported in Figure 15.3a;

(2) the minimum treatment cost (evidenced with an arrow in Figure 15.3b, which reports the results for centrifuge #1 as an example) does not correspond to the maximum achievable sludge TS concentration: this highlights a conflict between the 'technical-' and the 'economic-optimum'.

As the comparison among the devices is concerned, Figure 15.3c reports the best economic performance of each machine: the lowest cost was recorded for centrifuge #2 (506 € per ton of TS in the fed sludge), with a saving of 8 €/t_{TS} (1.6%) and 22 €/t_{TS} (4.2%) compared to centrifuge #3 and #1, respectively. These differences (lower than 5%) may be considered not enough relevant to drive into the choice of the most suitable machine. Here, another valuable feature of the proposed DSS can be emphasized: the final decision may be supported by collected technical data, the importance of which is strongly related to case-specific factors. For instance, key issues might be the contamination of the supernatant, the machine flexibility, power/reactants consumption, and so on. In the current case study, the plant manager looked upon the device flexibility as a core factor, thus addressing the choice towards centrifuge #2. In addition, a noticeable saving (around 50€/t_{TS}) with respect to the current dewatering unit installed at the plant was evidenced; this highlighted the urgency and convenience to substitute the existing machine.

15.4 CONCLUSIONS

This chapter presented a practical tool to assess and rank sludge dewatering devices. First of all, for the experimental procedure and the data processing, a well-established and standardized methodology, which can be easily followed whenever a device has to be evaluated, has been implemented. As model outputs, the tool provides both a technical (dewatering efficiency + process parameters) and an economic (treatment cost, split up for item) overview. Thanks to a comprehensive, objective and detailed representation of the costs, it is possible to drive the decision making process through an economic-criterion (as usual for sectors with limited economic resources, such WWTPs); anyway, the final decision on the most suitable device can be also supported by the other technical parameters analyzed by the DSS, the importance of which depends on the case-specific features.

In order to validate the model, it was applied to a real case study, where three industrial size devices (mobile centrifuges) were assessed and compared with the dewatering system installed in the WWTP. The latter was proven to be not convenient, with treatment costs higher (>10%) with respect to the tested centrifuges. Among them, the cheapest one was identified (#2), and the role of 'most suitable' was assigned to it, also weighing the technical factors (mainly machine flexibility). Moreover, as a general outcome, a conflict between the technical and the economic 'optimum' was highlighted: the minimum treatment cost may not correspond to the maximum achievable sludge TS concentration.

In conclusion, the plant manager entrusted to the model can identify the 'best technology' for this unit of treatment.

15.5 REFERENCES

Aziz A. A. A., de Kretser R. G., Dixon D. R. and Scales P. J. (2000). The characterisation of slurry dewatering. *Water Science and Technology*, **41**(8), 9–16.

Bertanza G., Papa M., Canato M., Collivignarelli M. and Pedrazzani R. (2014). How can sludge dewatering devices be assessed? Development of a new DSS and its application to real case studies. *Journal of Environmental Management*, **137**, 86–92.

Boran J., Houdkova L. and Elsaesser T. (2010). Processing of sewage sludge: dependence of sludge dewatering efficiency on amount of flocculant. *Resources Conservation and Recycling*, **54**(5), 278–282.

Chen C., Zhang P., Zeng G., Deng J., Zhou Y. and Lu H. (2010). Sewage sludge conditioning with coal fly ash modified by sulfuric acid. *Chemical Engineering Journal*, **158**(3), 616–622.

Dursun D., Turkmen M., Abu-Orf M. and Dentel S. K. (2006). Enhanced sludge conditioning by enzyme pre-treatment: comparison of laboratory and pilot scale dewatering results. *Water Science and Technology*, **54**(5), 33–41.

Emir E. and Erdincler A. (2006). The role of compactibility in liquid-solid separation of wastewater sludges. *Water Science and Technology*, **53**(7), 121–126.

Fytili D. and Zabaniotou A. (2008). Utilization of sludge in EU application of old and new methods – a review. *Renewable & Sustainable Energy Reviews*, **12**(1), 116–140.

Gratziou M., Ekonomou S. and Tsalkatidou M. (2005). Cost analysis and evaluation of urban sewage processing units. *Water Science and Technology: Water Supply* **5**(6), 155–162.

Hong J., Hong J., Otaki M. and Jolliet O. (2009). Environmental and economic life cycle assessment for sewage sludge treatment processes in Japan. *Waste Management*, **29**(2), 696–703.

http://epp.eurostat.ec.europa.eu/ (accessed 18 November 2014).

http://www.eea.europa.eu/data-and-maps/data/waterbase-uwwtd-urban-waste-water-treatment-directive-1 (accessed 18 November 2014).

http://www.end-o-sludg.eu/ (accessed 18 November 2014).

http://www.eu-routes.org/ (accessed 18 November 2014).

Kelessidis A. and Stasinakis A. (2012). Comparative study of the methods used for treatment and final disposal of sewage sludge in European countries. *Waste Management*, **32**(6), 1186–1195.

Kouloumbos V. N., Schaffer A. and Corvini P. F. X. (2008). The role of sludge conditioning and dewatering in the fate of nonylphenol in sludge-amended soils. *Water Science and Technology*, **57**(3), 329–335.

Mamais D., Tzimas A., Efthimiadou A., Kissandrakis J. and Andreadakis A. (2009). Evaluation of different sludge mechanical dewatering technologies. *Journal Residuals Science & Technology*, **6**(1), 27–34.

Mulder J. W., van Loosdrecht M. C. M., Hellinga C. and van Kempen R. (2001). Full-scale application of the sharon process for treatment of rejection water of digested sludge dewatering. *Water Science and Technology*, **43**(11), 127–134.

Neyens E., Baeyens J., Dewil R. and De Heyder B. (2004). Advanced sludge treatment affects extracellular polymeric substances to improve activated sludge dewatering. *Journal of Hazardous Materials*, **106**(2–3), 83–92.

Ozdemir B. and Yenigun O. (2013). A pilot scale study on high biomass systems: energy and cost analysis of sludge production. *Journal of Membrane Science*, **428**(1), 589–597.

Peng G., Ye F. and Li Y. (2011). Comparative investigation of parameters for determining the dewaterability of activated sludge. *Water Environment Research*, **83**(7), 667–671.

Qi Y., Thapa K. B. and Hoadley A. F. A. (2011). Application of filtration aids for improving sludge dewatering properties – a review. *Chemical Engineering Journal*, **171**(2), 373–384.

Ruiz-Hernando M., Labanda J. and Llorens J. (2010). Effect of ultrasonic waves on the rheological features of secondary sludge. *Biochemical Engineering Journal*, **52**(1), 131–136.

Saveyn H., Curvers D., Thas O. and Van der Meeren P. (2008). Optimization of sewage sludge conditioning and pressure dewatering by statistical modelling. *Water Research*, **42**(4–5), 1061–1074.

Scholz M. (2005). Review of recent trends in capillary suction time (CST) dewaterability testing research. *Industrial & Engineering Chemistry Research*, **44**(22), 8157–8163.

Uggetti E., Ferrer I., Molist J. and García J. (2011). Technical, economic and environmental assessment of sludge treatment wetlands. *Water Research*, **45**(2), 573–582.

Volcke E. I. P., Gernaey K. V., Vrecko D., Jeppsson U., van Loosdrecht M. C. M. and Vanrolleghem P. A. (2006). Plant-wide (BSM2) evaluation of reject water treatment with a SHARON-Anammox process. *Water Science and Technology*, **54**(8), 93–100.

Wang W., Luo Y. and Qiao W. (2010). Possible solutions for sludge dewatering in China. *Frontiers of Environmental Science & Engineering in China*, **4**(1), 102–107.

Weemaes M. P. J. and Verstraete W. H. (1998). Evaluation of current wet sludge disintegration techniques. *Journal of Chemical Technology and Biotechnology*, **73**(2), 83–92.

Yang Y., Zhao Y. Q., Babatunde A. O. and Kearney P. (2007). Co-conditioning of the anaerobic digested sludge of a municipal wastewater treatment plant with alum sludge: benefit of phosphorus reduction in reject water. *Water Environment Research*, **79**(13), 2468–2476.

Yu G. H., He P. J., Shao L. M. and He P. P. (2008). Stratification structure of sludge flocs with implications to dewaterability. *Environmental Science & Technology*, **42**(21), 7944–7949.

Chapter 16

Short-cut enhanced nutrient removal from anaerobic supernatants: Pilot scale results and full scale development of the S.C.E.N.A. process

Daniele Renzi[1], *Stefano Longo*[2], *Nicola Frison*[2], *Simos Malamis*[2], *Evina Katsou*[2] *and Francesco Fatone*[2,*]

[1]*Alto Trevigiano Servizi srl, Via Schiavonesca Priula, Montebelluna – Treviso (Italy)*

[2]*Department of Biotechnology, University of Verona, Via Le Grazie 15, Verona – Italy*

**francesco.fatone@univr.it – Tel/Fax: +39 045 8027965*

16.1 INTRODUCTION

Enhanced nutrient removal in municipal wastewater treatment plants (WWTPs) can be partly and efficiently carried out by treating the ammonium and phosphorus-rich reject water (other terms for reject water are 'return liquor', 'digester supernatant' or 'sludge digester liquid') produced from the dewatering of anaerobic digested sludge in order to meet more stringent effluent standards. In conventional plants the nitrogen flow from the reject water constitutes 10–30% of the total N-load (Cervantes, 2009; Gustavsson *et al.* 2011). As far as phosphorus is concerned, the concentration in reject water can be up to 130 mg L^{-1} (Oleszkiewicz & Barnard, 2006; Pitman, 1999; Ivanov *et al.* 2008). High P concentrations may be reached when anaerobic co-digestion of sewage sludge and organic waste are applied (Malamis *et al.* 2014; Battistoni *et al.* 2005). Thus, the reject water is returned to the activated sludge process and accounts for 10 to 50% of the nutrients in the main stream of municipal WWTPs.

In addition, innovative schemes aiming at energy neutral-positive municipal WWTPs consider the anaerobic digestion as the core process for biogas recovery from sewage sludge. As a consequence, the enhanced nutrient removal from the digester supernatant is proposed to take place separately from the main stream and becomes a significant stage within the new-conceived WWTPs.

16.1.1 Removal or recovery?

Separate treatment of the digester supernatant requires a minimum extension because of the high temperature of the supernatant (25–35°C), potentially leading to short sludge retention times (SRTs) and high reaction rates. Furthermore, a high ammonium concentration and a low COD:N ratio favor high autotrophic ammonium reduction rates. The alkalinity content is often around 1.1 mol HCO_3 per mol NH_4^+-N, while 1.98 mol HCO_3 per mol NH_4^+-N is required for complete nitrification, so extra alkalinity is required if complete ammonium oxidation is needed. Among the different treatment technologies for digester supernatant (cfr Chapter 5), the innovative biological processes have been proved to be the most economically sustainable in terms of nitrogen removal. Compared to the physicochemical processes, these processes do not allow nitrogen recovery. However, the sustainability of the currently available and future nitrogen removal systems has been investigated by several authors (Mulder, 2003; STOWA, 2012). Ammonium can be stripped from ammonium rich side streams (e.g., rejection water) by means of air stripping. This is a well-known technique. In order to strip ammonium a high pH is required (pH 10 to 12). Usually NaOH or Ca(OH) are added as alkali to realize this pH increase. In the air stripping process the rejection water is led through a stripping column in reverse flow through an air stream. The ammonia is transferred to the air stream which is led to an absorber. The adsorbed substance contains acid (H_2SO_4 or HNO_3) in which the ammonia dissolves and ammonium salts are formed. The ammonium salts are drained from the absorber while the ammonia free air can be recycled to the stripper (STOWA, 2012).

In general, the energy demand of such a reference stripping varies from 100 to 150 MJ/kg N (aeration, heat, chemicals as well as their cost) and is significantly higher than the energy demand of the nitrogen producing Haber-Bosch process combined with Anammox (total 60 MJ/kg N).

This shows that nitrogen recovery is more expensive (1.9–3.2 €/kg N) than nitrogen removal using Anammox (0.8 €/kg N) because of the higher energy utilization, as well as the price and quantity of the chemicals required (NaOH or CaO and H_2SO).

In contrast to nitrogen, phosphorus is a limited resource which must be recovered and reused. It is estimated that the remaining accessible reserves of phosphate rock will run out in 50 years, if the growth of demand for fertilizers remains at 3% per year (Gilbert, 2009; Elser & Bennet, 2011). Reducing usage will help the reserves last longer, but the biggest gains will probably be derived from the recovery of phosphates, both from wastewaters and livestock waste (Gilbert, 2009).

Struvite ($MgNH_4PO_4 \cdot 6H_2O$) is generally considered as the optimal phosphate mineral for recovery as it contains 51.8% of P_2O_5 (based on $MgNH_4$ PO_2) and could potentially be used as a slow-release fertilizer. If the economic and life cycle costs are taken into account, however, phosphate recovery in the form of struvite may not be the best approach, for the following reasons: (1) production of P-mineral with a high content of struvite from real wastewater is a difficult and costly process; and (2) struvite is not superior to other phosphate based compounds in terms of fertilization efficiency. Hence, phosphate recovery could be aimed at any forms of phosphate-based compounds acceptable by the fertilizer industry, depending on the onsite economic and environmental circumstances, including the local regulations about recovered material. Accordingly, efforts should also be targeted towards the use of (composted) sludge for effective fertilization (Hao *et al.* 2013). The application of innovative short-cut nitrogen removal and via-nitrite enhanced phosphorus removal can optimize the treatment costs and resource recovery, taking into account the potential usability of the recovered materials.

16.2 SHORT-CUT NITROGEN REMOVAL AND VIA-NITRITE ENHANCED PHOSPHORUS BIOACCUMULATION: FUNDAMENTALS

Short-cut nitrogen removal (SCNR) through the ammonium oxidation to nitrite and its subsequent reduction to gaseous nitrogen has gained increasing attention over the last years. The adoption of nitritation/denitritation as opposed to conventional nitrification/denitrification has significant advantages, since it theoretically reduces the oxygen demand up to 25% and requires up to 40% less organic carbon source. Furthermore, it decreases sludge production by 20–40% and carbon dioxide emissions by 20% (Gustavsson, 2010). The completely autotrophic nitrogen removal process is even more economically attractive as it further reduces energy demand and has no external carbon source requirements. However, its operational and environmental sensitivity and the facts that neither enhanced biological phosphorus removal can be achieved nor nitrogen is completely removed are major drawbacks for its potential implementation (Malamis *et al.* 2014).

To accomplish SCNR the growth of ammonium oxidizing bacteria (AOB) must be favored against the growth of nitrite oxidizing bacteria (NOB). NOB can be inhibited/washed out by maintaining a significant free ammonia (FA) and/or free nitrous acid (FNA) concentration in the reactor (FA > 1 $mgNH_3 \cdot L^{-1}$, FNA > 0.02 $mgHNO_2\text{-}N \cdot L^{-1}$) (Anthonisen *et al.* 1976; Gu *et al.* 2012). AOB are also favored against NOB at alkaline pH (8–8.8) (Zhang *et al.* 2007), high temperature (>25°C) (Hellinga *et al.* 1998) and low dissolved oxygen (DO) concentration (0.3–1.5 $mg \cdot L^{-1}$) (Peng & Zhu, 2006; Blackburne *et al.* 2008).

SCNR has been examined for strongly nitrogenous effluents, such as landfill leachate, supernatant produced from the anaerobic digestion of activated sludge (i.e., reject water) and from the organic fraction of municipal solid waste (OFMSW) (Hellinga *et al.* 1998; Fux *et al.* 2006; Ganigué *et al.* 2007; Ganigué *et al.* 2012) as well as from low strength effluents such as domestic wastewater (Blackburne *et al.* 2008). Nitritation/denitritation has been examined for various environmental and operating conditions, including low and high DO concentrations in the reactor (Pollice *et al.* 2002; Blackburne *et al.* 2008; Guo *et al.* 2009a), high temperature (Hellinga *et al.* 1998), high salinity (Ye *et al.* 2009), different FA and free nitrous acid (FNA) concentrations (Park *et al.* 2010), different solids retention times (SRTs) (Pollice *et al.* 2002).

SCNR can be conveniently coupled with suitable bioprocesses for phosphorus removal through its accumulation in biomass and can be a sustainable option resulting in the production of high added value products. In fact, the mechanism of phosphorus uptake can be realized under anoxic conditions by denitrifying phosphorus accumulating organisms (PAOs) that can utilize nitrate or nitrite as electron acceptors (Kishida *et al.* 2006; Carvalho *et al.* 2007). Denitrifying PAOs require less carbon source compared to aerobic PAOs (Li *et al.* 2011; Peng *et al.* 2011). The rate of phosphate uptake can be higher in the presence of nitrite compared to nitrate (Lee *et al.* 2001).

16.3 CAPITAL AND OPERATING COST OF ANAEROBIC SIDESTREAM TREATMENT

16.3.1 Energy consumptions and costs of short-cut nitrogen removal from anaerobic sidestream

In conventional nitrification-denitrification in the main line of WWTPs, the electrical energy consumption is normally 3.5–5.7 kWh/kg $N_{eliminated}$ (Beier *et al.* 2008). Electrical energy consumption at the DEMON in Strass WWTP was reported to be 1.16 kWh/kg N (Wett, 2007) and in the case of the SBRs in Zürich, similar figures were reported: 1.0 kWh/kg N (Joss *et al.* 2009). The designed electric energy consumption for the DeAmmon in Himmerfjärden WWTP was 2.3 kWh/kg $N_{eliminated}$ (Gustavsson, 2010), which is close to the consumption for an SBR with nitritation-denitritation at Sjölunda WWTP in Sweden, 2.9 kWh/kg N (Gustavsson *et al.* 2011). The electrical energy consumption in the DeAmmon in Hattingen was reported to be as high as 5.6 kWh/kg $N_{eliminated}$ (Jardin *et al.* 2006) and was explained by the low ammonium load (Gustavsson, 2010).

Personnel requirements were estimated to be 0.25 man/year for the DeAmmon in Hattingen (Jardin *et al.* 2006). In all the marketing nitritation-anammox solutions some license or royalties costs are required. There should also be certain agreements on support, particularly in the event of failure.

Investments costs are very site-specific. The investment costs for a Sharon/Anammox installation with a capacity of 1,200 kg NH_4-N/day are estimated at €2 million (2001). The operating costs are linked to the costs for energy, methanol, and lye. In the Table 16.1 indicative economical parameters are given for a full-scale Sharon-Anammox process (van Dongen *et al.* 2001), where the prices refer to the year 2001 and to design guidelines and technical data given by the authors van Dongen *et al.* (2001).

Table 16.1 Full scale cost of Sharon-AnAmmOx process (van Dongen *et al.* 2001).

Parameter	Unit	Case 1	Case 2
N-load	kg N/d	1200	1200
Flow rate	m^3/d	2400	1000
NH_4-conc.	m^3/d	500	1200
Investment	€ (x1.000)	2260	1810
Depreciation	€/year (x1.000)	240	196
Maintenance	€/year (x1.000)	46	41
Personnel	€/year (x1.000)	11	11
Electricity	€/year (x1.000)	82	76
Total cost	€/year (x1.000)	374	325
Cost per kg $N_{removed}$	€	1.05	0.90

On the other hand, Siegrist *et al.* (2012) reported the cost of ammonia stripping versus Nitritation/Anammox in SBR pointing out the 50% cost savings using the biological processes (Table 16.2).

Table 16.2 Cost comparison ammonia stripping versus Nitritation/Anammox in SBR (Siergrist *et al.* 2012).

	NH3-Stripping WWTP Opfikon 19.2 t NH_4–N_{elim}/year	**Nitritation/Anammox WWTP St. Gallen-Au 46.6 t NH_4–N_{elim}/year**
Operating costs (chemicals, energy, sludge disposal)	2.50	0.60
Maintenance costs (spare parts)	1.50	0.20
Personnel costs (25–30% of site)	1.50	0.70
Proceeds of sale of fertilizer	0.60	–
Capital cost	3.50	2.70
Net cost	8.40	4.20

Volcke *et al.* (2007) reported economic evaluations on the basis of the operating cost index (OCI) for a simulated WWTP treating 21100 m^3/day, which resulted in reject water of 172 m^3/day (Table 16.3).

As opposed to nitrogen, phosphorus is a non-renewable resource. There is a wide range of technologies to remove and recover phosphorus from wastewater, including chemical precipitation, biological phosphorus removal, crystallization, novel chemical precipitation approaches and other wastewater and sludge-based methods (Morse *et al.* 1998).

Table 16.3 Economic evaluations of reject water treatment by SHARON-AnAmmOx (Volcke *et al.* 2007).

Costs (€/year)	No reject water treatment (BSM2)	Reject water treatment with SHARON-AnAmmOx
Effluent quality (EQ)	437400	359700
Aeration energy (AE)	194330	194530
Mixing energy (ME)	16200	16570
Pumping energy (PE)	57770	66620
Sludge production (SP)	239000	231000
External carbon addition (EC)	43900	2000
Methane production (MP)	−128800	−123000
TOTAL (=OCI)	859800	747400

Struvite crystallization (SC) is the most widespread. The SC process removes nitrogen and phosphorus from nutrient-rich wastewater by binding these two compounds together in the form of crystallized struvite which can be used as a slow-release fertilizer and can have commercial value depending on the local regulations on the recovered products. The theoretical composition of the so-called MAP (Magnesium-Ammonium-Phosphate) on a weight basis is 9.9% magnesium, 5.7% nitrogen, 12.6% phosphorus with the remainder being crystalline water. MAP satisfies a need for mineral slow-release fertilizers and has many potential uses in horticulture, for nurseries, golf courses, and so on. MAP is likely to be of most benefit to customers as a 'boutique' fertilizer. An alternative to supplying the product directly to end-users is to sell it in bulk to a fertilizer manufacturer for use as a raw ingredient in their products (Munch *et al.* 2001).

The sales price for the MAP has a significant impact on the economics of operating the SC Process. Until a separate business plan for the MAP has been completed, a sales price has to be estimated. Based on the nitrogen and phosphorus content alone, a sales price of $234/t is possible. This would be the price if the MAP was to be used in broad-scale agriculture. However, the intention is to use MAP as a 'boutique' fertilizer for specialized applications. For these applications, much higher sale prices can be achieved. In Japan, a sale price of $3800/t has been reported (Taruya *et al.* 2000). Dockhorn (2009) reported a price of struvite of 760 €/t MAP (6 €/kg P).

A business plan for the Australian market reported the operating cost of the SC process and concluded the following: expenditure is mainly centered on capital acquisitions (of SC Process units of 200–400 k€), the actual costs of which may be less than the current estimates used in these calculations, significantly increasing profits returned: (a) Growth in net asset value from zero to \$3.8 million in 5 years; (b) High returns on equity of 44% in Year 5; (c) Profits returned in Year 2, profits consistently increasing to \$1.5 million in Year 5; (d) Most of the equity is held in physical assets.

However, besides the reasons concerning the agronomic properties and the cost stated above (Hao *et al.* 2013), struvite may be not easily marketable due to legislation constraints in some EU countries which can influence a lot the economical sustainability of the scheme anammox + struvite.

16.4 S.C.E.N.A. SYSTEM

16.4.1 Pilot-scale results

16.4.1.1 S.C.E.N.A. system integrated in co-digestion of WAS and OFMSW for bio-hythane production

The first pilot scale S.C.E.N.A. (Short-Cut Enhanced Nutrients Abatement) system was applied and validated for the treatment of the supernatant of anaerobic co-digestion of sewage sludge and organic fraction of the municipal solid waste (OFMSW) (Fatone *et al.* 2011) within the pilot hall of the Treviso (northern Italy) municipal treatment plant. The authors discussed the start-up strategy and carbon source to enhance the short-cut nitrogen removal and via-nitrite enhanced biological phosphorus uptake from anaerobic supernatant (Frison *et al.* 2012, 2013). The first integrated scheme of (1) two-phase anaerobic digestion for the bio-hythane production (Cavinato *et al.* 2013) and (2) via-nitrite biological nutrients removal was proposed by Frison *et al.* (2013) (Figure 16.1, adapted from Malamis *et al.* 2013).

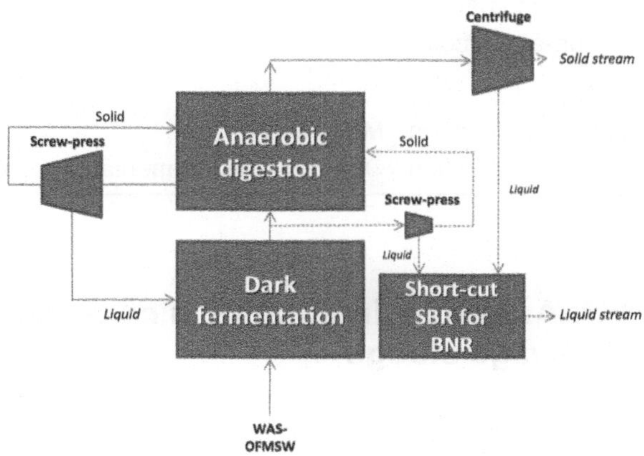

Figure 16.1 Biohythane production and scSBR integration in an anaerobic co-digestion plant (adapted from Malamis *et al.* 2013).

In this first scheme the hydrolysis reactor (dark fermentation) of the two-phase anaerobic digestion is used to provide short chain volatile fatty acids to the anoxic phase in the short-cut sequencing batch reactor (scSBR), which is treating the anaerobic supernatant. Starting from the conventional activated sludge inoculum, the start-up of the scSBR is carried out in two periods (Frison *et al.* 2012) and the stable via-nitrite route is achieved in 15–30 days. Then, nitritation-denitritation and significant phosphorus luxury anoxic uptake was observed. However, the dark acid fermentation did not optimize the contents of propionic and butyric acid that can enhance the via-nitrite (anoxic) biological phosphorus uptake.

On the basis of pilot scale trials, Frison *et al.* (2013) calculated the specific costs of via-nitrite nitrogen removal (Table 16.4). It was found that using the OFMSW fermentation liquid instead of methanol, the overall specific cost for nitrogen removal in the nitritation-denitritation decreased by 22%. In addition, the enhanced phosphorus biological removal was an important added value of the scSBR. Moreover, the added value of the contemporary via-nitrite anoxic phosphorus uptake was not considered, thus underestimating the advantages of the S.C.E.N.A. process with comparison to the complete autotrophic nitrogen removal which must be followed by struvite recovery to achieve the same nutrients removal from anaerobic supernatant.

Table 16.4 Specific costs comparison of nitritation with heterotrophic/autotrophic denitritation (adapted from Frison *et al.* 2013).

Process option	Type of carbon source	Specific costs
	–	€ kg $N^{-1}_{removed}$
Two reactor nitritation-anammox	–	2.5
One reactor nitritation-anammox	–	2.3
DEMON® full scale		
One reactor nitritation/denitritation	Methanol	3.24–3.64
One reactor nitritation/denitritation	OFMSW liquid fermentation	2.85

16.4.2 S.C.E.N.A. system integrated in conventional treatment of sewage sludge

The S.C.E.N.A. system (Figures 16.2 and 16.3) was applied at the conventional municipal WWTP of Carbonera (Italy), where a best available carbon source (BACS) was produced from the fermentation of sewage sludge. Therefore, it was called 'Best', for the enhancement potential on nutrient removal, 'Available' because it was recovered on-site from an available waste stream.

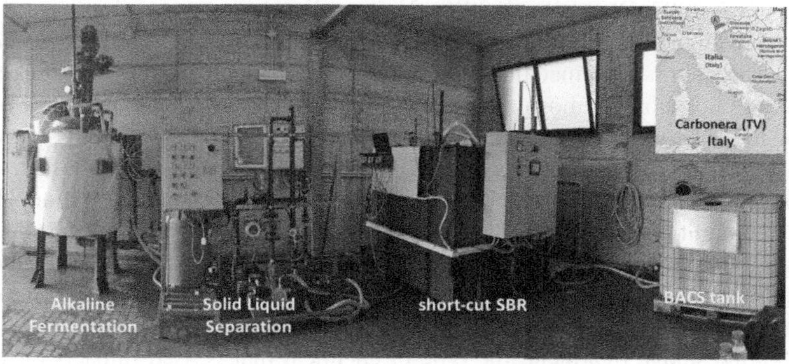

Figure 16.2 Pilot-scale S.C.E.N.A. system.

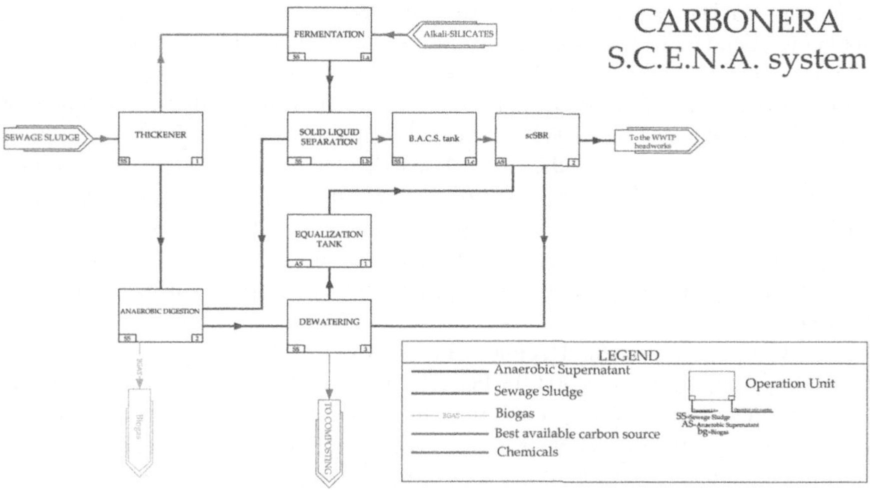

Figure 16.3 The S.C.E.N.A. system integrated in Carbonera municipal WWTP.

The S.C.E.N.A. system in municipal WWTP can be described according to the following key processes: (1) alkaline production of BACS from sewage sludge (or OFMSW); (2) nitritation in aerobic conditions (so as to also minimize N_2O emissions); (3) denitritation and via-nitrite biological phosphorus uptake achieved through the BACS dosage.

The biowaste derived alkaline fermentation liquid is rich in propionic and butyric acid and has been found to be a high added value carbon source. Several studies demonstrated that using sludge-derived Short Chain Volatile Fatty Acids (SCVFA) resulted in superior BNR performance than using synthetic acetate

(Tong *et al.* 2007; Zheng *et al.* 2010). Higher phosphorus removal efficiency was achieved with the use of SCVFA derived from WAS compared to acetate (Tong *et al.* 2007). The authors explained that the presence of propionate was probably the reason for better phosphorus removal, while the higher nitrogen removal efficiency might be due to the better use of exogenous denitrification pathway for nitrogen removal.

In the sludge fermentation process, the pH plays an important role on the hydrolysis of sludge and the production of SCVFAs from excess sludge in fermentation. Under alkaline conditions, the yield of SCVFAs can be significantly enhanced (Yuan & Weng, 2006). Recent studies have demonstrated enhanced SCVFAs production and inhibition of methanogenic activity (resulting in less SCVFAs consumption) under alkaline conditions (Wu *et al.* 2010). NaOH and $Ca(OH)_2$ are widely used for alkaline sludge treatment; the type of chemical that is used impacts on waste activated sludge (WAS) hydrolysis, acidification and dewatering ability (Kim *et al.* 2003). Thus, the type of reagent that is used for the pH adjustment in alkaline fermentation influences the effectiveness of the process. The optimum pH range 9–11 was reached using NaOH and $Ca(OH)_2$ (Su *et al.* 2013). This technique is not economically and environmentally sustainable and enhances the salinity of the carbon source, thus decreasing the rates of nitritation/nitrification. Furthermore, the sludge dewatering characteristics and the separation of the produced fermentation liquid from sludge can be adversely affected from the use of NaOH (Su *et al.* 2013; Longo *et al.* 2014). Recent studies have shown that the use of WAS fermented liquid as carbon source results in the reduction of nitrous oxide (N_2O) and nitric oxide (NO) production during the via nitrite processes (Zhu & Chen, 2011).

Therefore, the initial S.C.E.N.A. process was upgraded for application in a conventional municipal wastewater treatment plant and applied in the Carbonera WWTP.

The Carbonera (Veneto Region – Italy) WWTP plant has actual treatment potential of approximately 40,000 PE. The full scale wastewater treatment line is composed of the following operation units: screening and degritting, primary sedimentation, activated sludge process (+chemical P precipitation) and secondary clarifier; anaerobic digestion of the sewage sludge and dewatering (by centrifuge).

Currently, an S.C.E.N.A. pilot system is operating for the separate treatment of part of the reject water of the plant. The pilot unit consists of three main subunits: the alkaline fermentation unit, the membrane unit for the solid/liquid separation of the fermentation effluent and the sequencing batch reactor (SBR) for the via nitrite nutrient removal. The pilot scale fermentation unit (reaction volume 0.5 m^3) receives mixed (primary and secondary sewage sludge) from the full scale WWTP plant. The sewage sludge is fermented to produce an effluent that is rich is SCVFA. An ultrafiltration (UF) membrane filtration skid is employed for the solid/liquid separation of the fermentation effluent (MO P13U 1 m, Berghof, Germany), while alternative less energy-intensive microscreens are under investigation and validation at pilot scale. Fermented sludge is first screened through 50 mm to prevent clogging of the membrane modules. The sludge fermentation liquid is then directed to the short-cut via nitrite SBR process (3 m^3) that treats separately the anaerobic supernatant, removing N and P via nitrite.

According to the pilot-scale results, the best parameters for the alkaline fermentation of sewage sludge, the solid-liquid separation and the via-nitrite nutrients removal were found out (Longo *et al.* 2015). The production of SCVFA by alkaline fermentation has proved to be highly dependent on pH and temperature. The use of wollastonite was tested in order to avoid the addition of chemicals in the alkaline fermentation process. The fermentation liquid consisted mainly of acetic, propionic and butyric acid (37, 34 and 15% respectively). Under the presence of acids, the following silicate reaction can occur

$$CaSiO_3(s) + 2H^+(aq) \rightarrow Ca^{2+}(aq) + SiO_2(s) + H_2O \tag{16.1}$$

Through the H^+ consuming reaction, it is possible to maintain an alkaline pH (8–9). Besides, an increasing pH also shifts the carbonate equilibrium towards HCO^{3-} and CO_2^{3-} resulting in the precipitation of secondary carbonates as $CaCO_3$ and $Ca_3(PO_4)_2$.

The solid liquid separation is often the bottleneck of the process. The filtration performance of the fermented sludge by membrane ultrafiltration was analyzed. The addition of wollastonite decreased the CST and TTF (by 51% and 59% respectively), resulting in more favorable dewatering potentials.

The fermentation liquid produced was tested as a carbon source for nutrient removal into scSBR. The average influent, effluent characteristics together with the nutrient removal efficiency measured are reported in Table 16.5 while the nutrient removal rates are reported in Table 16.6.

Table 16.5 Average influent, effluent characteristics and nutrient removal efficiency.

Parameter	Influent ($mg \cdot L^{-1}$)	Effluent ($mg \cdot L^{-1}$)
TKN	480–520	26–50
NH_4-N	470–510	20–35
NO_2-N	<0.5	4–25
NO_3-N	<0.5	<0.5
PO_4-P	50–70	10–25
TP	80–100	15–25

Table 16.6 Specific nutrient removal rates.

sAUR (mgN/gVSS*h)	10–15
$sNUR_{BACS}$ (mgN/gVSS*h)	45–70
sPUR (mgP/gVSS*h)	4.5–8

The demonstration S.C.E.N.A. system was designed according to the best parameters derived from the pilot-scale experimentation for the alkaline fermentation of sewage sludge, the solid-liquid separation and the via-nitrite nutrients removal.

Considering a design treatment potential of 40,000 PE the summary of OPEX were preliminary outlined as in Table 16.7.

Table 16.7 Estimation of annual cost for reject water treatment in Carbonera WWTP.

		C.A.S.P.*	S.C.E.N.A.
ELECTICAL ENERGY	€/year	28,000	13,300
SLUDGE DISPOSAL	€/year	26,800	19,300
PolyAlluminiumChloride	€/year	4100	–
WOLLASTONITE	€/year	–	800
PERSONNEL ANNUAL COST	€/year	700	1700

*Current conventional activated sludge process.

As far as the environmental impact is considered, the S.C.E.N.A. system allows the biological uptake of the phosphorus in a form that can be recovered after composting of the S.C.E.N.A. sludge. The impact of the S.C.E.N.A. system was also preliminary evaluated in terms of quality of the secondary effluent as shown in Table 16.8.

Table 16.8 Quality of the secondary effluent in current and simulated future scenario.

	TSS ($mg \cdot L^{-1}$)	COD ($mg \cdot L^{-1}$)	NH_4–N ($mg \cdot L^{-1}$)	NO_3–N ($mg \cdot L^{-1}$)	TP ($mg \cdot L^{-1}$)	TN ($mg \cdot L^{-1}$)
Current scenario (without S.C.E.N.A.)	5.1	4.45	1.56	7.31	2.83	8.87
Future scenario (with S.C.E.N.A.)	5.04	4.67	1.33	3.59	2.32	4.92

16.5 CONCLUSIONS

Sludge reject water is a nutrient-rich flux which should be properly treated managed for the technical, economical and environmental optimization of the nitrogen removal and phosphorus recovery in WWTPs. The completely autotrophic nitrogen removal is the most attractive biological process for the treatment of sludge reject waters in municipal WWTPs with several full scale applications. This solution cannot enhance the phosphorus bioaccumulation and should be followed by struvite crystallization for sustainable phosphorus recovery.

On the other hand, efforts for the technical, economical and environmental sustainability of wastewater treatment plants should also address the novel denitrifying biological phosphorus removal via nitrite which offers the potential to integrate phosphorus and nitrogen removal in a robust single bioreactor in which ammonium is oxidized to nitrite under aerobic conditions, while under anoxic conditions the denitrification via nitrite and enhanced biological phosphorus uptake occur simultaneously by the denitrifying phosphorus accumulating organisms. Thus, the phosphorus could be recovered via the composted sludge.

The S.C.E.N.A. system achieved these objectives in the wastewater treatment plant of Carbonera (Veneto region, northern Italy). Here the real anaerobic supernatant was treated in a nitritation-denitritation short-cut SBR, where the best available carbon source to enhance the phosphorus bioaccumulation was in-situ recovered from the sewage sludge.

The OPEX estimation proved the economic viability of the system which led the water utility Alto Trevigiano Servizi srl to apply the S.C.E.N.A. system by retrofitting an existing tank. This will also minimize the CAPEX and demonstrate how this system can be applied to integrate existing WWTPs.

16.6 REFERENCES

Anthonisen A. C., Loehr R. C., Prakasam T. B. and Srinath E. G. (1976). Inhibition of nitrification by ammonia and nitrous acid. *Water Pollution Control Federation*, **48**, 835–852.

Battistoni P., Paci B., Fatone F. and Pavan P. (2005). Phosphorus removal from supernatants at low concentration using packed and fluidized-bed reactors. *Industrial & Engineering Chemistry Research*, **44**, 6701–7.

Beier M. and Schneider Y. (2008). *Abschlussbericht Entwickelung von Bilanzmodellen für die Prozesse Deammonifikation und Nitritation zur Abbildung gross-and halb-technischer Anlagen* (Final report development of balance models for deammonification and nitritation processes to illustrate full and half technical scale installations). Leibniz University Hannover, Hannover: 39.

Blackburne R., Zhiguo Y. and Keller J. (2008). Partial nitrification to nitrite using low dissolved oxygen concentration as the main selection factor. *Biodegradation*, **19**(2), 303–312.

Carvalho C., Lemos P. C., Oehmen A. and Reis M. A. M. (2007). Denitrifying phosphorus removal: linking the process the process performance with the microbial community structure. *Water Research*, **41**, 4383–4396.

Cavinato C., Bolzonella D., Pavan P., Fatone F. and Cecchi F. (2013). Mesophilic and thermophilic anaerobic co-digestion of waste activated sludge and source sorted biowaste in pilot-and full-scale reactors. *Renewable Energy*, **55**, 260–265.

Cervantes F. J. (2009). Environmental Technologies to Treat Nitrogen Pollution. IWA Publishing, London.

Dockhorn T. (2009). About the Economy of Phosphorus Recovery. International Conference on Nutrient Recovery from Wastewater Streams.

Elser J. and Bennett E. (2011). Phosphorus cycle: a broken biogeochemical cycle. *Nature*, **478**, 29–31.

Fatone F., Dante M., Nota E., Di Fabio S., Frison N., Pavan P. and Fatone F. (2011). Biological short-cut nitrogen removal from anaerobic digestate in a demonstration sequencing batch reactor. *Chemical Engineering Transactions*, **24**, 1135–1140.

Frison N., Lampis S., Bolzonella D., Pavan P. and Fatone F. (2012). Two-stage start-up to achieve the stable via-nitrite pathway in a demonstration SBR for anaerobic codigestate treatment. *Industrial & Engineering Chemistry Research*, **51**(47), 15423–15430.

Frison N., Di Fabio S., Cavinato C., Pavan P. and Fatone F. (2013). Best available carbon sources to enhance the via-nitrite biological nutrients removal from supernatants of anaerobic co-digestion. *Chemical Engineering Journal*, **215**, 15–22.

Fux C., Velten S., Carozzi V., Solley D. and Keller J. (2006). Efficient and stable nitritation and denitritation of ammonium-rich sludge dewatering liquor using an SBR with continuous loading. *Water Research*, **40**(14), 2765–2775.

Ganigué R., López H., Balaguer M. D. and Colprim J. (2007). Partial ammonium oxidation to nitrite of high ammonium content urban landfill leachates. *Water Research*, **41**(15), 3317–3326.

Ganigué R., Volcke E. I., Puig S., Balaguer M. D. and Colprim J. (2012). "Impact of influent characteristics on a partial nitritation SBR treating high nitrogen loaded wastewater. *Bioresource Technology*, **111**, 62–69.

Gilbert N. (2009). The disappearing nutrient. *Nature*, **461**(8), 716–718.

Gu S. B., Wang S. Y., Yang Q., Yang P. and Peng Y. Z. (2012). Start up partial nitrification at low temperature with a real-time control strategy based on blower frequency and pH. *Bioresource Technology*, **112**, 34–41.

Guo J. H., Peng Y. Z., Wang S. Y., Zheng Y. N., Huang H. J. and Ge S. J. (2009). Effective and robust partial nitrification to nitrite by real-time aeration duration control in an SBR treating domestic wastewater. *Process Biochemistry*, **44**, 979–985.

Gustavsson D. J. I. (2010). Biological sludge liquor treatment at municipal wastewater treatment plants – a review, *Vatten*, **66**, 179–192.

Gustavsson D. J. I. and Jansen J. la Cour (2011). Dynamics of nitrogen oxides emission from a full-scale sludge liquor treatment plant with nitritation. *Water Science and Technology*, **63**(12), 2838–2845.

Hao X., Wang C., van Loosdrecht M. C. M. and Hu Y. (2013). Looking Beyond Struvite for P-Recovery. *Environmental Science & Technology*, **47**, 4965–4966.

Hellinga C., Schellen A. A. J. C., Mulder J. W., van Loosdrecht M. C. M. and Heijnen J. J. (1998). The SHARON_ process: an innovative method for nitrogen removal from ammonium-rich waste water. *Water Science and Technology*, **37**, 135–142.

Ivanov V., Kuang S., Stabnikov V. and Guo C. (2009). The removal of phosphorus from reject water in a municipal wastewater treatment plant using iron ore. *Journal of Chemical Technology and Biotechnology*, **84**, 78–82.

Jardin N., Thöle D. and Wett B. (2006). Treatment of sludge return liquors: experiences from the operation of full-scale plants. In: Proceedings of the Water Environment Federation, WEFTEC 2006, Dallas, 21–25 October, 5237–5255.

Joss A., Salzgeber D., Eugster J., König R., Rottermann K., Burger S., Fabijan P., Leumann S., Mohn J. and Siegrist H. (2009). Full-scale nitrogen removal from digester liquid with partial nitritation and anammox in one SBR. *Environmental Science & Technology*, **43**(14), 5301–5306.

Kim J., Park C., Kim T.-H., Lee M., Kim S., Kim S. W. and Lee J. (2003). Effects of various pretreatments for enhanced anaerobic digestion with waste activated sludge. *Journal of Bioscience and Bioengineering*, **95**(3), 271–275.

Kishida N., Kim J., Tsuneda S. and Sudo R. (2006). Anaerobic/oxic/anoxic granular sludge process as an effective nutrient removal process utilizing denitrifying polyphosphate-accumulating organisms. *Water Research*, **40**(12), 2303–2310.

Lee D. S., Che O. J. and Jong M. P. (2001). Biological nitrogen removal with enhanced phosphate uptake in a sequencing batch reactor using single sludge system. *Water Research*, **35**(16), 3968–3976.

Li X., Chen H., Hu L. F., Yu L., Chen Y. G. and Gu G. W. (2011). Pilot-scale waste activated sludge alkaline fermentation, fermentation liquid separation, and application of fermentation liquid to improve biological nutrient removal. *Environmental Science & Technology*, **45**, 1834–1839.

Longo S., Katsou E., Malamis S., Frison N., Renzi D. and Fatone F. (2015). Recovery of volatile fatty acids from fermentation of sewage sludge in municipal wastewater treatment plants. *Bioresource Technology*, **175**, 436–444.

Malamis S., Katsou E., Di Fabio S., Bolzonella D. and Fatone F. (2013). Biological nutrients removal from the supernatant originating from the anaerobic digestion of the organic fraction of municipal solid waste. *CRC Critical Reviews in Biotechnology*, **34**(3), 244–257.

Morse G. K., Brett S. W., Guy J. A. and Lester J. N. (1998). Review: Phosphorus removal and recovery technologies. *Science of The Total Environment*, **212**, 69–81.

Mulder A. (2003). The quest for sustainable nitrogen removal technologies. *Water Science and Technology*, **48**(1), 67–75.

Münch, Elisabeth V. and Keith Barr. (2001). Controlled struvite crystallisation for removing phosphorus from anaerobic digester side streams. *Water Research*, **35**(1), 151–159.

Oleszkiewicz J. A. and Barnard J. L. (2006). Nutrient removal technology in North America and the European Union: a review. *Water Quality Research Journal of Canada*, **41**, 449–462.

Peng Y. and Zhu G. (2006). Biological nitrogen removal with nitrification and denitrification via nitrite pathway. *Applied Microbiology and Biotechnology*, **73**, 15–26.

Peng Y.-Z., Wu C.-Y., Wang R.-D. and Li X.-L. (2011). Denitrifying phosphorus removal with nitrite by a real-time step feed sequencing batch reactor. *Journal of Chemical Technology and Biotechnology*, **86**, 541–546.

Pitman A. R. (1999). Management of biological nutrient removal plant sludges – change the paradigms? *Water Research*, **33**, 1141–1146.

Pollice A., Laera G., Saturno D. and Giordano C. (2008). Effects of sludge retention time on the performance of a membrane bioreactor treating municipal sewage. *Journal of Membrane Science*, **317**(1), 65–70.

Seongjun P., Bae W. and Rittmann B. E. (2010). Multi-species nitrifying biofilm model (MSNBM) including free ammonia and free nitrous acid inhibition and oxygen limitation. *Biotechnology and Bioengineering*, **105**(6), 1115–1130.

STOWA (2012). Explorative research on innovative nitrogen recovery, Rapport 51.

Su G., Huo M., Yuan Z., Wang S. and Peng Y. (2013). Hydrolysis, acidification and dewaterability of waste activated sludge under alkaline conditions: combined effects of NaOH and Ca(OH) 2. *Bioresource Technology*, **136**, 237–243.

Taruya T., Ueno Y. and Fujii M. (2000). Development of phosphorus resource recycling process from sewage, Poster Paper, First World Water Congress of the IWA, Paris, 3–7 July.

Tong J. and Chen Y. (2007). Enhanced biological phosphorus removal driven by short-chain fatty acids produced from waste activated sludge alkaline fermentation. *Environmental Science & Technology*, **41**, 7126–7130.

van Dongen U., Jetten M. S. M. and van Loosdrecht M. C. M. (2001). The Sharon – Anammox process for treatment of ammonium rich wastewater. *Water Science and Technology*, **44**(1), 153–160.

Volcke E. I. P., van Loosdrecht M. C. M. and Vanrolleghem P. A. (2007). Interaction between control and design of a SHARON reactor: economic considerations in a plant-wide (BSM2) context. *Water Science and Technology*, **56**(9), 117–125.

Wett B. (2007). Development and implementation of a robust deammonification process. *Water Science and Technology*, **56**(7), 81–88.

Wu H., Gao J., Yang D., Zhou Q. and Liu W. (2010). Alkaline fermentation of primary sludge for short-chain fatty acids accumulation and mechanism. *Chemical Engineering Journal*, **160**, 1–7.

Ye L., Tang B., Zhao K. F., Pijuan M. and Peng Y. Z. (2009). Nitrogen removal via nitrite in domestic wastewater treatment using combined salt inhibition and on-line process control. *Water Science and Technology*, **60**(6), 1633–1639.

Yuan H., Chen Y., Zhang H., Jiang S., Zhou Q. and Gu G. (2006). Improved bioproduction of short-chain fatty acids (scfas) from excess sludge under alkaline conditions. *Environmental Science & Technology*, **40**, 2025–2029.

Zhang S., Peng Y., Wang S., Zheng S. and Gou S. (2007). Organic matter and concentrated nitrogen removal by shortcut nitrification and denitrification from mature municipal landfill leachate. *Journal of Environmental Sciences*, **19**, 647–651.

Zheng X., Chen Y. and Liu C. (2010). Waste activated sludge alkaline fermentation liquid as carbon source for biological nutrients removal in anaerobic followed by alternating aerobic–anoxic sequneching batch reactors. *Chinese Journal of Chemical Engineering*, **18**, 478–485.

Zhu X. Y. and Chen Y.G. (2011). Reduction of N2O and NO generation in anaerobic-aerobic biological wastewater treatment process by using sludge alkaline fermentation liquid. *Environmental Science & Technology*, **45**, 2137–2145.

Chapter 17

Investigation of the potential energy saving in a pilot-scale sequencing batch reactor

Luca Luccarini[1], Dalila Pulcini[2], Davide Sottara[3] and Alessandro Spagni[4]

[1]Enea – Technical Unit Models, Methods and Technologies for the Environmental Assessment, Via Martiri di Monte Sole 4, 40129 Bologna, Italy

[2]Department of Civil and Environmental Engineering, Politecnico di Milano, Piazza L. da Vinci 32, Milano, Italy

[3]Biomedical Informatics Department, Arizona State University, 13212 East Shea Boulevard, Scottsdale, AZ 85259

[4]Enea – Technical Unit Models, Methods and Technologies for the Environmental Assessment Water Resource Management, Via Martiri di Monte Sole 4, 40129 Bologna, Italy

17.1 INTRODUCTION

Energy is one of the three highest costs for wastewater treatment facilities besides personnel and sludge disposal (Lazarova *et al.* 2012; Zessner *et al.* 2010). Most of the middle and large-scale wastewater treatment plants (WWTPs) are designed to leverage some kind of activated sludge process; however, such processes are highly demanding in terms of energy, since more than 50% of the total energy consumed in WWTPs is normally used for aeration and mixing of the activated sludge tank (Lazarova *et al.* 2012). Since the aeration systems are fundamental to support the biological activity in activated sludge and, at the same time, the most energy demanding process, much research and technological innovation has been funded to improve their efficiency (Amand & Carlsson, 2012; Jeppsson *et al.* 2013).

WWTPs are designed to manage peak flowrates or at least a large fraction of the worst-case scenario flow rates (Tchobanoglous *et al.* 2003). This leads to the design of oversized tanks if compared to the volumes actually required to treat the average flowrate, so WWTPs usually work at low loading rates. The design of oversized tanks (or 'safety margins') together with large plant dynamics could cause large energy waste if plants are not properly managed (Olsson & Newell, 1999).

Nowadays, an efficient management of any WWTP cannot ignore the support offered through automation and information technology, in order to reach the effluent quality standards required by legislation and process sustainability and to limit the energy costs. To this aim, optimizing oxygen addition through the aeration systems by using control logic or automatic policies is fundamental; however, the choice of the most effective control logic or policy strictly depends on the configuration of the evaluated plant.

This chapter provides an overview of a pilot-scale study carried out to evaluate the potential energy saving for aeration in wastewater treatment by using sequencing batch reactors.

17.1.1 Sequencing batch reactors

Sequencing batch reactors (SBRs) are based on a fill-and-draw activated sludge technology with a timed sequence of processes (phases) that take place in the same tank. SBRs present high operational flexibility due to the possibility of modifying the duration of each individual phase (aeration, settling, etc.) depending on the influent characteristics (flow and concentration) (Wilderer *et al.* 2001).

Among the advantages associated with this technology we can mention:

- elimination of the secondary settling tank and of sludge recirculation;
- good tolerability to flow rate and pollution loading variations;
- good clarification conditions due to the capability to control filamentous bulking;
- simple and compact construction, producing appreciable savings in terms of civil engineering works.
- Applicability to simple automation;

Nevertheless, the drawbacks include the following:

- air system has to be over-designed because of the reduced aeration time per cell and to cover the peaks in oxygen demand occurring at the beginning of each aerobic phase;
- an elaborate and high performance decant system is essential.

Previously used by small and medium capacity plants, the SBR technique is now applied to major urban areas too (over 1 million p.e.) (Degremont, 2007). The largest SBR plants of the world have been working as secondary treatment in the WWTP of Dublin City, Ireland, which serves 1.7 million p.e., in Yannawa (Bangkok, Thailand), with size 500,000 p.e. (Kirkwood, 2004) and in Le Havre (France), 415,000 p.e. The SBR is considered a 'state of the art' technology in

Germany and, up to 1999, 138 SBR plants for domestic sewage were built, while about 50 SBRs were installed for industrial applications. The size of most of the plants is less than 5,000 p.e., but the number of plants with a capacity of more than 10,000 p.e. is increasing. Among these, the largest SBR plant is located in the City of Neubranbenburg and serves 140,000 p.e. Nowadays, SBR appears to be the most appropriate treatment for wastewater management in sensitive coastal tourist areas, where a good flexibility to seasonal fluctuations in wastewater quantity and quality is very important (Tasli *et al.* 2001).

A SBR and a conventional (i.e., continuous-flow) activated sludge (CAS) plant can be compared. Both the systems are designed to treat easily a wide range of highly variable influent conditions. When the hydraulic loading increases, the hydraulic retention time (HRT) in the continuous flow automatically decreases. In SBRs, the same relationship is applied simply by changing the cycle time, adjusting the time set for the idle phase. Indeed, this phase thus serves to buffer peak influent loads while the react phase is kept in the required range (Wilderer *et al.* 2001). Most WWTPs include a CAS process, where wastewater is continuously fed to the plants. In these plants, aeration systems management is normally performed by one or more cascaded controllers belonging to the relatively simple PID family (Wahab *et al.* 2009) or by more advanced systems such fuzzy controllers (Baroni *et al.* 2006), which are designed to continuously optimise the aeration of the aerobic tank (Olsson *et al.* 2005).

Biological nitrogen removal from wastewater is usually achieved via nitrification and denitrification processes. During the nitrification process, ammonia is oxidized to nitrate, which is then reduced to nitrogen gas using organic matter as electron donor during the denitrification process. Therefore, while conventional activated sludge plants require two tanks operating under different conditions (i.e., aerobic and anoxic), in SBRs nitrogen removal is accomplished in the same tank by simply managing the switch on and off of the blowers. However, if the process is not monitored, the length of the different phases is usually fixed considering the time required to process the maximum load. Assuming that the plant works properly, this conservative choice guarantees that the concentration of pollutants in the effluent will generally be low, but, at the same time, it causes a significant waste of energy. In order to save energy and increase the overall performance of the treatment system, instead, the duration of the phases should not be set to a fixed-time control strategy based on a worst-case, but, on the contrary, should be managed according to the actual duration of the reactions, which, in turn depend on wastewater load; in particular, aeration should only be maintained on as long as the time necessary to complete nitrification.

17.1.2 Automation of sequencing batch reactors

Over the last decade, several studies have demonstrated that monitoring and control infrastructures can be designed for SBRs management, allowing to optimise the operating conditions and to achieve the best cost/performance ratio (Pat *et al.* 2011;

Shaw *et al.* 2009). Moreover, SBR management should also be automated in order to allow the plant to work continuously and autonomously (Marsili-Libelli *et al.* 2008; Spagni *et al.* 2008). Recently, the automation of SBR processes has been performed using 'intelligent control systems', which rely on AI-based tools, such as neural networks (Aguado *et al.* 2009; Luccarini *et al.* 2002; Luccarini *et al.* 2010; Sottara *et al.* 2007) or fuzzy logic controllers (Marsili-Libelli, 2006; Ruano *et al.* 2010). These tools have been employed to analyse time series acquired by probes installed in SBR tanks, trying to detect the end of the process reactions, in order to optimize the length of the phases.

Recently, much interest has also been shown in the development of remote control infrastructures, which should include a data acquisition system, a data storage facility to store the data, a remote control channel to issue commands to the actuators such as pumps and blowers, and a user interface. Such infrastructures are desirable since plant management includes activities, such as the diagnosis of malfunctioning or the regulation of control parameters, where the experience of expert operators is fundamental. Moreover, Dürrenmatt and Gujer (2012) investigating on the applicability of various data-driven modelling techniques to support WWTP management, concluded that a high degree of expert knowledge was available for long-term operation. All this available knowledge could potentially be transferred to automated Environmental Decision Support Systems (EDSS) and integrated with remote control infrastructures for plant running optimisation.

This work shows the potential advantages of the application of control systems to wastewater treatment. Specifically, this study compares the amount of energy effectively consumed with that theoretically required by aeration of a pilot-scale SBR fed with real municipal wastewater. The former was estimated as the total consumed energy by the blower considering a fixed length of the oxidation phase according to the worst-case scenario. The latter, instead, was estimated as the required energy considering a variable length of the aerobic phase, according to the real duration of the nitrification process. The analysis has been executed off-line, using the data collected in the course of seven days (28 complete cycles) of uninterrupted running of the pilot plant. We focused, especially, on the monitoring of the state of the nitrification process, manageable and controllable by a knowledge-based EDSS. This infrastructure, presented as case-study able to control the pilot-scale SBR plant in Sottara *et al.* (2014), allowed the application of control policies, recognizing some anomalous operating conditions in order to ensure a safe operation of the plant.

17.2 DESCRIPTION OF THE CASE STUDY

17.2.1 Pilot plant

The experimental activity was carried out on a pilot scale SBR plant (Figure 17.1) which has been located side stream to the Trebbo di Reno (Bologna, Italy) full-scale municipal WWTP, managed by the multi-utility company HERA SpA, designed for 2,000 p.e. The raw wastewater was collected after an initial screening treatment.

The reactor had a working volume of 500 L and was fed with 150 L of screened municipal wastewater 4 times per day. The plant has been equipped with a mechanical mixer, a blower connected to a membrane diffuser set to provide a dissolved oxygen (DO) concentration during the nitrification of approximately 1 mgO_2/L, two peristaltic pumps for influent loading and effluent discharge (flow rate = 6 L/min) and a pump for sludge wastage (flow rate = 1 L/min). Besides, it had been equipped with a digital modular multi-parameter system for the measurement and the on-line acquisition of pH, oxidation-reduction potential (ORP), DO and temperature. All signals were acquired in current (4–20 mA) as analog inputs by a multi-function data acquisition device (National Instruments 6052E), while the electrical components were actuated by a PLC (Omron Sysmac CJ series). The equipment was located on a car trailer.

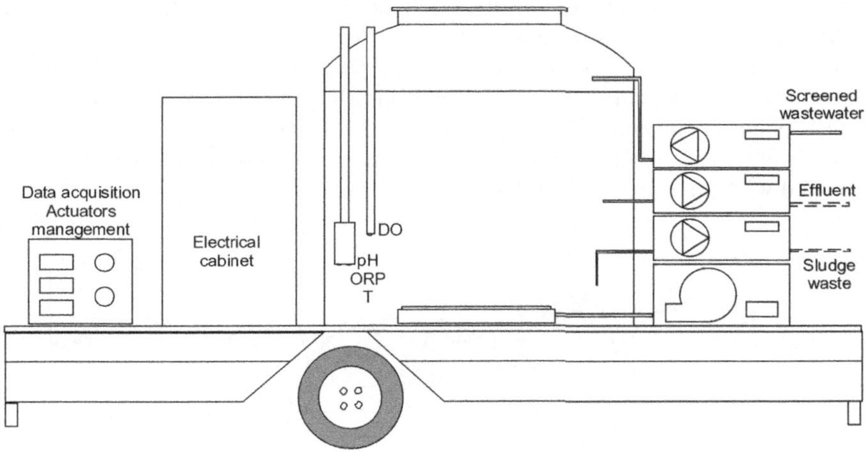

Figure 17.1 Schematic diagram of the pilot-plant.

17.2.2 Process monitoring

Preliminary experiments were performed without the use of EDSS, hardcoding a fixed-phase duration policy, performing four 6-hour cycles per day. During the fill phase, 150 L of sewage was loaded in the reactor. The anoxic phase lasted 90 min while the aerobic phase was ensured by the activation of a blower for 3 hours. A 30-minute settling phase was necessary before discharging the effluent during the draw phase. Sludge waste took place at the end of the aerobic phase to manage the solids retention time (SRT) at approximately 20 d. The plant was operated under similar conditions for a 6 month period, during which the influent, the effluent and the content of the tank were sampled approximately three times per week. Weekly track studies were also carried out to monitor the trend of the biological processes within a complete cycle.

The most relevant values of wastewater composition, measured during the experimental period, were: total COD of 250–400 mg/L, soluble COD of 200–350 mg/L, BOD_5/COD (0.4–0.7), pH of 8.0–8.2, NH_4^+-N of 45–75 mgN/L, and total Kjeldahl nitrogen (TKN) of 55–85 mgN/L. During the whole experimental runs, the plant usually showed good removal efficiency, always higher than 80%, both for ammonia and organic compounds. Concentrations of ammonia in the effluent of a few mgN/L were occasionally observed, having been caused by unusually high nitrogen loads occurring sporadically. Moreover, the fairly low COD/TKN ratio and the variable characteristics of the treated wastewater caused some variability of the nitrogen-oxidized forms in the effluent. The plant also showed good settling characteristics and the total suspended solids (TSS) content in the effluent usually remained below 50 mg/L. Figure 17.2 shows an example of continuous on-line acquisition of pH, ORP and DO signals, which display both high variability of the influent and the repetitiveness of the processes. Figure 17.3 confirms that it is possible to observe a relationship between the indirect but easily measurable signals (such as pH, ORP and DO) and the biological processes in SBR plants (Peng *et al.* 2004; Spagni *et al.* 2001, 2007). Figure 17.3 shows an example of a cycle, where denitrification and nitrification processes take place as expected with the relative pH, ORP and DO trends.

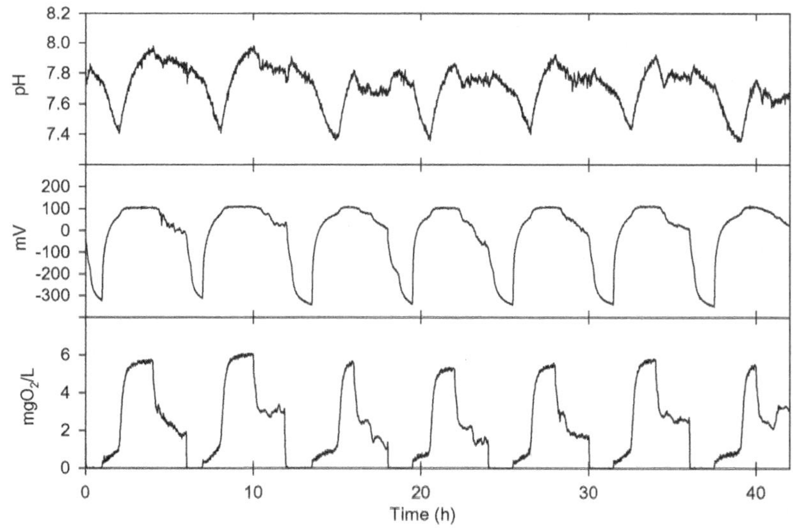

Figure 17.2 pH, ORP and DO trends in the reactor.
Source: (Luccarini *et al.* 2010).

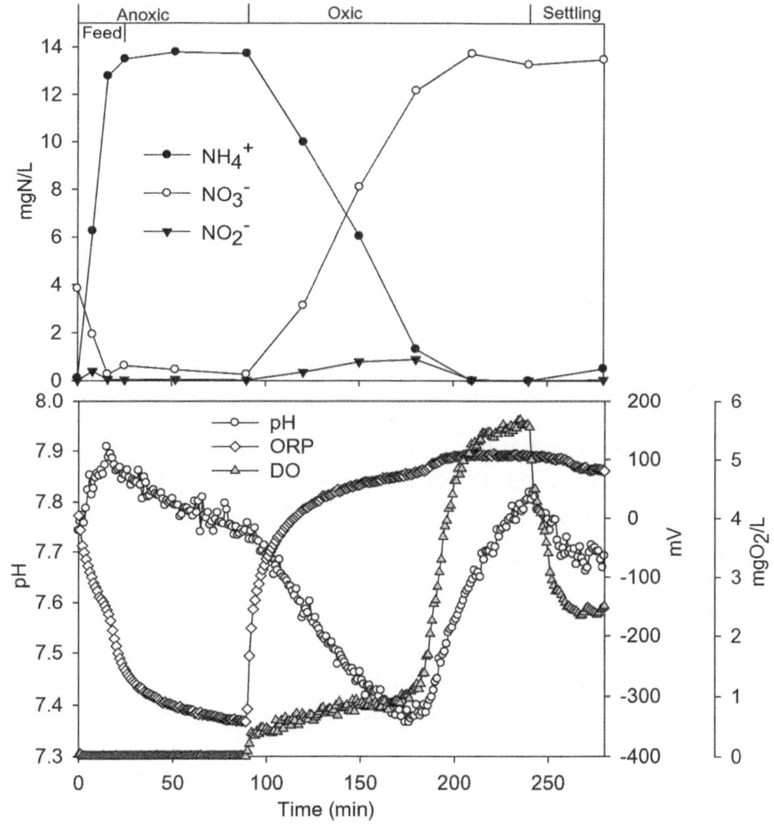

Figure 17.3 Example of a SBR cycle: nitrogen forms above, pH, ORP and DO below. *Source:* (Luccarini *et al.* 2010).

17.2.3 EDSS architecture

An effective optimization of the treatment process requires constant monitoring, in order to adapt the configuration of the plant actuators to the actual state of the process. Instrumentation, control and automation (ICA) is a reasonably cost-effective solution to enable the continuous management of one or more plants, even when the scale or the location of the plant would not justify the employment of human personnel (Olsson & Newell, 1999). The control policies, however, are context-dependent and involve some non-trivial decisions. The decision to switch from one operational phase to a different one – typically the one that follows in the operating cycle – is generally based on a trade-off between the attempt to ensure the correct completion of the current process phase and the desire to minimize its duration.

However, an optimal choice depends on the complete knowledge of the state of the process, which to this date is not (economically) feasible. Since only an indirect estimate is usually available, any decision should take into account the inherent uncertainty about the real operating conditions. More generally, the automated controller should also be able to detect, handle and possibly recover from failures both in the process and the plant instrumentation, so that their impact is minimized. This complex scenario suggests the adoption of an automated Environmental Decision Support System (EDSS) (Poch *et al.* 2004): the EDSS will constantly monitor the plant's state, estimate the current operating conditions and apply the appropriate policies to optimize the cost/effectiveness ratio of the treatment process, actively involving the operators only when the circumstances mandate it.

The EDSS we propose is based on the architecture shown in Figure 17.4. It is based on a hybrid Service Oriented (SOA) and Event-Driven (EDA) Architecture. Rather than building a dedicated, monolithic software, we have partitioned the common functionalities of the EDSS into self-contained modules with different responsibilities, including (but not necessarily limited to):

- A data acquisition service, that collects and pre-processes the signals acquired in real time from the probes installed on the plant;
- A control interface, which commands the plant actuators (pumps, blowers, stirrers, etc.) in order to enforce the desired operational conditions, according to the policy recommendations;
- A data store, where historical time series are persisted for short- and long-term usage;
- A user interaction module, to enable (remote) communication between the EDSS and the plant operators;
- One or more modules implementing the decision support and decision making policies, as described later in this section;
- Modules dedicated to functionalities supporting the architecture itself, such as enabling security and authentication, or a registry of the available modules.

The modules are exposed as services: the core of the architecture is based on an Enterprise Service Bus (ESB) that enables the communication between servers and clients. The SOA, then, allows to implement more complex, higher level functionalities by orchestration of the different services. SOAs, however, are inherently request-driven: any process must be explicitly initiated by some client. An EDSS with monitoring and control responsibilities should also be reactive, to respond promptly to external stimuli such as the changes observed in the plant. To combine the best of both approaches, (Luckham, 2007), we have created a hybrid architecture where external events can trigger the execution of one or more services. During their execution, services can generate additional events which, in turn, might trigger additional computations, effectively creating virtual

Event Processing Networks (Luckham, 2001). The events themselves are filtered and routed using a dynamic content-based routing service (Enterprise Patterns), ensuring that events are delivered only to those subsystems for which they are relevant. More details on the architecture can be found in (Sottara *et al.* 2009a).

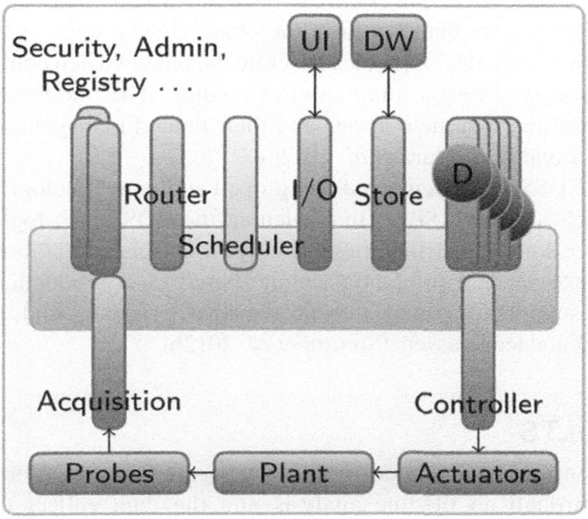

Figure 17.4 EDSS architecture.

The core modules of the EDSS implement the analysis, decision and control logic: in their development, we have adopted a model-driven, knowledge-based approach, based on hybrid artificial intelligence (AI) technologies. A combination of predictive models such as neural networks, rules, ontologies and business processes has been used to model and execute the management and optimization policies. When compared to 'ad-hoc' software, such an approach is considered more robust and flexible from several perspectives. As in a model-driven architecture, the separation of the model from the runtime platform allows the two components to be developed independently, to the point that the latter is usually considered commodity, while the real value lies in the former. The more declarative nature of the knowledge base facilitates the interaction with stakeholders and subject matter experts and improves maintainability.

The complexity of the application justifies the hybrid combination of modelling techniques. Quantitative approaches such as neural networks are more suitable for signal analysis and prediction, while declarative, logic-oriented techniques are more appropriate to define the business logic and the operational policies to manage the plant. In our EDSS, we have deployed an ensemble of different predictors to estimate the process state (and thus the progress in each phase) using the indirect signals (pH, redox potential and DO concentration). We have applied self-organizing maps to determine the advancement of the process and feed-forward neural networks to estimate the concentration of pollutants (ammonia, nitrates) in the tank (Sottara *et al.* 2009b) as well as to detect the state changes directly (Luccarini *et al.* 2010). However, alternative techniques such as fuzzy logic (Marsili-Libelli, 2006) have also proven to be successful and could easily be integrated in the EDSS.

The results of the predictors inform the phase control logic, which is defined using a set of business rules. The rules generate the phase switch commands which, in turn, are processed by the main control module. This module uses business process to model the treatment cycles and their related management procedures: details can be found in (Sottara *et al.* 2012a).

The entire EDSS is implemented using open source technologies with liberal licensing models such as ASLv2. In particular, the EDSS core logic is run using the open source knowledge integration platform Drools (http://jboss.org/drools), a flexible, object-oriented inferential engine based on a production rule system extended with support for complex events, business processes and, more recently, other types of knowledge assets (Sottara *et al.* 2012b).

17.3 RESULTS

The energy consumption, related to the oxygen used in the aerobic phase, was investigated through an off-line analysis and the data collected by the data acquisition system of the plant in May 2008 during seven continuous days. The real duration of each nitrification phase (which determines the minimum, necessary duration of the aerobic phase) was estimated considering ten minutes after the identification of the ‘ammonia valley’ in the pH (point A in Figure 17.5), the ‘ammonia break-point’ in the ORP (point B in Figure 17.5) or the ‘ammonia knee’ in the DO time series (point C in Figure 17.5) (Al-Ghusain *et al.* 1994; Spagni *et al.* 2001). Figure 17.5 presents an example where all these events (ammonia valley, break-point and valley) are clearly visible; moreover, the oxygen added during the aerobic phase which was not used for biological processes is highlighted in the same figure (Figure 17.5). Figure 17.5 confirms a significant, unnecessary energy consumption in the aerobic phase; in fact, we can easily evaluate in the specific example that the effective duration of the nitrification process is approximately 40% of the overall time the plant has been operating in the aerobic phase. However, due to the variability of the influent wastewater, this time changes continuously and, therefore, the relative length of the aerobic phase should be adjusted accordingly on a case-by-case basis.

Using the signal profiles of the pilot-plant, the results were divided as follows:

- Percentage of nitrification time;
- DO consumption comparison for different cycles;
- Evaluation of energy saving.

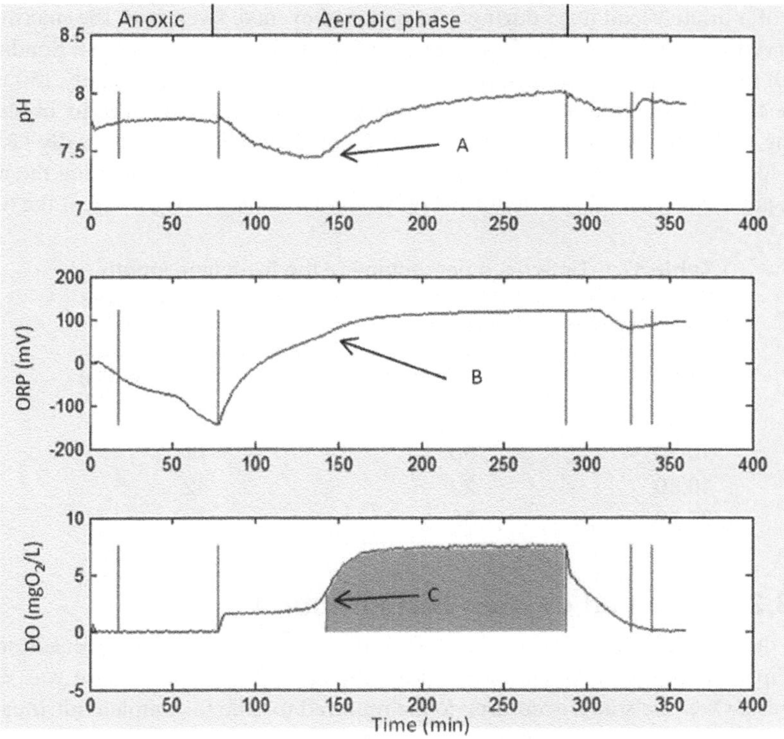

Figure 17.5 Acquired signals and wasted oxygen (filled area).

17.3.1 Nitrification time

The effective duration of the nitrification process evaluated in two different days of the first week of May, Friday and Sunday, is reported in Table 17.1. On Friday, the required time to complete the nitrification in the cycles at 4:30, 16:30 and 22:30 was 30% of the fixed time, while at 10:30, the time was longer (55% of fixed time) than others, which was expected since it corresponds to the maximum load of the day. On Sunday, the required time to complete the nitrification process was always longer than Friday, being 70% for the cycle at 10:30 and approximately 40% for the others. This longer time is related to life style of the habitants of the specific residential area: in fact, the sampled wastewater was collected from a sewer which serves a typical residential area, whose inhabitants move to the City of Bologna for working activities during the week. It is noteworthy that the effective nitrification time was, in average, approximately 40% of the total aerobic phase, confirming that the length of that phase (chosen as a precautionary value) could be significantly reduced when the plant is not fed with a peak load, and even more in case of a diluted load (e.g., during rain or storm events). Even with the maximum observed load during the weeks reported in this study (i.e., the cycle on Sunday at 10.30) the time required to complete the nitrification was approximately 150 min, while the total aerobic phase duration was set at 195 min. However, as observed during the 6 month study, several cycles presented aerobic phases actually lasting even longer than the worst-case pre-fixed duration (195 min), confirming the need of an automatic control system for the proper and reliable management of the SBR.

Table 17.1 Time (as a percentage of the fixed time length for aerobic phase) required to complete nitrification process on Friday and Saturday.

Cycle	Friday (%)	Sunday (%)
04:30	26	35
10:30	56	71
16:30	29	42
22:30	30	44

17.3.2 Dissolved oxygen consumption

The duration of the aerobic phase was designed according to precautionary principles to assure complete nitrification; therefore, the aerobic phase was set at 195 min. Over the study, according to the required to time to complete nitrification (see Section 17.3.1), the required oxygen to actually complete the process was always much lower. Moreover, the required oxygen followed significant daily and weekly trends. In fact, the 10:30 cycle only (Table 17.2) turned out to be adequately designed, because it treats the highest load of the day; nevertheless, although the oxygen wasted was significantly lower than those during the other cycles, it was still of approximately 20–40% confirming the potential energy saving by better management of the SBR phase. On the contrary, the cycles at 4:30 (i.e., those during the night) always resulted to be the ones with the highest unused oxygen (over 70%). In general, the amount of oxygen that could have been saved was always rather high, being comprised between 20 and 70% (Table 17.2).

Table 17.2 Saveable oxygen for the aerobic phase (unit as percentage).

Cycle	Monday (%)	Tuesday (%)	Wednesday (%)	Thursday (%)	Friday (%)	Saturday (%)	Sunday (%)
04:30	64	64	68	75	73	71	63
10:30	34	43	27	37	36	29	21
16:30	67	64	74	65	69	56	55
22:30	66	58	68	69	68	70	54

17.3.3 Cost analysis

The electricity consumption shows large variation according to the applied processes and the size of the plant. The cost analysis, thus, is evaluated on the basis of a small (about 2,000 p.e.) full scale municipal WWTP. The energy consumption can be estimated for small plants to be between 30 and 50 kWh per p.e. per year (a) (Becker & Hansen, 2013; Kampschreur & van Loosdrecht, 2012).

According to the data reported in Table 17.2, approximately 50–55% of energy related to aeration could be saved. Therefore, for the hypothetical full-scale plant, a saving of 30,000 to 50,000 kWh/a corresponding to approximately 4,500 to 7,500 E/a at an electricity cost for aeration of 0.15 E/kWh (assumed as an average value in Italy) could be achieved. Similarly, assuming a wastewater production of 200 L/d per p.e., a saving equivalent to 0.03–0.05 E/m^3 could be achieved.

Reaching an equivalent energy savings with a CAS plant is feasible, but the required equipment is higher and much more expansive. First, a PI controller to maintain DO concentration to a fixed set-point is essential and, consequently, an inverter to regulate the air flow insufflated by the compressor in the oxidation tank. Second, since the oxygen consumption depends on many factors, the best value of the set-point is variable, but it can be estimated from ammonium measurements. Today, this is a proven technology and the energy savings by DO control can be significant. In effect, controlling the DO concentration to a constant set-point can save 30–50%, while to base the set-point on ammonium measurement can save another 10–15% (Olsson, 2012). Nevertheless, the investment costs to equip a plant with this technology are probably too high for small and medium plants, for which the SBR solution appears to be more sustainable, at least for plants which serve up to 20,000 p.e.

17.4 CONCLUSIONS

The study confirms that small residential communities produce wastewater with highly variable concentrations and flow rates, which could greatly affect the WWTP processes.

Moreover, a continuous monitoring of the processes seems to be essential to achieve an optimal cost/performance ratio, especially for those cases where high influent variability occurs.

For this kind of very variable sewers, SBRs may be an effective and economic solution; SBRs also seem to facilitate the monitoring of the processes.

For discontinuous processes, the use of cheap and reliable sensors, such as pH, ORP and DO allows to observe the organic matter and nitrogen removal processes. In particular, this study demonstrates that the aerobic phase in SBRs can be easily and effectively monitored and controlled by using proper tools of analysis and management policies. In fact, the results show that an energy saving up to 55% can be achieved on the aeration energy consumption, when compared to a static management, based on safety parameters used at design time.

These signals may be used by a proper EDSS to deploy several predictors and estimate the process state in real time. In particular, self-organizing maps to determine the advancement of the process, feed-forward neural networks to estimate the concentration of pollutants (ammonia, nitrates) in the tank and to detect the state changes directly may be applied and be easily integrated in the EDSS. Business rules and processes can then leverage the estimated values to apply optimal management policies.

17.5 ACKNOWLEDGEMENTS

The authors thank HERA S.p.A for having partly supported the activities reported in this chapter.

17.6 REFERENCES

Aguado D., Ribes J., Montoya T., Ferrer J. and Seco A. (2009). A methodology for sequencing batch reactor identification with artificial neural networks: A case study. *Computers & Chemical Engineering*, **33**, 465–472.

Al-Ghusain I. A., Huang J., Hao O. J. and Lim B. S. (1994). Using pH as a real-time control parameter for wastewater treatment and sludge digestion processes. *Water Science and Technology*, **30**, 159–168.

Amand L. and Carlsson B. (2012). Optimal aeration control in a nitrifying activated sludge process. *Water Research*, **46**, 2101–2110.

Baroni P., Bertanza G., Collivignarelli C. and Zambarda V. (2006). Process improvement and energy saving in a full scale wastewater treatment plant: air supply regulation by a fuzzy logic system. *Environmental Technology*, **27**, 733–46.

Becker M. and Hansen J. (2013). Is the energy-independence already state-of-the-art at NW-European wastewater treatment plants? In: Proceedings of Asset Management for Enhancing Energy Efficiency in Water and Wastewater Systems. Marbella, Spain, 24–26 April.

Degremont (2007). Water Treatment Handbook. 7th edn, Degremont, France.

Dürrenmatt D. J. and Gujer W. (2012). Data-driven modeling approaches to support wastewater treatment plant operation. *Environmental Modelling and Software*, **30**, 47–56.

Jeppsson U., Alex J., Batstone D. J., Benedetti L., Comas J., Copp J. B., Corominas L., Flores-Alsina X., Gernaey K. V., Nopens I., Pons M.-N., Rodríguez-Roda I., Rosen C., Steyer J.-P., Vanrolleghem P. A., Volcke E. I. P. and Vrecko D. (2013). Benchmark simulation models, quo vadis? *Water Science and Technology*, **68**, 1–15.

Kampschreur M. J. and van Loosdrecht M. C. M. (2012). Finding the balance between greenhouse gas emission and energy efficiency of wastewater treatment. In: Water-energy Interactions in Water Reuse, V. Lazarova, K. H. Choo and P. Cornel (eds), IWA Publishing, London.

Kirkwood S. (2004). Yannawa wastewater treatment plant (Bangkok, Thailand): design, construction and operation. *Water Science and Technology*, **50**(10), 221–228.

Lazarova V., Peregrina C. and Dauthuille P. (2012). Toward energy self-sufficiency of wastewater treatment. In: Water-Energy Interactions in Water Reuse, Lazarova V., Choo K. H. and Cornel P. (eds), IWA Publishing, London.

Luccarini L., Porrà E., Spagni A., Ratini P., Grilli S., Longhi S. and Bortone G. (2002). Soft sensors for control of nitrogen and phosphorus removal from wastewaters by neural networks. *Water Science and Technology*, **45**, 101–107.

Luccarini L., Bragadin G. L., Colombini G., Mancini M., Mello P., Montali M. and Sottara D. (2010). Formal verification of wastewater treatment processes using events detected from continuous signals by means of artificial neural networks. case study: SBR plant. *Environmental Modelling and Software*, **25**, 648–660.

Luckham D. The Power of Events – An Introduction to Complex Event Processing in Distributed Enterprise Systems, Addison-Wesley Longman Publishing Co., Inc. Boston, MA, USA © 2001, ISBN: 0-201-72789-7.

Luckham D. (2007). SOA, EDA, BPM and CEP are all Complementary. http://www.complexevents.com/wp-content/uploads/2007/05/SOA_EDA_Part_1.pdf; http://complexevents.com/wp-content/uploads/2007/07/Soa_EDA_Part2.pdf

Marsili-Libelli S. (2006). Control of SBR switching by fuzzy pattern recognition. *Water Research*, **40**, 1095–1107.

Marsili-Libelli S., Spagni A. and Susini R. (2008). Intelligent monitoring system for long-term control of sequencing batch reactors. *Water Science and Technology*, **57**, 431–438.

Olsson G. (2012). Water and Energy. Threats and Opportunities. IWA Publishing, London.

Olsson G. and Newell B. (1999). Wastewater Treatment Systems: Modelling, Diagnosis and Control. IWA Publishing, London.

Pat A. M., Vargas A. and Buitron G. (2011). Practical automatic control of a sequencing batch reactor for toxic wastewater treatment. *Water Science and Technology*, **63**, 782–788.

Peng Y. Z., Chen Y., Peng C. Y., Liu M., Wang S. Y., Song X. Q. and Cui Y. W. (2004). Nitrite accumulation by aeration controlled in sequencing batch reactors treating domestic wastewater. *Water Science and Technology*, **50**, 35–43.

Poch M., Comas J., Rodriguez-Roda I., Sanchez-Marrè M. and Cortes U. 2004. Designing and building real environmental decision support systems. *Environmental Modelling and Software*, **19**(9), 857–873.

Ruano M. V., Ribes J., Sin G., Seco A. and Ferrer J. (2010). A systematic approach for fine-tuning of fuzzy controllers applied to WWTPs. *Environmental Modelling and Software*, **25**, 670–676.

Tasli R., Artan N. and Orhon D. (2001). Retrofitting SBR systems to nutrient removal in sensitive tourist areas. *Water Science and Technology*, **44**(1), 121–128.

Shaw A., Watt J., Fairey A. W. and Iler M. (2009). Intelligent sequencing batch reactor control from theory, through modelling, to full-scale application. *Water Science and Technology*, **59**, 167–173.

Sottara D. (2010). Integration of Symbolic and Connectionist AI Techniques in the Development of Decision Support Systems Applied to Biochemical Processes. PhD dissertation, University of Bologna.

Sottara D., Luccarini L. and Mello P. (2007). AI techniques for waste water treatment plant control case study: denitrification in a pilot-scale SBR. *Lecture Notes in Computer Science*, **4692**, 639–646.

Sottara D., Manservisi A., Mello P., Colombini G. and Luccarini L. (2009a). A CEP-based SOA for the management of waste water treatment plants. IEEE Workshop on Environmental, Energy, and Structural Monitoring Systems, 2009. EESMS 2009, doi: 10.1109/EESMS.2009.5341314.

Sottara D., Colombini G., Luccarini L. and Mello P. (2009b). A pool of experts to evaluate the evolution of biological processes in SBR plants. Lecture Notes in Computer Science (including subseries Lecture Notes in Artificial Intelligence and Lecture Notes in Bioinformatics) 5572 LNAI, pp. 368–375.

Sottara D., Mello P., Morlini G., Malaguti F. and Luccarini L. (2012a). Modelling SBR Cycle Management and Optimization Using Events and Workflows. Proceedings of 6th International Congress on Environmental Modelling and Software (iEMSs) 1–5 July, Leipzig, Germany.

Sottara D., Mello P., Bragaglia S., Pulcini D., Giunchi D. and Luccarini L. (2012b). Ontologies, Rules, Workflow and Predictive Models: Knowledge Assets for an IEDSS. Proceedings of 6th International Congress on Environmental Modelling and Software (iEMSs), 1–5 July, Leipzig, Germany.

Sottara D., Bragaglia S., Pulcini D., Mello P. and Luccarini L. (2014). A hybrid, integrated IEDDS for the Management of Sequencing Batch Reactors. In: Proceedings of the 7th International Congress on Environmental Modelling and Software, D. P. Ames, N. W. T. Quinn and A. E. Rizzoli (eds), June 15–19, San Diego, California, USA. ISBN: 978-88-9035-744-2.

Spagni A., Buday J., Ratini P. and Bortone G. (2001). Experimental considerations on monitoring ORP, pH, conductivity and dissolved oxygen in nitrogen and phosphorus biological removal processes. *Water Science and Technology*, **43**, 197–204.

Spagni A., Lavagnolo M. C., Scarpa C., Vendrame P., Rizzo A. and Luccarini L. (2007). Nitrogen removal optimization in a sequencing batch reactor treating sanitary landfill leachate. *J. Environ. Sci. Heal. A.*, **42**, 757–765.

Spagni A., Marsili-Libelli S. and Lavagnolo M. C. (2008). Optimisation on sanitary landfill leachate treatment in a sequencing batch reactor. *Water Science and Technology*, **58**, 337–343.

Tchobanoglous G., Burton F. L. and Stensel H. D. (2003). Wastewater Engineering: Treatment and Reuse. Metcalf and Eddy, Mc Grw Hill, Boston.

Wahab N. A., Katebi R. and Balderud J. (2009). Multivariable PID control design for activated sludge process with nitrification and denitrification. *Biochemical Engineering Journal*, **45**, 239–248.

Wilderer P. A., Irvine R. L. and Goronszy M. C. (2001). Sequencing Batch Reactor Technology. Scientific and Technical Reports N. 10. IWA Publishing, London.

Zessner M., Lampert C., Kroiss H. and Lindtner S (2010). Cost comparison of wastewater treatment in Danubian countries. *Water Science and Technology*, **62**, 223–230.

Chapter 18

Economic impact of upgrading biogas from anaerobic digester of sewage sludge to biomethane for public transportation: Case study of Bekkelaget wastewater treatment plant in Oslo, Norway

Rashid Abdi Elmi[1] *and G. Venkatesh*[2]

[1]*Senior Advisor, Oslo Vann og Avløpsetaten, Oslo, Norway.*
E-mail: rashid.elmi@vav.oslo.kommune.no

[2]*Researcher, Department of Hydraulic and Environmental Engineering, Norwegian University of Science and Technology, S P Andresensv 5, Trondheim, Norway 7491.*
E-mail: venkatesh.govindarajan@ntnu.no

18.1 INTRODUCTION

Mitigation of global warming is a top priority in the energy and environmental policies of the city of Oslo. The city of Oslo targets a 50% reduction in greenhouse emissions by year-2030 compared to 1991 levels (*Byrådsavdeling for klima og samferdsel, Oslo Kommune* (2013)). One of the sectors being focused on, is public transport. Meeting this ambitious target the city's administration has set for itself, entails the quick adoption of innovative approaches to optimisation of energy consumption as well as alternate modes of renewable energy generation. Oslo has stepped up to this challenge by effecting a switch from diesel to renewable energy in its public transport system, since the beginning of year-2010. A wonderful synergistic relationship has been uncovered in the process between one public service (transport) and another (sanitation). The Bekkelaget wastewater treatment plant (WWTP) in Oslo – more about which will be discussed later in this chapter – in addition to performing its primary function of treating wastewater and ensuring that the nutrient loading of the receiving water body (the Oslo fjord) is curtailed, has doubled up as a site for renewable energy production.

Anaerobic digestion of sewage sludge is the oldest and the most-prevalent method for decomposing and stabilizing organic matter in an oxygen-free environment. In most cases, the essential and often the only purpose of digesting sewage sludge anaerobically has been (and still is) the partial sterilization and mass-reduction of the sludge generated. Biogas is generated in the process. While it can be flared and released into the atmosphere (as happens in many WWTPs around the world, where sludge is anaerobically digested), it can also be valued for its energy-content. In the 21st century, with climate change (read global warming) emerging as a key global concern, biogas from sludge digesters ought to be looked upon as a source of renewable energy (Speece, 1996; Lemaand Omil, 2001; McCarty, 2001). Biogas in general – and sewage sludge digester biogas in particular – has been a subject of interest for several researchers in the fields of energy and water/wastewater engineering, with the water-energy-carbon nexus gaining increasing importance as a subject of research and analysis (Venkatesh & Dhakal, 2008; Venkatesh, 2012). Figure 18.1 below (Venkatesh & Elmi, 2013) is a simplified schematic sketch which illustrates the various ways in which energy can be recovered (and used) from biogas in WWTPs.

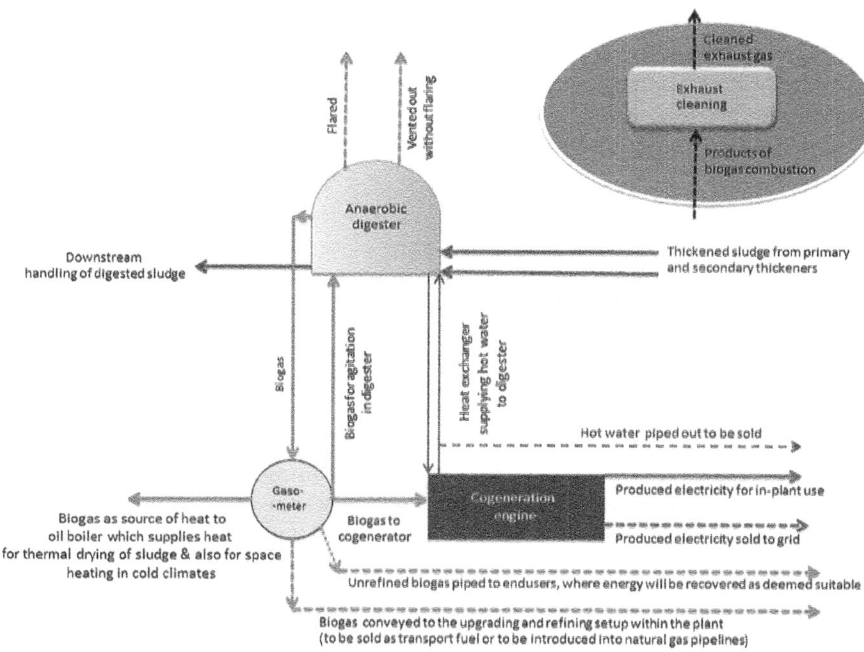

Figure 18.1 Options of utilization of biogas from sewage sludge digesters in wastewater treatment plants, excluding the possible production of chemicals (Venkatesh & Elmi, 2013). (The pre-treatment process for biogas cleaning and/or upgrading have not been included).

As stated in Starr *et al.* (2012), biogas from anaerobic digestion process is mainly a mixture of methane (35–65%) and carbon dioxide gas (15–50%). It may also contain traces of hydrogen sulphide, water vapour, oxygen, and siloxanes. The concentration of each compound in the biogas depends on the type of substrate digested, and its composition. Upgrading the biogas implies increases the content of methane and removing carbon dioxide and other impurities from it. A study from Finland stated that biogas from sewage sludge digestion has the highest methane content and is least contaminated with benzene, hydrogen sulphide and nitrogen (Rasi *et al.* 2007). This Finnish study also concluded that if biogas has to be utilized as a transportation fuel (in addition to being a possible source of heat and electricity), preference should be given to sewage sludge or manure given that both have lower content of nitrogen, halogenated and silicon compounds; and so the biogas generated therefrom is easier to upgrade. Appels *et al.* (2008), had put the annual potential of biogas production in Europe at over 200 billion cubic metres, and had stated that upgraded biogas would be an important future contributor to the energy supply of Europe. While noting that biogas-use in Combined Heat and Power plants in WWTPs was quite well-entrenched, this paper also observed that its use as transport fuel was becoming more and more common in Europe.

At the time of writing, technologies to upgrade biogas to natural-gas standards recommended for vehicle fuels, are in vogue. As listed in Zhao *et al.* (2010), the main technologies used are water and polyethylene scrubbing (costs put at 0.13 Euro/Nm^3), pressure swing adsorption using molecular sieves (0.4 Euro per Nm^3), chemical absorption (0.17 Euro per Nm^3), bio-filter, cryogenic separation (0.44 Euro per Nm^3) and membrane separation (0.17 Euro per Nm^3). Ryckebosch *et al.* (2011) conclude in their paper that most of the choices are determined by the presence or absence of suppliers for the technology in the country; Sweden prefers water scrubbers, Germany has a penchant for pressure swing adsorption, while in the Netherlands, both these, as well as membrane technology is in vogue. Kaparaju (2011) has noted that in year-2011, the water scrubber technology was used in 48 biogas upgrading facilities in Europe, physical absorption using organic solvent in 10, chemical absorption using organic solvent in 31, membrane technology in 6, cryogenic separation in 1 and pressure swing adsorption in 41. The total installed upgrading capacity in 2011 was approximately 115,155 Nm^3 per hour. As far as the capital investment in biogas upgrading plants are concerned, Urban (2009) has put it at 1 million Euros on average, for plants treating 500 Nm^3 biogas per hour, and close to 3 million Euros for plants treating 2000 Nm^3 biogas per hour; implying economies of scale.

Bekkelaget WWTP – the WWTP in Oslo, has been using the chemical absorption technology for upgrading biogas to biomethane, since the beginning of 2010. This waste water treatment plant was built in 2001 with two thermophilic anaerobic digesters for sludge treatment. The two digesters, during the period 2001–2010, produced approximately 20 GWh-equivalent of biogas, annually (with a methane content of 60–65%). Till 2010, most of the biogas (16.5 GWh) was used onsite to deliver heat to the sludge digester (12 GWh) and for sludge drying (4.5GWh). The rest – equivalent to 3.5 GWh – was flared. In year-2007, on the grounds of high operation and maintenance expenses, the sludge-drier was decommissioned. Consequently, more biogas had to be flared. However, in the 10th year of the lifetime of the Bekkelaget WWTP, a major change was implemented – the plant evolved into a supplier of energy for the public transportation sector.

The objective of this chapter is to dwell on the economic impact of this change, after a short description of the processes at the Bekkelaget WWTP.

18.2 WASTEWATER TREATMENT AND SLUDGE HANDLING AT BEKKELAGET WWTP

The Bekkelaget WWTP is fully owned by the City of Oslo but managed and operated by a private company, Bekkelaget Vann AS (BVAS). The plant is located about 3 kilometres inland, to the east of the Oslo fjord. The treatment process setup (Figure 18.2) is situated inside a rock cavern. When the plant was constructed in year-2001, it had a capacity of 270,000 PE (Population-equivalents). At the time of writing, the capacity has raised to 320,000 PE. The plant handles an average daily flow of 125,000 m³/d; the peak wet weather flow reaching over double this average value. A capacity expansion project will get underway in year-2014, and by 2021, the WWTP will be able to handle a load of 490,000 PE.

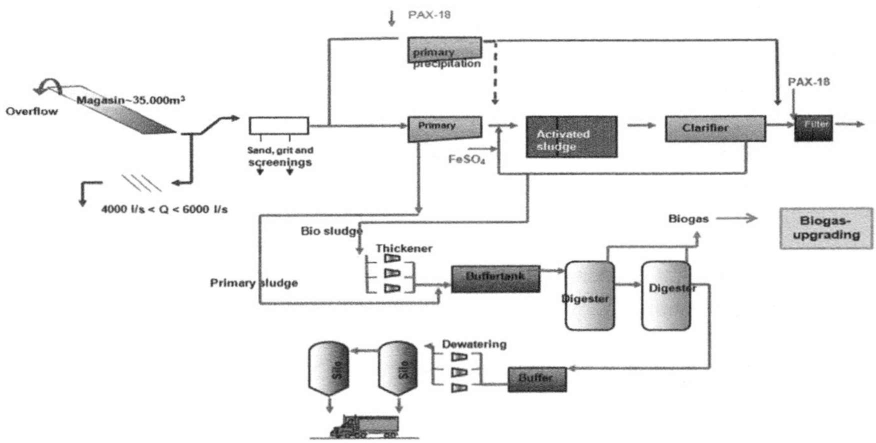

Figure 18.2 Bekkelaget WWTP Processes Flow chart (picture courtesy: Bekkelaget Vann AS (BVAS)).

The influent wastewater is subjected to preliminary treatment (screening, sand and grease separation) which removes *inter alia*, objects which may damage pumps and other equipment down stream. This is followed by primary settling, by which heavy sediments and fine grit are separated and pumped to the sludge handling process. The influent from this stage flows into the denitrification zone in the activated sludge process. The activated sludge system consists of two zones – anoxic and aerated. Nitrogen removal occurs in two steps – through alternating biological processes. Denitrification occurs in the anoxic phase, while nitrification in the aerobic phase. In this approach, ammonia is oxidised to nitrate which is then denitrified to N_2 gas. Ferrous sulphate is added to the excess recycled activated sludge for phosphorus removal (simultaneous precipitation). The aerated wastewater enters the secondary settler (also called the clarifier in Figure 18.2) wherein the activated sludge flocs are removed, resulting in the outflow of a clear treated effluent. This secondary sludge – also called biological sludge – is pumped to the centrifugal thickener. The dual media filter (prior to which polyaluminium chloride is added) is used as the final polishing step before the final treated effluent is discharged to the Oslofjord.

The volume of the primary sludge is reduced in a gravity belt thickener and polymer is added to it for conditioning. The biological sludge undergoes a thickening process in a centrifugal thickener as mentioned earlier. The thickened sludge, prior to entry into the two thermophilic anaerobic digesters, needs to be heated to the temperature of the digesters. Post-digestion, the sludge is dewatered by centrifuges, before being despatched for use as fertiliser in agricultural farmland close to Oslo. The other important output from the digesters is, of course, biogas.

18.3 BIOGAS HANDLING AT BEKKELAGET WWTP

The WWTP produced approximately 3.4–3.6 million Nm^3 of biogas annually over the period 2001–2006; the equivalent energy value being 20 GWh. About 16.5 GWh was utilized in-house to provide the heat for the thermophilic digestion process and sludge drying; the former accounting for nearly 12 GWh. The remaining biogas – equivalent to 3.5 GWh- was flared. The electricity consumption within the plant was about 11 GWh annually.

In 2007, the sludge dryer was decommissioned owing to the high maintenance expenses incurred, poor air quality in the rock cavern and the preference of farmers- who were the main recipient of the treated sludge – for dewatered rather than dried sludge. This is due to the fertilizer value of the dewatered sludge, particularly its Nitrogen content. The total amount of biogas flared rose thereby, from an equivalent of 3.5 GWh to 8 GWh. It was realised that the loss of usable energy could be avoided. Several application options were evaluated. The plant zeroed in on the following option – Upgrading most of the biogas generated to vehicle fuel; installing a heat pump (consuming 2.6 GWh of electricity) and new heat exchangers to provide the 13 GWh required for sludge heating prior to digestion. Some biogas however would still have to be flared. In 2012, for instance, out of 3.99 million cubic metres of biogas produced, 14.7% was used for heating needs, 8% was flared and the remaining (3 million cubic metres) was sent to the upgrading plant.

At the upgrading plant (Figure 18.3), activated carbon is used to adsorb and remove hydrogen sulphide, siloxanes and other undesirable constituents of the biogas, which if present may corrode/abrade gas storage tanks, compressors and the engines in which the biomethane will eventually be combusted (Ryckebosch *et al.* 2011). The biogas upgrading facility at Bekkelaget, uses LP COOAB (**L**ow **P**ressure **CO_2** Absorption by an amine) technology. The amine solution used for absorption of the carbon dioxide from the biogas, is regenerated by heating. The heat energy required for the regeneration is provided by a pellet-fuelled boiler; the heat requirement being 2.6 GWh annually.

Figure 18.3 The biogas upgrading plant at Bekkelaget, Oslo (picture courtesy: Oslo Water and Wastewater Authority, Norway).

The digester biogas is upgraded to natural gas quality by this process. It meets the vehicle fuel specifications which mandate that the combined CO_2-N_2 composition should be between 1.5%–4.5%, the rest being methane (Ryckebosch *et al.* 2011). The annual production of biomethane is estimated at 2.1 million Nm^3, sufficient to replace the diesel consumed by 80 public transport buses in the city of Oslo. The capacity of production of upgraded biomethane (over 97% methane by volume; with methane loss less than 0.1%) is in the range of 195 and 488 m^3 (measured at normal temperature and pressure – NTP – conditions) per hour. Oslo Water and Sewerage Works (VAV), Waste-to-Energy (EGE) are publicly run utilities. Oslo Water and Sewerage Works is responsible for the supply and treatment of drinking water and sewage. EGE on the other hand is responsible in producing environmentally friendly energy from waste. Both are owned by Oslo municipality. Ruter AS is a common management company for public transport in Oslo and Akershus which is owned by Oslo and Akershus County authority. Oslo Water and Sewerage Works (VAV) has an agreement with EGE to facilitate the sale of biomethane and, in return, EGE is paid by 0.10 NOK per m^3 to cover its administrative expenses. The EGE in turn sells it to AGA AS (an industrial gas company belonging to the Munich-headquartered Linde Group), which in turn supplies it as transport fuel to the 80 buses operated by RUTER AS (Figure 18.4).

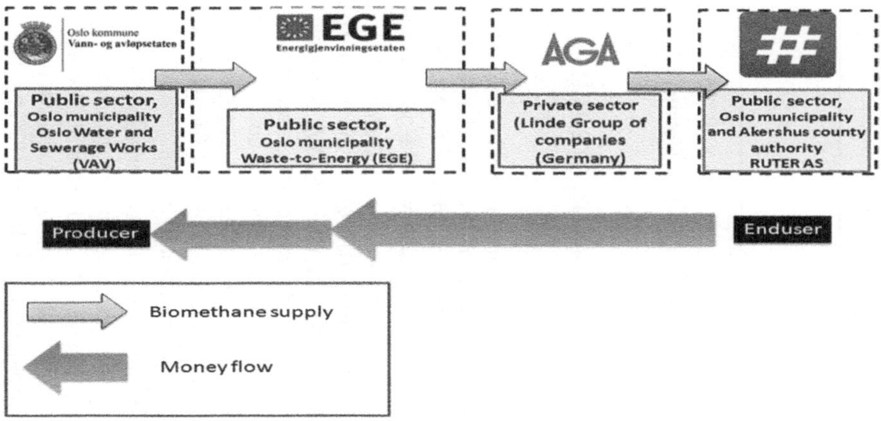

Figure 18.4 Biomethane value chain – from producer to end-user (the increasing thickness of the green arrows indicates the price increment from left to right, along the supply chain).

18.4 THE ECONOMICS OF THE UPGRADING FACILITY

Table 18.1 lists the various cost elements for year-2012, and also explains how these costs are expected to change in the future, annually, till year-2024. The period 2010–2024 is considered in this analysis, as the lifetime of the upgrading plant is set at 15 years. The costs incurred in years 2010 and 2011 are depicted in Figure 18.5. In a life-cycle-costing analysis, the capital investment is committed in a so-called 'year-0'; which is the one in which there are no other cash-flows into or out of the facility. In this case, this is Year-2009, as indicated in the second column of Table 18.1. All the cash flows during the period 2010–2024 are discounted to year-2009, using a discount rate of 3.1%. The annual inflation rate from 2012 onwards is assumed to be constant at 2.07% (Data sourced from www.inflation.eu, in 2013). All other assumptions of changes over time have been included in the last column of Table 18.1. The values are reported in Norwegian kroner (at the time of writing, 1 NOK equals 0.123 Euro or 0.17 USD).

Table 18.1 The cost elements and cash-flows explained.

Cost Elements	In 2012 (million NOK)	Comments
Annual capital expenses		
Fixed annual depreciation, after discounting to year-2009	2.53	Linear depreciation assumed over the 15-year lifetime
Interest payments	0.93	Constant payment of 930,000 NOK annually (at 2.44%)
Operation and maintenance expenses		
Electricity	0.599	Consumption of 1 GWh assumed to remain constant till 2024. The tariff rate in Norway fluctuates a lot, and is highly dependent on a host of external factors – precipitation, temperature and demand from other countries on the Nordic grid (Eika, 2013). Tariff rate in nominal NOK is assumed to rise in tandem with the general inflation rate – 2.07% per annum.
Pellets	1.436	In 2012, pellets with an energy content of 4.49 GWh were consumed (2.47 GWh and 3.32 GWh in 2010 and 2011) From 2013–2024, the primary pellet energy consumption is assumed to decrease from 2.6 GWh to 2.4 GWh uniformly, courtesy efficiency improvement in the combustor. The cost of efficiency improvement is accounted for in the rising maintenance expenditure. The unit cost of pellets (nominal NOK per kWh primary energy content), in 2012 and 2013 is 0.33 NOK per kWh. This is assumed to rise in tandem with the general inflation rate of 2.07% for Norway (the inflation rate in 2013, which can be assumed to be constant till 2024).
Amine	0.259	Amine consumption in litres assumed to be constant over time (1400 litres from 2013 onwards, needed to replenish losses during regeneration. The facility was equipped with amine for operation in 2010 and 2011 and the costs thereof were factored in into the capital investment. The unit cost of amine assumed to rise in tandem with the general inflation rate of 2.07%, from a value of 185.56 NOK per litre in 2013.

Activated carbon	0.03	Assumed to constant over time, at 1000 kilograms per year. The unit cost assumed to increase in tandem with the general inflation rate of 2.07%, from a value of 33 NOK per kg in 2013. The facility was equipped with activated carbon for operation in 2010. In 2011 however, 1000 kilograms were purchased.
Salaries to employees	0.39	This is assumed to increase from 2013 onwards, at a rate of 3% per year (above the general inflation rate). In 2010 and 2011, the salaries were 0.21 and 0.35 million NOK respectively.
General maintenance	0.126	Assumed to be 3.5% of the capital costs in 2010 (Starr, 2013); and thereafter rise uniformly at a rate of 2% annually. As a plant gets older, maintenance expenses are bound to increase. This also factors in the cost of phased efficiency improvement of the pellets-combustor.
Income to upgrading facility		
Sale of biomethane	6.57	The energy equivalent of the biomethane sold annually assumed to remain constant at 19.9 GWh from 2013 onwards. If capacity of the wastewater treatment plant increases and excess biogas is generated, we assume that this would be flared. This is taken into account into the cashflows. The selling price of biomethane (per cubic metre at NTP conditions), is 3.3 NOK in 2012, assumed to rise in tandem with inflation of 2.07% annually.
Compensation from EGE for flared biogas	0.27	The compensation paid by EGE for the flared biogas in 2012 was 0.27 million NOK. This cashflow started in 2012. It is assumed that the total revenue (sale + compensation) in every year after 2012, is the average of the revenues of the two previous years. This results in an annual rise in the compensation paid of 2%. However, it can be assumed that the compensation is not indexed to the general inflation rate. A rise of 30% in the compensation paid, over the 13-year period thereby implies that the excess biogas which may be generated when more wastewater is handled, would also be flared.

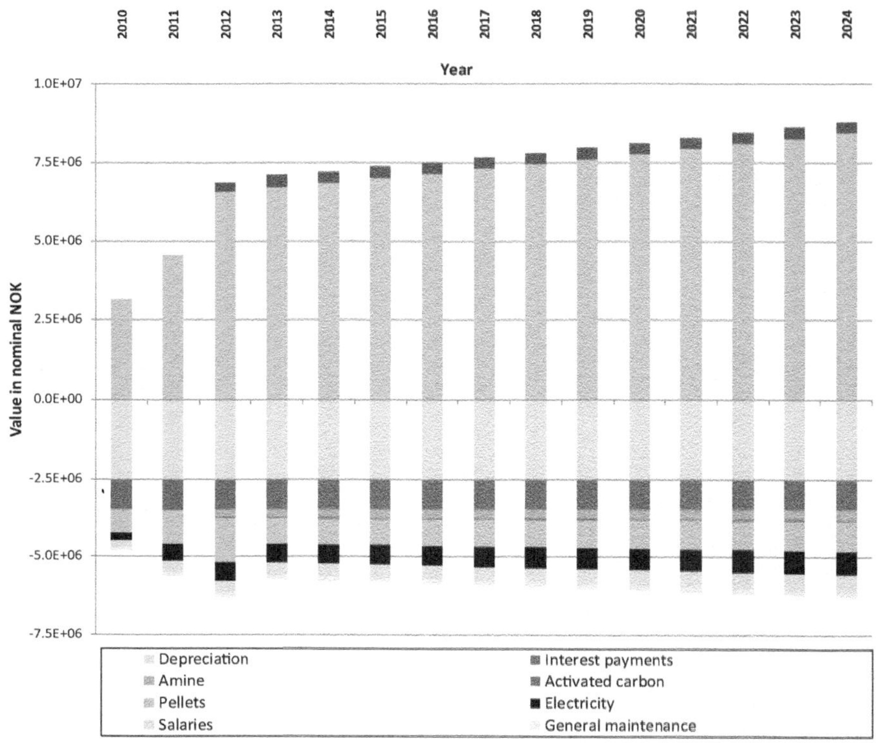

Figure 18.5 Life-cycle costing of the biogas upgrading facility (from 2009–2024) (the X-axis labels, which are the calendar years from 2009 to 2024 have been omitted for the sake of clarity).

Figure 18.5 depicts the cash-flows over time. The *X*-axis labels have been omitted for clarity. The capital investments are committed to year-2009, and the income (positive values) and expenses (negative values) for years 2010 to 2024 follow thereafter. Using a discount rate of 3.1%, the cash flows are discounted back to year-2009, in order to estimate the net present value (NPV). The NPV at this discount rate is 13.63 million NOK. While 3.1% is the discount rate adopted by Oslo VAV in its accounting, the authors have tested for a series of discount rates ranging from 2% to 10%, in Figure 18.6. As long as the discount rate is less than approximately 7.7%, the NPV would be positive and the investment economically feasible.

A sensitivity analysis can be conducted by assuming different unit cost scenarios for electricity, pellets, amine and activated carbon, various selling price scenarios for biomethane. Assumptions made about salaries and maintenance expenses, as well as about the quantities of resources consumed, could be changed. However, the authors contend that these exercises would be beyond the scope of this chapter. It has also been assumed that the capacity of the upgrading plant remains constant and about 20 GWh of biomethane are produced annually. The possibility that investments could be committed before 2024 to enhance its capacity and handle more biogas is ruled out (the biogas flow rate handled by the plant, at the time of writing is around 450 cubic metres per hour; considering uninterrupted operation throughout the year).

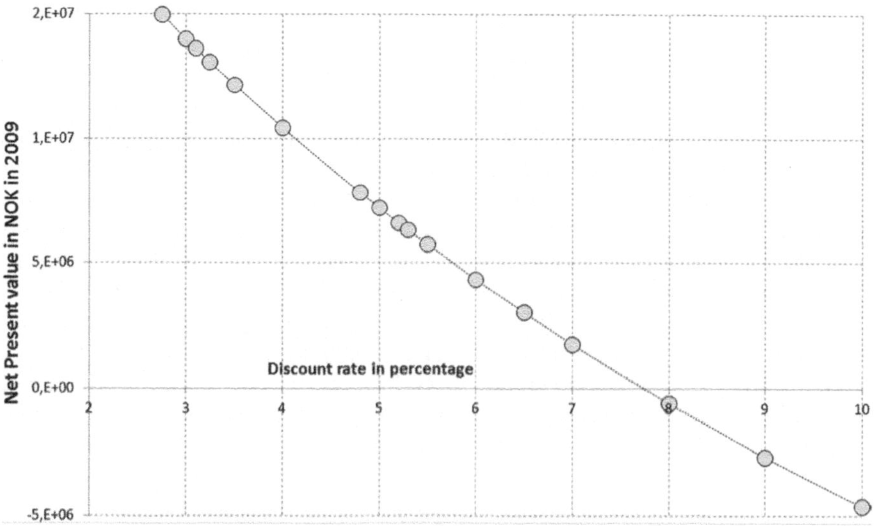

Figure 18.6 Effect of different discount rates on the Net Present Value.

The prevalent confidentiality requirement dictated that the exact unit price paid by the end-user RUTER AS (the last link on the right in Figure 18.4) for the biomethane could not be disclosed to the authors. However, as per an analysis done by Xynteo (Oslo, Norway), the price more than triples en route; and the final selling price of biomethane is almost as much as that of the fossil-diesel it replaces. Thus, RUTER AS does not really profit economically by switching over from fossil-diesel to sludge-derived-biomethane; though environmentally, it succeeds in 'greening' its operations. On the one hand, the carbon dioxide emitted by the combustion of biomethane is biogenic, and on the other, emissions of other pollutant gases (which would have occurred had diesel been used) are reduced considerably.

18.5 CONCLUSION

Whether the investment in the facility is profitable or not would strongly depend on the assumptions made about the cashflows and the discount rate chosen for the analysis. As mentioned earlier, the authors have only investigated the effect of different discount rates for a fixed set of assumptions about the costs and price. For this given set of assumptions, at the discount rate adopted by Oslo VAV (3.1%), the investment in the facility turns out to be profitable. At the end of its 15-year lifetime, though economically, the capital value of the plant would have been depreciated down to zero, there would be a 'technical' salvage value, which along with the net benefits incurred (as indicated by the positive NPV) over the 2010–2024 period, could be carried forth into the next spell of operation, with the aid of a fresh infusion of capital, and possibly an expansion in capacity to meet both the rising supply of biogas and a likely rise in demand for biomethane.

In addition, the biomethane supply provides the Oslo municipality and the public transport company RUTER AS with a 'green profile' and improves their status in the eyes of the inhabitants of Oslo city who avail of the sanitation and transportation services. Environmentally as well, the biogas-to-biomethane-to-transport-fuel project contributes to the truncation of the carbon footprint of Oslo city, while expanding its 'green-print' a little (refer Venkatesh & Elmi (2013) for the environmental assessment which is beyond the scope of this chapter).

Diesel consumption by RUTER AS is supplanted by biomethane sourced from Bekkelaget's sewage sludge digester biogas, and the bus company benefits economically in the process (Figure 18.7). In a green economy, after all, it is a zero-sum game within the energy sector. What is good for the goose (read renewable energy sub-sector) is not good for the gander (read fossil energy sub-sector).

Figure 18.7 A RUTER AS bus running on biomethane from Bekkelaget (picture courtesy: Oslo Water and Wastewater Authority, Norway).

18.6 REFERENCES

Appels L., Baeyens J., Degreve J. and Dewil R. (2008). Principles and potential of the anaerobic digestion of waste-activated sludge. *Progress in Energy and Combustion Science*, **34**, 755–781.

Byrådsavdeling for klima og samferdsel, Oslo Kommune (Department of Transport and Environmental Affairs, City of Oslo) (2013). *Miljø og Klimarapport 2013* (Environment and Climate Report 2013). http://www.enoketaten.oslo.kommune.no/oslo_kommunes_miljo_og_klimarapport_2012 (accessed 18 November 2014).

Eika T. (2013). Statistics Norway. Personal Communication with Rashid Elmi (one of the authors) in November.

Inflation. eu. (2013). Available at www.inflation.eu/inflation-rates/norway/historic-inflation/cpi-inflation-norway.aspx (accessed 30 October 2013).

Kaparaju P. (2011). 2011: D5.1 Evaluation of Potential Technologies and Operational Scales Reflecting Market Needs for Low-cost Gas Upgrading Systems. Valorgas. Available as http://www.valorgas.soton.ac.uk/Deliverables/111129_VALORGAS_241334_D5-1_Final_version.pdf. (accessed 17 October 2013).

Lema J. M. and Omil E. (2001). Anaerobic treatment: a key technology for a sustainable management of wastes in Europe. *Water Science and Technology*, **44**, 133–140.

McCarty P. L. (2001). The development of anaerobic treatment and its future. *Water Science and Technology*, **44**, 149–156.

Rasi S., Veijanen A. and Rintala J. (2007). Trace compounds of biogas from different biogas production plants. *Energy*, **32**, 1375–1380.

Ryckebosch E., Drouillon M. and Vervaeren H. (2011). Techniques for transformation of biogas to biomethane. *Biomass and Bioenergy*, **35**(5), 1633–1645.

Speece R. E. (1996). Anaerobic Biotechnology for Industrial Waste Waters. Archae Press, Nashville, Tennessee, USA.

Starr K. (2013). Environmental and Economic Assessment of Carbon Mineralisation for Biogas Upgrading. Thesis submitted in September 2013, in Fulfillment of the Requirements for the PhD Degree in Environmental Sciences and Technology. Universit at Autonoma de Barcelona, Barcelona, Spain.

Urban W., Girod K. and Lohmann H. (2009). Technologienund Kosten der BiogasaufbereitungundEinspeisung in dasErdgasnetz. Ergebnisse der Merkterhebung 2007–2008. 123s. Fraunhoefer-InstitutfuerUmwelt-, Sicherheits-und Energietechnick, Oberhausen (German).

Venkatesh G. (2012). The water-energy-carbon nexus. *Asian Water*, **28**(2), Published by SHP Media Sdn. Bhd. Kuala Lumpur, Malaysia.

Venkatesh G. and Dhakal S. (2012). An international look at the water-energy nexus. *Journal of American Water Works Association*, **104**(5), 93–96.

Venkatesh G. and Elmi R. A. (2013). Economic-environmental analysis of handling biogas from sewage sludge digesters in wastewater treatment plants for energy recovery: case study of Bekkelaget wastewater treatment plant in Oslo (Norway). *Energy*, **58**(10), 220–235.

Zhao Q., Leonhardt E., MacConnell C., Frear C. and Chen S. Purification technologies for biogas generated by anaerobic digestion. Chapter 9, Climate Friendly Farming, CSANR Research Report 2010–001. http://csanr.wsu.edu/publications/researchreports/CFF%20Report/CSANR2010-001.Ch09.pdf (accessed 10 August 2012).

Chapter 19

A wind PV hybrid system for power supply of a sewage treatment plant in a small town in Southern Brazil

Giuliano Daronco[1] *and Alexandre Beluco*[2]

[1]*Universidade Regional do Noroeste do Rio Grande do Sul (UNIJUÍ)/ Companhia Riograndense de Saneamento (CORSAN), Santa Rosa, RS, Brazil*

[2]*Universidade Federal do Rio Grande do Sul (UFRGS), Porto Alegre, RS, Brazil*

19.1 INTRODUCTION

The initial concern of mankind, with respect to hygiene conditions, was to remove the wastewater from inhabited areas without worrying about the effect of wastewater on the environment. The problem induced by the wastewater generation was intensified when large cities started to develop. The first public sewer system was built in Rome (Cosgrove, 1909). It was a huge channel called *Cloaca Maxima* that had the purpose of transporting the waste water away from the people of the city to the Tiber River. Few other cities had systems for removal of sewage before the Middle Ages and even before the Industrial Revolution.

The gradual increase of population and population concentration in large cities made the infrastructure for waste removal inadequate. At the same time, concerns about the capacity of the receiving bodies to accept increasing amounts of sewage without being affected grown (Science Channel, 2013). A method to reduce costs in wastewater treatment and seek better overall performances is to reduce the one of the highest costs: energy consumption (Marco *et al.* 2013).

The increase in demand for sewage treatment will naturally lead to the increase of energy consumption. The increase in energy consumption will eventually supersede the increase in power supply to consumers. A survey carried out by Silveira *et al.* (2010) concluded that the energy consumption in sewage treatment plants is in the order of 0.050% of total energy consumption in Brazil. This percentage will be doubled in ten years and quadrupled in twenty years, according to projections in non worst-case scenarios.

There is a gradual evolution in the methods of treatment, in terms of higher performance and lower energy consumption. One very interesting aspect would be to install and use autonomous systems for power generation, such as wind turbines and photovoltaic modules supported by diesel generator sets, possibly connected to the grid. This would allow the installation of treatment plants in places far from urban centers and close to the point of disposing the treated sewage to water bodies, without requiring the extension of distribution lines for electricity.

Renewable energy is more suitable for decentralized use, with the energy converters located close to consumers and providing supplies at concentrations far lower than those obtainable with non-renewables. Among the renewable resources with technical and economic viability to meet the typical demands of sewage treatment plants, are mainly micro hydro, solar photovoltaic and wind power. The gases released in sewage treatment plants, when in sufficient quantity and adequate heat capacity, can often be exploited to recover a part of the heat energy they carry.

The photovoltaic modules can be used to meet the demands of sewage treatment plants in places where sunlight is sufficient to produce a reasonable annual amount of energy. However, the costs of PV modules are still quite high and equivalent to the annual energy cost of a typical treatment plant. But it is possible to design systems that have the support of a diesel generator or that are connected to the energy system and are able to meet this energy demand.

Photovoltaic systems provide a power supply that has its own characteristics and availability of energy concentration. A greater penetration of photovoltaic components in hybrid systems necessarily require a reduction in their costs. The increase in the world production of photovoltaic modules and a greater number of incentive programs for their installation, among some other factors, could contribute to cost reduction.

This chapter presents the results of a study on up-to-date alternatives for energy supply to a sewage treatment plant that is under construction. The plant is located near the town center and therefore connected to the grid. Among the alternatives considered, wind turbines and photovoltaic modules were included. Generation systems thus obtained were analyzed with respect to the best composition of the energy system with regard to the energy cost using the software Homer.

19.2 THE SEWAGE TREATMENT PLANT CONSIDERED IN THIS STUDY

The sewage treatment plant considered in this study is located in the municipality of Alto Alegre, in the northern center of the State of Rio Grande do Sul, the southernmost state of Brazil. The city of Alto Alegre can be seen in the image goo.gl/maps/JBCkm (Google Maps, 2014) and its location in the State appears in the image goo.gl/maps/jlnTj (Google Maps, 2014).

The treatment plant will serve the population of the main city of the municipality, built 370 meters above the sea level at the highest part of the watershed of Lake Guaiba, near the lake formed by the dam of Passo Real. This dam [which can be seen in the image goo.gl/maps/aT3zB (Google Maps, 2014)] is the first in a sequence of hydraulic structures along the river Jacuí built by the electric state utility company.

The municipality had a total population of 2137 inhabitants in 2000, with 607 residents in the urban core (190 homes) that will be served by the plant. Figure 19.1a shows a satellite image of the city. Sewage is currently eliminated through individual solutions. The treated effluent will be disposed in a river that appears in the upper left corner of Figure 19.1a, according to the current environmental legislation.

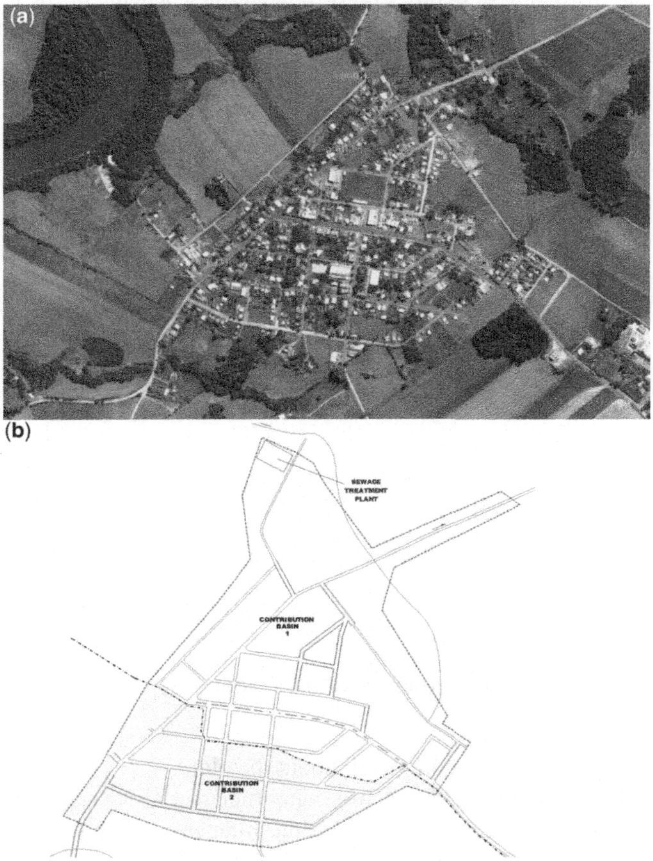

Figure 19.1 (a) Satellite image and (b) map of the city of Alto Alegre. On the map, the city appears divided into two basins, one to the north, in light gray, and the other to the south, in gray.

The city was divided into two basins, because of the topography of the region. The sewage from the southernmost part should be pumped to the north, where the treatment plant is located. These two basins appear on the map of Figure 19.1b.

In Figure 19.1, the map in (b) is covering the region that appears in the image of (a). On the map, the pumping station appears represented in the central region. The map also shows the location of the treatment plant in the north, represented as a rectangle, in a region that does not appear in (a).

The facility is under construction and it will consist of septic tanks followed by an anaerobic filter. The total area occupied by the septic tanks will be 84 m^2, while the area devoted to the anaerobic filter is 108 m^2. The septic tank should be 14 m long and 6 m wide, with a depth of 1.80 m. The filter should be 18 m long and 6 m wide, also with 1.80 m depth. Figure 19.2 shows a scheme of the plant, septic tanks first and then the anaerobic filter. The effluent will be released respecting environmental laws.

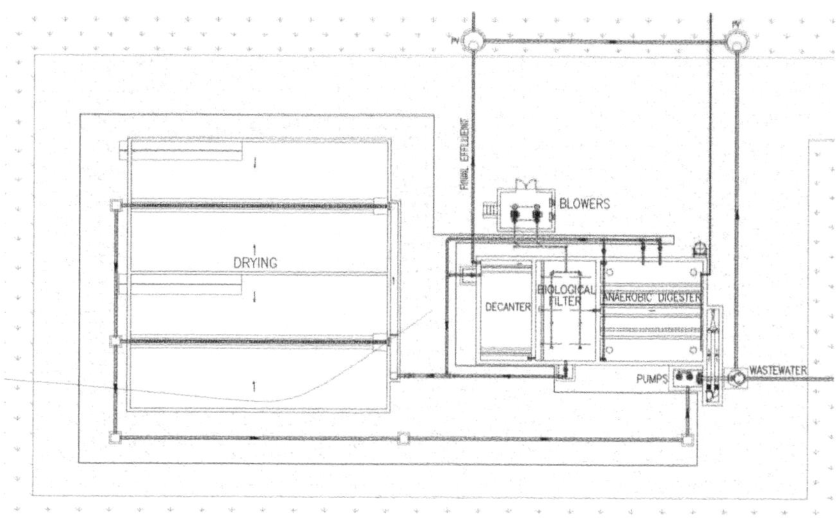

Figure 19.2 Scheme of the sewage treatment plant.

Daronco (2013) provides more detail about the plant.

The plant has been designed to achieve an efficiency of 30% and 80% removal of the organic load in the septic tank and the filter respectively. The removal of fecal coliforms has been designed to be 50% and 90% in the septic tank and the filter respectively. The design flow rate is 7.40 l/s (estimated value in a 20-year lifetime of the plant) and up-to-date flow rate is 5.54 l/s (2.94 l/s from the south and 2.60 l/s from the northern basin).

The main consumption of electrical energy in the treatment plant is due to the pumping station and blowers used in the filter. The pumping station will consist of two pumps, one consuming 1475 kWh/h and the other consuming 0737 kWh/h. The blowers consume 15.77 kWh/h. The average consumption is 539.46 kWh per month. There is also an estimated consumption of 40.00 kWh per month for auxiliary loads such as lighting and security system of the components of the treatment system.

19.3 COMPONENTS OF THE ENERGY SYSTEM

The power supply to the sewage treatment plant in this study can be obtained with a combination of photovoltaic modules, small wind turbines and diesel generators. A system with such components needs buses in DC and AC. As energy supplies must be provided without fail, the connection to the grid is necessary. Storing energy in batteries can help to overcome periods of reduced availability of renewable resources, reducing the energy uptake from the grid.

The diesel system consists of a set of generators used for power supply, aimed to support the electrification of the sewage plant if needed. The average cost of diesel is about USD$ 1.10 per liter and the nominal power of the diesel system is 5kW installed power. The acquisition cost for this equipment is USD$ 500 and the replacement cost is US$ 400, with operating and maintenance costs estimated at USD$ 0.05 per hour and lifetime estimated at 15,000 hours.

The diesel oil consumption is associated with the consumption of 0.72 g of hydrocarbons per liter of diesel oil and the emission of 6.5 g of carbon monoxide per liter of diesel oil, 0.49 g of particulate matter per liter of diesel oil and 58 g of nitrogen oxides per liter of diesel oil. Emission limits, and even limit oil consumption, may be established in order to prevent significant increases in its use.

The software HOMER (Homer Energy, 2009) allowed the acquisition of data from solar availability in Alto Alegre, where the sewage treatment plant is located. Figure 19.3 shows the solar radiation incident on a horizontal plane. The acquisition cost for PV modules is about USD$ 6000/kW, with replacement cost of USD$ 5400/kW and maintenance cost of US$ 300 per year and lifetime estimated at 12.5 years.

Figure 19.3a shows the average incident solar radiation on a horizontal plane for each month as well as the deviation from the average, the maximum and minimum values. The maximum insolation occurs in January, while the minimum occurs in June. Figure 19.3b shows the variation of sunlight available within the day, with the lowest values appearing in the first and the last hour of the daylight and the peak near midday. The variation of sunlight hours throughout the year is also evident.

The system consists of wind turbines with a diameter of 13 meters placed on towers of 28 meters. The area around the treatment plant is quite open and suitable for the installation of wind turbines of medium size. The acquisition cost is

USD$ 90,000, with replacement cost of USD$ 72,000, operating and maintenance cost of USD$ 4500 per year and lifetime estimated at 12.5 years. The acquisition of these turbines in pairs reduces costs for USD$ 162,000, USD$ 129,600 and USD$ 8100 for each two turbines.

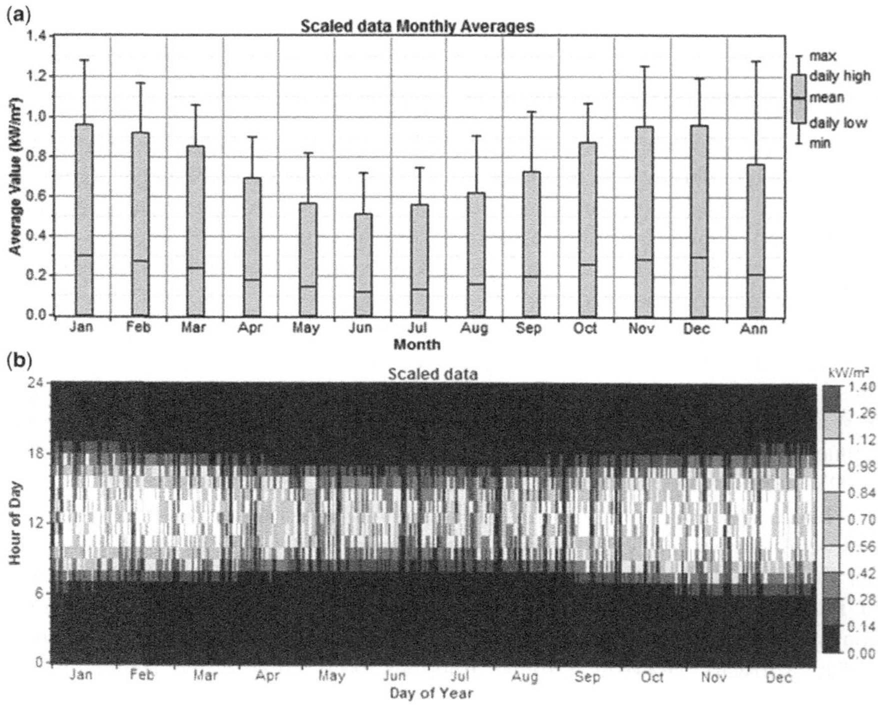

Figure 19.3 Incident solar radiation on a horizontal plane, obtained with software Homer. In (a), monthly averages; and in (b) daily availability.

The series of wind data is synthetic, derived from characteristic parameters of the region (mainly the Weibull shape parameter, equal to 2.5), shown in Figure 19.4: the average wind speed, the deviation from the average as well as the maximum and minimum values. Figure 19.4a shows the typical variability of the wind characteristics. Figure 19.4b shows the enhanced variability of wind over the hours and days of the year as indicated by the intense variation in the gray scale.

Batteries 6FM200D model were adopted in the simulation (Vision Battery, 2013). This model is part of the database of the software and has features very similar to the batteries available in the local market. Automotive batteries were selected due to their availability in the market and their low cost. The model operates with 12 V, with nominal capacity of 200 Ah, equivalent to 0.66 kWh, and a lifetime throughput of 256 kWh. The acquisition cost is USD$ 100, with replacement cost of USD$ 90.

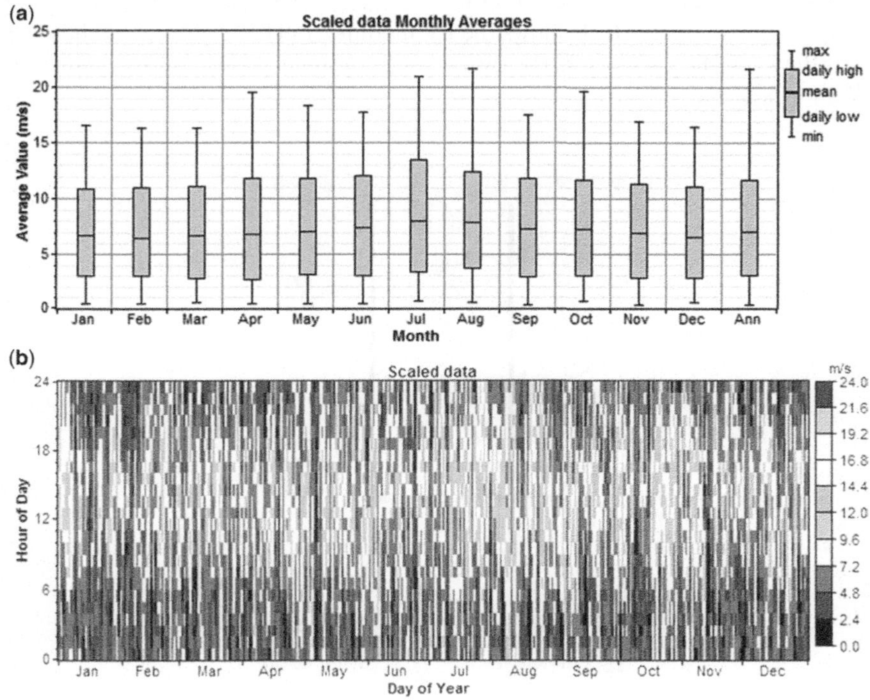

Figure 19.4 Synthetic series of wind speed at the site of the treatment plant, at a height of 50 meters. In (a), monthly averages; and in (b) daily availability.

In this study, converters, which carry out the functions of both inverters and rectifiers in a single component, were considered. The device can operate as rectifier and inverter with 100% of total capacity, with performance of 85% as an inverter and 90% as a rectifier. The lifetime is estimated at 12.5 years.

The energy provided by the network has a cost of $ 0.162 during off peak and $ 0.80 at peak times. The sale occurs with values of $ 0.08 during off peak and $ 0.45 at peak times. The installed power is 22 kW, but the sale to the grid is limited to 10 kW, at no charge for the contracted power or operations d interconnection. The peak time is from 19 hours to 22 hours.

Electric loads, all in AC, are divided into two sets. The equipment for sewage treatment has an average consumption of 432 kWh per day and peak consumption of 18 kW. This set of loads can not experience interruptions in the power supply. Lighting loads and safety, among other auxiliary loads, have an average consumption of 50 kWh per day and peak consumption of 4 kW.

Figure 19.5 shows a schematic of the system considered for power supply to the treatment plant. This figure was extracted from Homer interface. It shows the DC and AC buses, as well as wind turbine, diesel generator and the connection to the grid on the left side. It shows the PV modules and battery bank on the right side. Between the two buses, the two sets of consumer loads and the converter are shown.

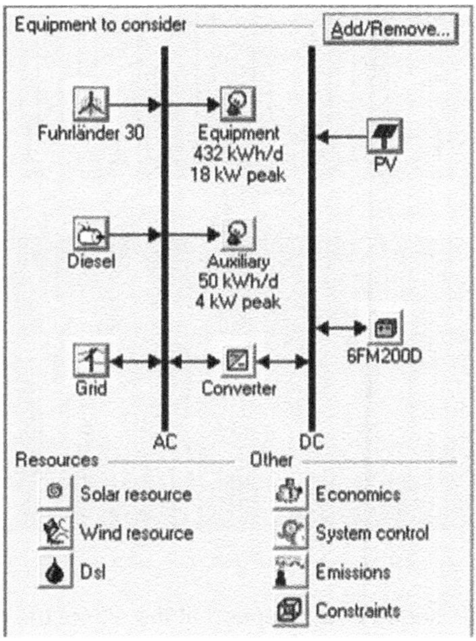

Figure 19.5 Wind PV diesel hybrid system, connected to the grid and storing energy in batteries, supplying energy to the sewage treatment plant considered in the study.

This figure was taken from Homer and shows yet another of its commands. Figure 19.5 left, the commands for insertion of the availability of energy resources. Figure 19.5 right, the parameters for the simulation related to economic data, with the control system and greenhouse gas emissions.

19.4 SIMULATIONS WITH HOMER

The software Homer (Homer Energy, 2009), The Micropower Optimization Model, was developed by National Renewable Energy Laboratory (NREL),

U.S. Department of Energy, and is available for universal access in its version 2.68 beta. Homer simulates a system for power generation over the time period considered in the project at intervals of 60 minutes, presenting the results for a period of one year (Lilienthal, 2004; Lambert, 2005). Homer interface facilitates the composition of the system being simulated and the use of input data for the simulations.

The simulations were performed for an operation period of 25 years, with 12% annual interest and 6% internal rate of return. The different generators and converters could operate simultaneously and in parallel and that can be adopted generators with power less than the maximum value of consumer demand. The generators will be triggered in order to maintain the batteries at a minimum of 80% of their maximum capacity and not just meet the demand.

The hybrid system designed to supply power to the sewage treatment plant considered in this case study is shown in Figure 19.5. It is a wind PV diesel hybrid system connected to the grid and storing energy in a battery bank. This system is also simulated without PV modules, batteries and converter and without grid connection, but it was always simulated with the maximum annual capacity shortage equal to zero.

Simulations were performed for the following values for the optimization variables: 0 kW, 8 kW, 12 kW, 16 kW and 20 kW for PV modules; 0, 2, 4 and 6 wind turbines; 0 kW, 5 kW, 10 kW and 15 kW for diesel generation set; 0, 8, 16, 24 and 32 batteries; 0 kW, 4 kW, 8 kW, 12 kW and 16 kW for the converter; 22 kW for the purchase capacity and 10 kW for the sell capacity from the grid, when considered.

Simulations were performed for the following values for the sensitivity inputs: 2 m/s, 4 m/s, 6 m/s, 8 m/s, 10 m/s and 12 m/s for the scaled annual average velocity of the wind and US$ 0.80, US$ 0.95, US$ 1.10, US$ 1.25 and US$ 1.40 per liter of diesel.

A set of 2,000 simulations, with 25 different values for the variables of sensitivity, were performed. The operation was repeated without PV modules, batteries and converter and without connection to the grid. The results are presented and discussed in the next section.

19.5 RESULTS AND DISCUSSION

The results of the simulation of the complete system of Figure 19.5, connected and selling excess power to the grid, indicate the optimal combinations of the energy system components (Figure 19.6). The purchase and sale of power from the grid have been taken into account in all cases. In the case of the lower wind speeds, the solution of combining the diesel generators and batteries should be preferred. On the other hand, in the cases of the highest wind speeds and lower diesel costs, the combination of diesel gen set and wind turbines seen to be optimal, eliminating the energy storage in batteries.

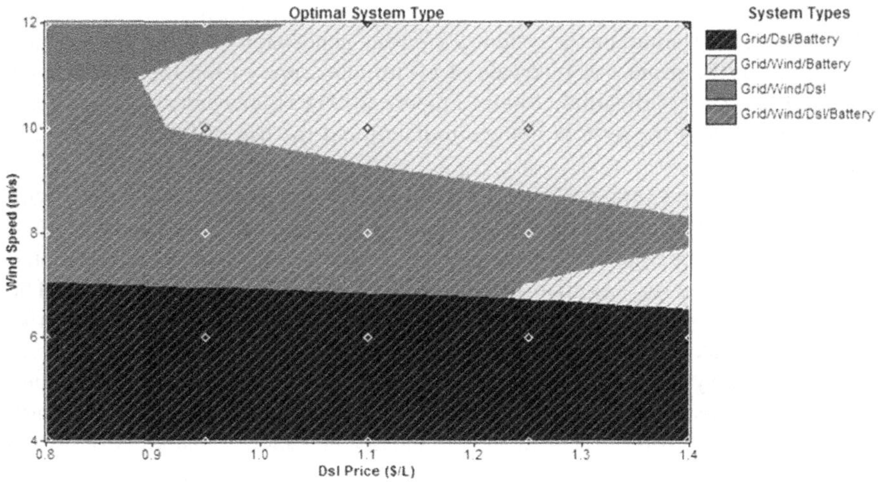

Figure 19.6 Optimization space for the system of Figure 19.5, with the ability to sell excess power to the grid.

The solutions of combining the wind turbines and batteries, including diesel generators are more favourable in the range of lowest price of the fuel. The polygonal line that divides gray and light gray hatched areas shows the costs in these two areas are very similar. Few solutions do not require the use of batteries. PV modules do not appear in the solution space.

The solution corresponding to the average values of wind speed and cost of diesel observed in the region is obtained with a combination of three 5 kW diesel generators, 16 batteries with 200 Ah each, two wind turbines and a converter with 4 kW. This solution represents a total investment of US$ 467,374, an upfront investment of US$ 173,100 and an energy cost of US$ 0.195 per kWh.

Levelized costs of the energy required in all cases are shown in Figure 19.7. Obviously, the higher costs correspond to systems based on diesel consumption and lower costs correspond to the solutions associated with higher wind speeds. In the intermediate region, the costs are distributed in horizontal bands, meaning that they depend more on the wind speed than the cost of fuel.

Similar simulations were performed assuming that the system is connected to the grid but without the ability to sell the excess energy (Figure 19.8). As expected, the combination of diesel generator and batteries is preferable at higher wind speeds compared to Figure 19.6. Similarly, the solutions corresponding to higher wind speeds and lower costs of diesel also occupy a region slightly larger in the optimization space. A consequence of the inability to sell excess power is the increase in the black-stripped region, or a larger number of solutions in the intermediate region without diesel fuel consumption and with the energy storage in batteries.

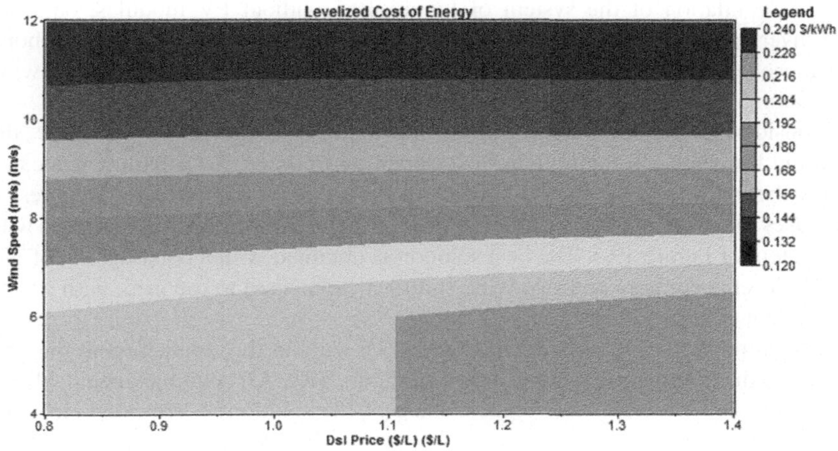

Figure 19.7 Levelized costs of energy for the system of Figure 19.5, selling excess power to the grid.

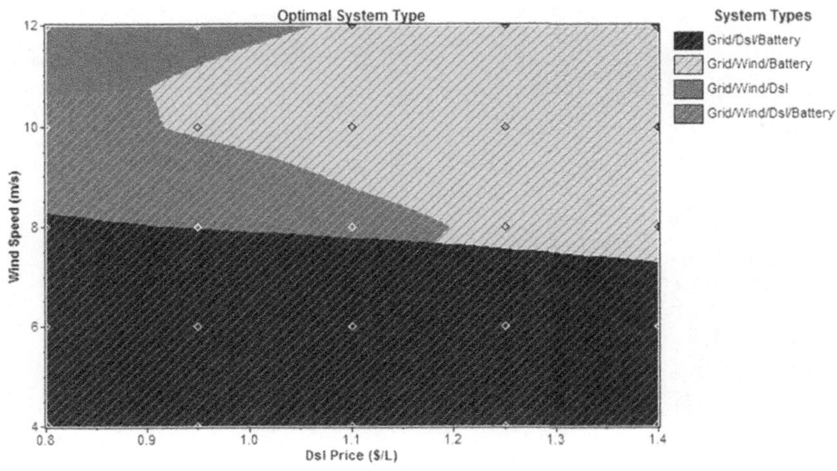

Figure 19.8 Optimization space for the system of Figure 19.5, without the ability to sell excess power.

In this case, the solution corresponding to the average values of wind speed and cost of diesel observed in the region is obtained with a combination of three 5 kW diesel generators, 16 batteries with 200 Ah each, no wind turbines and a converter with 4 kW. This solution represents a total investment of US$ 477,381, an capital investment of US$ 11,100 and a cost of energy of US$ 0.209 per kWh.

The simulation of the system of Figure 19.5 without PV modules, batteries and inverter will result in a space optimization with only two solutions. In both, the diesel generator and the connection to the network, with or without wind generators, will lead to increased costs.

Considering the simulation without PV modules, batteries and inverter, the dividing line between the regions corresponding to these two solutions is located below the top line of the region in black in Figure 19.6. These solutions represent increases in the cost of energy of the order of 10% to 20% compared to results with the system of Figure 19.5. The best solution is obtained with a combination of one 5 kW diesel generators and two wind turbines, connected to the grid, with a total investment of US$ 537,389.

The simulation of the system of Figure 19.5 without the connection to the grid results in the optimization space shown in Figure 19.9. All solutions include diesel generators and batteries. The biggest share of the solutions corresponds to wind diesel systems with energy storage in batteries. The PV modules appear only in solutions with low wind speeds. The solutions to this system represent increases in capital investment from 50% to 400% and in the cost of energy from 50% to 400%.

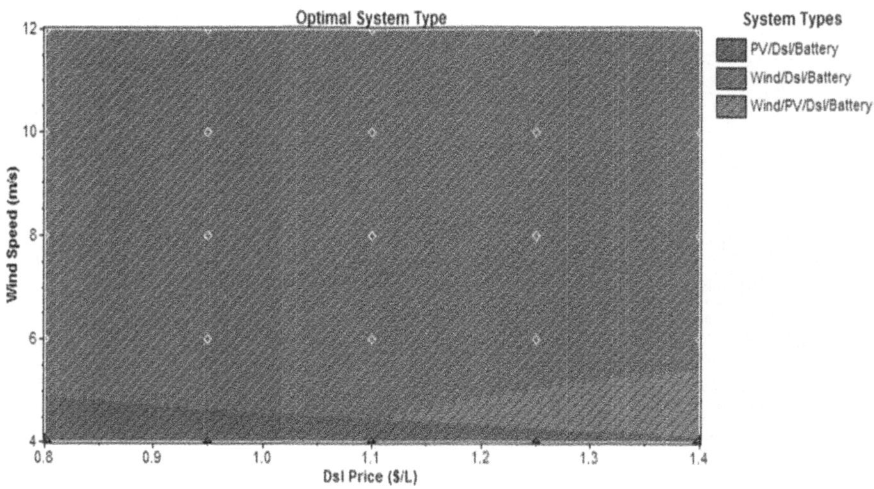

Figure 19.9 Optimization space for the system of Figure 19.5, without connection to the grid.

In this case, the solution corresponding to the average values of wind speed and cost of diesel observed in the region is obtained with a combination of three 5 kW diesel generators, 16 battery with 200 Ah each, two wind turbines and a converter with 4 kW. This solution represents a total investment of US$ 867,023, an initial investment of US$ 197,000 and a cost of energy of US$ 0.350 per kWh.

The solution which presents the lowest cost was the first, involving a combination of diesel generators, wind turbines and energy storage in batteries, besides the connection to the network. The costs of these solutions were high, higher than those obtained with the simple power supply of the interconnected system. The need for plant operation without failure of the power supply contributes to higher costs. Unfortunately, the PV modules do not appear among the feasible solutions.

Table 19.1 summarizes the main results. Three basic configurations of the hybrid system were simulated. The three settings were composed of 16 200 Ah batteries, a converter with 4 kW and three 5 kW diesel generators. One of the differences between the first configuration and the other two is that the former was simulated including the connection to the grid. Another difference is that the first and third configurations include three wind turbines, while the second configuration did not have wind energy.

Table 19.1 Summary of the main results of the simulations.

Configuration	Optimization space	US$/ kWh	US$ capital	US$ total	System components
# 1	Figure 19.6	0.195	173,100	467,374	two wind turbines, three 5 kW diesel gen sets, 16 batt w/200Ah, 4 kW converter, grid-connected
# 2	Figure 19.8	0.209	11,100	477,381	no wind turbines, three 5 kW diesel gen sets, 16 batt w/200Ah, 4 kW converter, no sale of excess energy
# 3	Figure 19.9	0.350	197,000	867,023	two wind turbines, three 5 kW diesel gen sets, 16 batt w/200Ah, 4 kW converter, isolated

The lower capital cost of the second configuration is due to the absence of wind turbines. The small difference compared to the cost of energy of the first configuration, combined with limited financial availability, can enable this solution.

The difference of the total cost over the initial cost is mainly due to fuel consumption. The third configuration involves a high annual fuel consumption, to maintain the supply of electricity for 100% of the time. Better operation strategy can yield better results.

Beluco and Daronco (2013) provide more details on the results.

19.6 FINAL REMARKS

This chapter evaluated alternatives for energy supply at a sewage treatment plant located in a small town in southern Brazil. The plant is located near the center of the city and therefore connected to the grid. Even so, some alternatives were analyzed, including wind turbines and PV modules, using the software Homer.

The combination of wind turbines and a diesel generator with PV modules, with the ability to sell excess power to the grid, results in a cost of US$ 0.195 per kWh. This cost will raise to US$ 0.209 per kWh if there is no opportunity to sell the excess energy to the grid. The cost of energy would raise further to US$ 0.350 per kWh if no PV modules and storage batteries are included.

This feasibility study is only the first step in designing a hybrid system based on renewable resources for power supply to the sewage treatment plant considered in this chapter. The large sunny area suggests the installation of PV modules and the typical wind potential in the region suggests the installation of wind turbines, but the cost for connection to the grid will certainly represent the threshold to be reached.

19.7 REFERENCES

Beluco A. and Daronco G. (2013). Simulation Results with Homer on a Wind PV Hybrid System for Power Supply of a Sewage Treatment Plant in a Small Town in Southern Brazil. Internal Report, IPH, UFRGS. Available at www.beluco.net/reports/homer-alto-alegre.pdf (accessed 22 August 2013).

Beluco A., Souza P. K. and Krenzinger A. (2008). A dimensionless index evaluating the time complementarity between solar and hydraulic energies, *Renewable Energy*, **33**, 2157–2165.

Beluco A., Souza P. K. and Krenzinger A. (2012). A method to evaluate the effect of complementarity in time between hydro and solar energy on the performance of hybrid hydro PV generating plants. *Renewable Energy*, **45**, 24–30.

Caetano Branco Company (2012). Branco Generators Set, www.branco.com.br/us (accessed 28 December 2012)

Cosgrove J. J. (1909). History of Sanitation. Pittsburgh, p. 124.

Daronco G. (2013). Wastewater Treatment Plant – Alto Alegre. Internal Report, CORSAN. Available at www.beluco.net/reports/wwtp-alto-alegre.pdf (accessed 22 August 2013).

Google Inc. (2014). Google Maps. Available at maps.google.com (accessed 8 August 2014).

Homer Energy (2009). Software HOMER, version 2.68 beta, The Micropower Opyimization Model. Available at www.homerenergy.com (accessed 9 September 2009).

Lambert T. W., Gilman P. and Lilienthal P. D. (2005). Micropower system modeling with Homer. In: Integration of Alternative Sources of Energy, F. A. Farret and M. G. Simões (eds), John Wiley & Sons, pp. 379–418, ISBN: 0471712329.

Lilienthal P. D., Lambert T. W. and Gilman P. (2004). Computer modeling of renewable power systems. In: Encyclopedia of Energy, Cleveland C. J. (ed.), Elsevier, **1**, pp. 633–647, NREL Report CH-710-36771.

Marco R., Menendez P. E. and Black & Veatch. (2013). How to Use Energy at Waste Water Plants and How We can Use Less. Available at www.ncsafewater.org/Pics/Training/AnnualConference/AC10TechnicalPapers/AC10_Wastewater/WW_T.AM_10.30_Menendez.pdf (accessed 12 December 2013).

Science Channel (2013). What is the history of sewage treatment systems? Available at curiosity.discovery.com/question/what-history-sewage-treatment-systems (accessed 25 August 2013).

Silveira D. A., Beluco A. and Wartchow D. (2010). Projeção do consumo energético no tratamento de esgotos sanitários. 2. Congresso Internacional de Tecnologias para o Meio Ambiente.

Vision Battery (2013). Vision 6FM200D Model. Available at www.vision-battery.com (accessed 2 April 2013).

IWA Publishing's authorised EU representative for General Product Safety Regulations is Diane D'Arras, 15 rue Duret, 75116 Paris, France, e-mail: safety@iwap.co.uk.

Printed and bound by CPI Group (UK) Ltd, Croydon, CR0 4YY
03/08/2026
02184432-0004